Frontispiece

This group of silver gilt and enamel miniature carriage clocks was sold by Christie's in April, 1980. The prices realised are indicated below. The influencing factors here would have been that they were good examples of miniature carriage clocks, and that costly decorative finishes were employed.

2½ins. £1,260	2¼ins. £1,320	2¼ins. £1,080
2¼ins. £1,920	2¼ins. unsold at £1,620	2⅜ins. £1,080
2⅜ins. £1,440	3ins. £1,920	2½ins. unsold at £1,440

These clocks are still holding their price and are sought after, but condition must be perfect — no chips, made up enamel, etc. It must also be remembered that a standard miniature timepiece in a silver case would be approximately £400—£600 depending upon decorative finish etc., and that these were exceptional examples.

COLLECTABLE CLOCKS

1840 – 1940

Reference and Price Guide

Alan & Rita Shenton

ANTIQUE COLLECTORS' CLUB

First published 1977
Reprinted 1981
2nd revised edition 1985
Reprinted 1987
3rd edition (updated prices and introduction) 1994
Reprinted 1996, 1998

ISBN 1 85149 195 3

British Library Cataloguing-in-Publication Data
A catalogue record for this book is available from the British Library

Printed in England
by the Antique Collectors' Club Ltd., Woodbridge, Suffolk IP12 1DS
on Consort Royal Era Satin from Donside Mill, Aberdeen, Scotland

The Antique Collectors' Club

The Antique Collectors' Club was formed in 1966 and quickly grew to a five figure membership spread throughout the world. It publishes the only independently run monthly antiques magazine, *Antique Collecting*, which caters for those collectors who are interested in widening their knowledge of antiques, both by greater awareness of quality and by discussion of the factors which influence the price that is likely to be asked. The Antique Collectors' Club pioneered the provision of information on prices for collectors and the magazine still leads in the provision of detailed articles on a variety of subjects.

It was in response to the enormous demand for information on 'what to pay' that the price guide series was introduced in 1968 with the first edition of *The Price Guide to Antique Furniture* (completely revised 1978 and 1989), a book which broke new ground by illustrating the more common types of antique furniture, the sort that collectors could buy in shops and at auctions rather than the rare museum pieces which had previously been used (and still to a large extent are used) to make up the limited amount of illustrations in books published by commercial publishers. Many other price guides have followed, all copiously illustrated, and greatly appreciated by collectors for the valuable information they contain, quite apart from prices. The Price Guide Series heralded the publication of many standard works of reference on art and antiques. *The Dictionary of British Art* (now in six volumes), *The Pictorial Dictionary of British 19th Century Furniture Design, Oak Furniture* and *Early English Clocks* were followed by many deeply researched reference works such as *The Directory of Gold and Silversmiths,* providing new information. Many of these books are now accepted as the standard work of reference on their subject.

The Antique Collectors' Club has widened its list to include books on gardens and architecture. All the Club's publications are available through bookshops world wide and a full catalogue of all these titles is available free of charge from the addresses below.

Club membership, open to all collectors, costs little. Members receive free of charge *Antique Collecting*, the Club's magazine (published ten times a year), which contains well-illustrated articles dealing with the practical aspects of collecting not normally dealt with by magazines. Prices, features of value, investment potential, fakes and forgeries are all given prominence in the magazine.

Among other facilities available to members are private buying and selling facilities and the opportunity to meet other collectors at their local antique collectors' clubs. There are over eighty in Britain and more than a dozen overseas. Members may also buy the Club's publications at special pre-publication prices.

As its motto implies, the Club is an organisation designed to help collectors get the most out of their hobby: it is informal and friendly and gives enormous enjoyment to all concerned.

For Collectors — By Collectors — About Collecting

ANTIQUE COLLECTORS' CLUB
5 Church Street, Woodbridge, Suffolk IP12 1DS, UK
Tel: 01394 385501 Fax: 01394 384434
————————— or —————————
Market Street Industrial Park, Wappingers' Falls, NY 12590, USA
Tel: 914 297 0003 Fax: 914 297 0068

Contents

Colour Plates

Introduction

The majority of books written for the horologist have tended to refer to the rare and exotic pieces made by the famous makers of the past. These are invaluable sources of information for connoisseurs with ample means with which to finance their pastime, but of little practical use to the present-day collector. Prior to the Second World War the collecting of clocks and watches together with the study of their principles and technical differences had been the pleasure of an enlightened few. These pioneers had, therefore, been fortunate in having a virtually untrodden hunting ground. Magnificent collections were built up, many of which are now part of the extensive displays in our specialist museums, the fine collection at the British Museum being but one example. The bulk of the items displayed had been acquired by Mr. Courtney Ilbert after a life-long diligent search both in this country and abroad. Through his acumen and insight many unique and rare pieces were recognised, acquired and preserved in his collection. When Mr. Ilbert died in 1957, through funds provided by the Museum, together with monies raised from public subscription, his collection of clocks and watches was acquired by the Museum, Mr. Gilbert Edgar having donated £60,000. Additional material to that already on display in the public gallery can be seen in the Ilbert Room by serious students of horology upon application in writing to The Director, Department of Medieval and Later Antiquities, British Museum, Bloomsbury, London, W.C.1.

After the Second World War interest grew in the collecting of antiques of all kinds and prices began to rise. These prices have so escalated that most of the traditional collectors' items have become impossible to find at attainable prices. Even if the funds are available these pieces are not finding their way on to the open market, remaining in private hands to play the dual role of being a hedge against inflation, but at the same time providing a decorative feature for the home. The natural sequence to this is that those who wish to acquire, research or merely enjoy old timepieces are having to turn to areas that were previously spurned. Fortunately many of the formerly rejected examples have a great deal of merit and are well worth a second or third glance. The Industrial Revolution may have brought with it new factory methods of production to supersede the established handmaking and finishing methods, but this led in turn to the swift introduction of many new ideas and designs. The era of the gimmick and novelty clock had arrived, with quality taking second place. Nothing demonstrates this better than the huge sales in this country of the clocks manufactured by the American factories: cheap, reliable and abreast with the latest furnishing trends. The Black Forest clockmakers in Germany swiftly relinquished their older traditional methods and followed on their heels. They also introduced standardisation of parts, the use of machinery for the slow laborious procedures and assembling of workers under one roof to work. These simple lessons were ones that the English clockmakers could not and would not accept. They doggedly kept to their beliefs that the customer wanted above all things a quality clock and that they would pay for it. They refused to accept that there was a different clientele — the working man. Admittedly there was still the rich land or factory owner who wanted the best and could pay for it, but such people were vastly outnumbered and now only formed a small part of the potential market. Several English companies did see the problem and attempted to compete — the British United Clock Company being the first, but they went out of business through still attempting to produce too good a movement at a competitive price. As early as 1747 there had been in the English horological trade diversity of labour, with each man specialising in one part of a clock or watch, but this was only a cottage industry with each man working in his own home as an outworker on piece rates. It can readily be seen that this method of working in no way lent itself to an industry where the various components were of standard size or pattern, and while in some circumstances the method is highly desirable it is not conducive to producing any item in quantity at a competitive price. There was the added problem of distributing raw materials and collecting the finished products from scattered workers.

Slowly realisation dawned, but by this time the market had been taken over by foreign goods and so the trade limped on through the 1900s still producing good quality movements but being undercut as regards price by competitors. The dependency upon foreign goods was highlighted during and at the end of the First World War by the almost total absence in the shops of the simple alarm clock. Although many of the cheaper clocks had been cased in this country to avoid customs duty as well as complying with the average Englishman's preference for our case designs, the movements had been shipped from Germany. Between the two World Wars endeavours to revitalise the industry were made. Several English factories opened up and appeals were made to 'Buy British' which were emphasised by exhibitions at Wembley. However, the repeal of the various trade tariffs did little to help home industry, with the final blow coming in the form of the Anglo-German Agreement in 1933 when the tariff was lowered from 33⅓ per cent to 25 per cent. The refusal of some to purchase German clocks because of the maltreatment of the Jews did little to help, and it is now realised that it was part of Hitler's overall plan to dump cheap clocks in this country and thereby kill our industry. Eventually after the Second World War, during which the dire need for clockwork mechanisms on timing devices, shells, etc., had highlighted the plight of our horological trade, large sums of Government money were used to finance expansion and update machinery.

This brief résumé of the background of the trade during the period 1840 to 1940 explains why so many of the clocks described in this book are not of English manufacture. They are, however, typical of the period and it is hoped that the following pages provide some intimation of the overall picture of the spectrum of clocks on the market during those years. Although many months have been spent in researching and documenting the examples shown and described here, it is appreciated only too well that a great deal has of necessity been omitted. In some instances this has been through lack of space, and in others unavailability of sources of information. Hopefully this the third edition will further stimulate interest and provoke others to undertake similar studies.

The first edition of this book appeared in 1977 and was extensively enlarged in 1985 and reprinted in its present format. This edition – the third – has little change apart from the revision in the text of the suggested valuations and this brief addition to the introduction.

What has happened in the collecting world since 1977? Over the past sixteen years we have seen more awareness of the interest to be found in timepieces of this particular period. A mixed blessing! Interest does generate preservation and whatever the traditional collectors might feel these clocks are part of our horological history and should be taken seriously. There is little doubt that mechanical clocks will eventually become a thing of the past. There is a strong argument for collecting (and preserving) not just the pre-1940 examples but also those of the 1960s, 1970s and 1980s. Such awareness does however have a down side. Prices rise. In the 1980s prices escalated rapidly – quite unrealistically in some instances. This was universally true of course. Almost anything purchased today was worth more tomorrow. Prices have dropped back but even so those who were selective and took heed of the advice to purchase good undamaged examples have still seen a huge return on their original outlay. This is an area of collecting where demand will continue to expand with each new generation of collectors. Admittedly these clocks were produced in large numbers but as many were discarded by their original owners the supply is not always as plentiful as could be expected. In many cases only limited numbers of certain models ever came on the market. An instance where it is essential to have read around your subject sufficiently to recognise a rarity when you see it! There is always the unarguable fact that there is something of interest whatever the size of your pocket!

Of the many various types of clocks discussed in this book there have been two areas where the prices have remained strong and are continuing to rise. The electric clocks and the novelty clocks. With the first mentioned this is hardly surprising. It was a hitherto totally ignored area of collecting and whilst most of these clocks cannot be regarded as aesthetically pleasing for the drawing room they demonstrate many complex and varied technical features. Sufficient documentation of the history of the manufacturers and the range of models, etc. produced is available to make identification and study of the variants possible. Much valuable work in this area has been accomplished by members of the Electrical Section of the Antiquarian Horological Society.

It is not difficult to form a representative collection whatever one's spending capacity may be – another bonus point. A good example of a Eureka is now over four figures but it is possible to find an encased movement for less – the technical interest is still there. Partly through the diligent searching of specialist dealers there is a steady trickle of Continental examples appearing on the market, thus such a collection can be quite expensive. All these and many more reasons indicate that this will be an area of collecting that unlike the Swatch watch is not based on fashion or whim and will continue to expand and provide a rewarding area of interest for future generations. Perhaps the young collector could do worse than look towards a representative collection of clocks with synchronous movements? These can be found now for a few pounds in boot fairs etc. What will they be worth in thirty or forty years time?

Novelty clocks do, of course, hold a universal appeal and continue to attract high prices. Usually most attractive and by their very nature eye-catching and interesting. The fact that the use of inferior materials and poor finish does not prevent relatively high prices being realised proves that as their designers intended the novelty factor is of more consequence than quality. Perhaps we are now entering the age of collecting where technical interest provides the premium?

CHAPTER I

Guide to Evaluation

Although an attempt has been made by means of the following illustrations and text to demonstrate the factors that make one clock a superior collector's item to another (or for that matter of a higher value) this is not information which can be readily learned. It is only by using common sense and constant observation backed up by textbook knowledge that this is acquired. The bibliography has been compiled with this in mind. Once a collector has decided in what field he wishes to specialise, his mind turns to cost of good examples. The horologist is no exception. The answer is not simple especially as there are always two aspects to consider — the aesthetic appeal and the mechanical interest. Some tentative guidelines have been given on the following pages, but it must be remembered that they are only guidelines and are not intended to be memorised and used blindly. If this is done there is a strong possibility of paying too high a price or missing a bargain. It is far more important to study the comparative prices and interpret the reasons behind them. Many factors govern price. Apart from the more obvious ones of condition, authenticity, rarity, etc., there is also a geographical element — some clocks fetch more money in various parts of the country but as these areas and types are constantly changing it is not possible to tabulate them! Even the time of year when purchasing or selling can contribute to a raised or lowered price. Continental and other foreign buyers tend to have set times for coming to this country to purchase and they obviously have a great influence on the current price. Problems can be created by changing exchange rates. Unfortunately the final decision must rest on the purchaser's own common sense, perhaps tempered by the depth of his pocket and desire for a particular piece! There is always the consolation that, with the coming of the digital quartz crystal clocks, any mechanical timepiece will eventually become a novelty; even if it was not a tremendous bargain at the time of purchase the value must be caught up with one day. Not all pleasure can be calculated in pounds and pence and many of the examples covered in this book do not need hours of cogitation. The only consolation to any lack of concrete price guidance is that, without any standardisation of prices for many of the hitherto considered uncollectable items, bargains are still to be found. Possibly the following general comments will be of assistance.

Buying a damaged or non-working timepiece is a risky action by any but the well-equipped, reasonably skilled clockmaker. Naturally the cost of the item in question plays an extremely relevant part in this decision. It is one thing purchasing a marble-cased clock for a few pounds and enjoying the challenge of attempting to repair and clean the case and movement, but quite a different matter buying a skeleton clock or carriage clock for several hundreds of pounds and then deciding to dabble. Personal limitations must be acknowledged and also the fact that repairs carried out by a professional usually attract a higher charge than the customer expects. This is not because restorers overcharge but because amateurs frequently neither appreciate the time necessary to carry out some repairs satisfactorily, nor that in some extreme situations parts may have to be specially made. There is also the problem that arises when any mechanical object needs repair — more than one part may need attention but this is not always apparent until the movement is dismantled. Most restorers will be only too happy to give a customer an estimate and comment on the work that needs to be carried out if only to clarify the situation and avoid confrontation at the time of settling the bill.

Some concept of what is available must be gained in order to be able to judge whether a particular example is worth acquiring or whether to wait for something a little better to appear. So long as the finances are available there are many complicated carriage clocks to be had in preference to a simple timepiece, but you would search for ever for a Black Forest cuckoo clock in an ormolu case with porcelain panels. Quality must only be looked for where it can be found. Quality in this context can be explained as being the use of any material or process that increased the cost of producing the article in the first instance. This also includes additional time needed to embellish or finish cases or movements, as extra labour also increases the initial cost.

If a clock is extremely rare, or the chances of finding an undamaged example are negligible, it is feasible to contemplate purchasing one with a few minor defects. However, if the clock being offered is in the cheaper price bracket it needs to be virtually perfect to qualify as a collector's piece. Any prices mentioned on the following pages are intended to refer to perfect working examples.

Some attempt has been made in the following chapters to give a basic price and then an estimation of the extra value placed on the various features. Although this is accepted as being the era of mass production it is remarkable to what extent the exceptions occur. Personal avarice also plays a great deal in deciding how much to pay for these!

CHAPTER II

Marble-Cased Clocks

If asked which clock most typified the Victorian drawing-room with its air of middle class respectability and solidarity, it would not be necessary to look further than the polished black marble mantel clocks that appeared in abundance throughout the reign of Queen Victoria and well into the Edwardian period. Although marble and onyx had always been popular, especially on the Continent, as a material in the making of decorative clock cases there had not been a widespread vogue for these solid cases until the mid-nineteenth century. There would appear to have been a number of contributory factors. As the quantity of clock movements manufactured was increased by the new factory methods of production, so the need grew for cases in which to house them. Advancements had also been made in the cutting of marble and similar materials. During the eighteenth century the marble had to be rough sawn and it had not been possible to obtain thin sheets. With the coming of the mechanical saw at the beginning of the nineteenth century, it became possible to obtain sheets as thin as 4mm, and it can be noted that many of the later Belgian marble clock cases have only thin facings of marble over a cement framework. Possibly these technical achievements, together with the then prevalent French fashion for ebonised furniture, provided the necessary impetus. The tragic death of Prince Albert and the Queen's subsequent mourning may have played their roles in the prolonged appeal of this style of case in England; more likely the vogue for neoclassical furniture and the prize-winning cabinet by Wright and Mansfield at the Paris Exhibition of 1867 was an influence.

That these clocks were imported in large numbers is indisputable. Smith & Sons Ltd. of The Strand, London, lists them in their 1900 catalogue amongst their 'Foreign Made Clocks' and with few exceptions they had French movements.

The source and nature of the material used to case them is less certain. One writer describing the clocks at the French International Exhibition of 1889 refers to "a show of those cheap black marbles found in Belgian quarries and cut, polished, and finished for the French market on French soil, as the quarries are not far from the frontier". He goes on to say that "the white and grissotte marbles are drawn from Italy, yellow marbles from the Pyrenees, red from Greece, malachite and lapis lazuli from Russia, onyx from Mexico". Strictly speaking the term 'marble' when used in this context is not accurate. Geologically marble is a calcite that has changed its physical appearance due to metamorphosis, but the term is frequently used in the trade in a broader sense. The material referred to as 'Belgian marble' is not a marble at all. The Ardennes district of Belgium possesses large deposits of a hard limestone with a wide range of colours which includes a dark grey/black variety. It is this that is referred to as 'black Belgian marble' but for the sake of clarity this term will continue to be used rather than the more accurate 'black Belgian limestone'.

T.D. Wright when reporting on the French clockmakers in 1889 states "nearly all the marble cases are made in Belgium; sometimes imported complete, but more often all the finished pieces are separate, and are put together by the clockmaker". This only clarifies the *French* sources. The questions that arise are: Did the English importer also import the complete cases? Or the finished pieces from Belgium for our casemakers to assemble? Or did we use local stone? Some of the cases are slate (as were many of the cheap 'marble' chimney-pieces), in which event this would have come from either Caithness or North Wales. Curiously neither area can provide any information as to whether they supplied slate specifically for the clock casemakers. From information generously supplied by Trevor Ford it appears that there was a source of suitable material in Derbyshire. A dark grey limestone that polished up black was quarried around Ashford-in-the-Water from the sixteenth until the beginning of the twentieth century. This, as with the Belgian limestone, is usually referred to as 'marble'. It was quarried, sawn and polished locally with much of it being supplied for the making of chimney-pieces, tables, vases, jewellery, etc., and was an extremely thriving industry. Again no specific references can be found of any local casemaking on a commercial basis, although the odd case is known to have been made. Although not conclusive, it is felt that sufficient evidence has been assembled to demonstrate that suitable

Figure 1

An exploded view of a marble clock case. Note that while the shaped pieces are of solid marble, the side 'pillars' are hollow, with only the outer facing being of marble. The carcase is of cement. A zinc tube normally encased the movement to exclude dust falling from the interior of the case. The base is enclosed by a wooden plank which provides support for the gong and also acts as a sounding board. The cases examined to date of clocks with English movements and possibly English made cases were of far more solid construction. For example the side pillars that are hollow in this illustration were of solid marble.

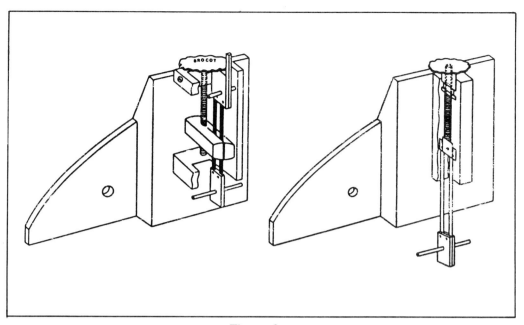

Figure 2a

The mechanism on the left was patented by Achille Brocot, for adjusting the timing of a clock by altering the length of the suspension. A small arbor protrudes through the bezel or dial (see Figure 2b) and by means of a watch key or the smaller end of a double ended key the length of the spring can be altered. 1865 stamped BROCOT Proprieté; 1874 BROCOT A PARIS: 1877 L. BROCOT — Bve SGDG. The similar mechanism on the right is attributed to Vallet (c.1873).

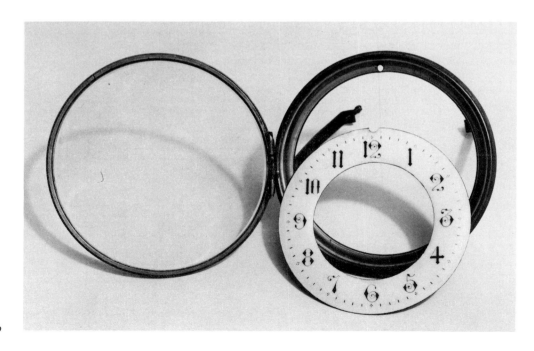

Figure 2b

material for making these cases was available in this country, but as one writer explained early this century "the manufacture of marble clock cases gives employment to a large number of hands between France and Belgium. The work is specialised, the workmanship and finish of the articles being particularly good. Clock cases are produced abroad at such exceptionally low prices as to make competition impossible in this country".

There are a few marble casemakers listed in the Trade Directories for the London area for the period (1852-90) among whom appear the Henson family. They were first at premises in 70 Strand (1852) as Robert Henson, then at 113a Strand (1860) while ten years later the business is under the name of Mrs. E. Henson — possibly Robert's widow. A further move to 277 Strand (1880) was accomplished while Samuel Henson was the owner. A patent was taken out in 1855 by Moses Poole for 'Marble Sculpturing' which was a method of embossing clock pillars, but in all probability cases made here would have been executed by the chimney-piece manufacturers of whom there were a great number.

As can be seen from the illustrations following, the cases range from monumental edifices embellished with bronze reliefs to simpler cases with coloured marble or brass pillars or even small severely plain cases. Accompanying side ornaments were common — many catalogues carried pages of examples — urns, equestrian figures, candelabra etc., from which customers could make their personal choice. Alternatively the complete sets would be advertised. For example "Presentation Suites with 14 day movements, Cathedral Gongs" appeared in a 1906/9 trade catalogue at prices ranging from £18.0s.0d. to £23.0s.0d. Not a cheap item. It was usual for both clocks and side ornaments to be enclosed in glass shades standing on ebonised bases.

The construction of the cases is not without interest. As can be seen from Figure 1, they were usually built up from these pre-shaped pieces — some solid marble, others hollow. The base, main structure and side 'pillars' comprising a fine cement carcase veneered with thin sheets of black marble. The very large cases were reinforced with metal rods that were bolted through the top and bottom sections to give added strength. If feeling sufficiently strong to up-end a clock, it is possible to see the retaining washers and nuts. The smaller, lighter cases, Figures 18 to 21 merely had metal rods keyed into the cement across the corners. According to Britten the cement used by the marble casemakers was "composed of Russian tallow, brick dust and resin melted together, and it sets as hard as stone at ordinary temperatures".

To reconstruct one of these cases is no simple matter if the basic carcase is damaged. If it is intact, it should be possible to remove the old adhesive and replace facings or solid shaped pieces of the case. Lime and white of egg were used for adhering closely fitting surfaces, but a modern epoxy or contact adhesive could replace the older recipe. To attempt more than relatively minor repairs can lead to disappointment, so it is as well not to purchase a clock that requires major renovation to the case on the assurance that it just needs a little work carried out on it. Considerable difficulties can be experienced with even the minor task of tightening the threaded rods holding ornamental columns in place as they are frequently rusted. Prevention is better than cure, so never lift a large heavy clock by the top alone; always support the full weight by lifting from the bottom to avoid the total weight of the case pulling the segments apart.

Problems can also be experienced when attempting to restore the original polish, which has been dulled through exposure to dust and heat from a coal fire, etc. Originally putty powder (oxide of tin) was used on a felt pad. It is important before attempting to repolish to remove any greasy marks by cleaning with benzine (not petrol). If the colour has faded, a good quality black shoe polish softened by gentle warming is ideal. For the final gloss it is difficult to improve upon a good wax polish, without any silicone, as this removes the organic impurities in the marble and dulls the surface. Proprietory polishes such as Gilbert's and Wilkin's were sold in their day; sometimes concoctions of bullocks' gall, soap lees, turpentine and pipeclay were used. In response to present day demand several manufacturers of horological solvents and polishes have now included a marble polish and staining compound in their range of products. These are advertised in current horological periodicals. For light coloured marbles a mixture of quicklime and soap lees spread over and left on the case for twenty four hours, cleaned off and polished with fine putty powder and olive oil was said to produce a good result, though it would be necessary to ascertain first that the case was made of true marble and not some other ornamental stone. Donald de Carle in his book *Watch and Clock Encyclopedia* lists some hundred and seven "Marbles and Ornamental Stones suitable for the manufacture of clock cases". A simple test for marble

Courtesy of Sotheby's

Courtesy of Phillips

Colour Plate 1

An example of a monumental marble and onyx clock (see also Figure 33). This was also noted in an early 1900s catalogue of Smith & Sons Ltd. The estimated price when it appeared in the saleroom was £2,500–£4,000 but it actually realised the higher figure of £5,800. A somewhat intriguing occurrence in view of the general state of the market at the time (1983) when many clocks were having difficulty in realising a sensible valuation.

Colour Plate 2

4ft. high. Marble clock with ormolu mounts inlaid with cloisonné enamel. As several examples of this particular style have appeared on the market over the past few years it was obviously part of a production run and not just made to fill a specific order.

is to place a few drops of acid on a freshly cut surface; if bubbles form the case is made of marble. Spons' *Workshop Receipts for Manufacturing Mechanics and Scientific Amateurs* (any edition) provides a rich source of information for anyone wishing to indulge in pursuing the older recipes for polishes, etc.

Finally, while on the subject of cases, it is necessary to mention the American 'marbleised' cases and those with an 'Adamantine Finish'. The former was achieved by paints and enamels — Edward Ingraham of Bristol, Connecticut, patented a method of 'Japanning Wooden Clock Cases' in 1885, while the latter was a coloured celluloid applied as a veneer. The American manufacturers were obsessed with complying with the latest fashions and furnishings but at the same time needing to keep their overall costs competitive. Although they did import some genuine marble or onyx cases, they usually imitated them with other cheaper materials. They achieved this by making iron or wooden cases in appropriate styles and then copied the veining or colouring of marble by the two methods just mentioned. In a 1906 catalogue of one manufacturer, the comparative prices are eight to eleven dollars for a marbleised wooden case, twelve to twenty dollars for a marbleised iron case and thirty dollars for a genuine marble one. One of the advertisements of Wm. L. Gilbert Clock Co. (showroom in Chicago and factories at Winsted, Connecticut) showed cases made of "plastic marble onyx inlay". One can only assume the word 'plastic' is being used here in its original sense.

Figures for placing on the top of the case or a pair of side ornaments could then be chosen from another section of the catalogue according to personal taste. Most of the large manufacturers produced these cases between 1867 and 1914 as can be ascertained from contemporary catalogues, including those of W. Gilbert Clock Co., Seth Thomas Clock Co., Terry Clock Company and the Ansonia Clock Company. Some of the cases were so well simulated that it is necessary to touch them in order to determine by texture and temperature that they are not in fact made of marble. A second look reveals a poor quality paper dial, thin pressed hands and an American movement. These movements being very similar in finish and appearance to those made by the German factories. The name of the company making them is usually stamped on the back plate.

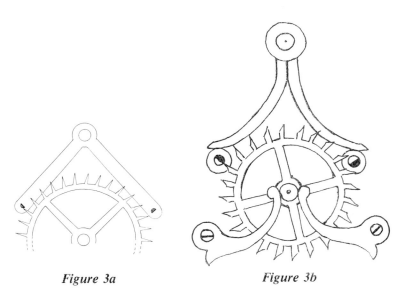

Figure 3a   *Figure 3b*

The two escapements shown here are both attributed to the Brocot family of Paris. It would appear from studying the information readily available in this country that the example in Figure 3a was patented by the father Louis-Gabriel Brocot and later perfected to become the version seen in Figure 3b by his son Achille Brocot (1817-1878). However, a further reference mentions the grandson Paul Brocot (1846-1882) as having perfected an escapement! It is known that he was in business with his father, Achille, in 1873 with addresses in 6, rue de Parc Royal, Paris and 8 Red Lion Square, London, W.C. It is sufficient to say they were a talented family.

Figures 4a and 4b

Achille Brocot was responsible for introducing a perpetual calendar of which this is one type. The advantage of a perpetual calendar is that the need to reset the mechanism at the beginning of each month is removed. This dial shows Month, Day of the Week, Date, Phase of the Moon and Equation of Time. The latter being the difference between mean and solar time, which agrees four times a year, as marked by small sunbursts on the dial, c. 15th April, 14th June, 31st August, and 25th December.

The majority of the black Belgian marble-cased clocks have French or German movements, although there are some interesting exceptions as shown by the example in Figure 25. The French movements are all of a relatively high standard but those made by Japy Frères being of special note. This family was one of the first in France to tool-up in order to mass produce clocks and parts of clocks as well as an amazing variety of items of hardware. They soon became one of the largest suppliers of *blancs roulants* (just the two plates of the movement, separated by the pillars and with the mainspring barrel in place) for many clocks, especially carriage clocks. In the Exhibition of 1855 they were awarded a Grande Medaille d'Honneur for the volume of their output (60,000 *blancs roulants* in a year) while still maintaining a high standard with regards to quality. The five brothers traded under the name of 'Japy Frères' from about 1837 onwards, but after 1854 'Japy Frères et Cie' appeared, followed in 1928 by 'Société Anonyme des Etablissements Japy Frères'. Although nominally eight-day movements, many of the French clocks happily carry on for fourteen days before needing to be rewound. Variations occur — prior to 1880 there was a tendency for the strike to be on a locking plate rather than a rack. There is one extremely simple way of ascertaining which method was used, as the mechanism incorporating a locking plate (alternatively referred to as a count wheel) is readily visible on the backplate of the movement, whereas that for a rack is on the front plate and therefore hidden from view unless the clock is disassembled. The alternatives of a gong or bell have no dating significance in this context. It is interesting to note from contemporary advertisements that a gong cost fractionally more than a bell. The opposite would be true today. It is worth remembering that if purchasing a clock minus its bell or gong it is not possible to replace one with the other, and that the best source of supply for either is another clock of the same type! At the risk of stating the obvious, do check if there are two winding apertures to see if the gong or bell is in place and not missing. A friend omitted to note and realise the significance of the two holes and is still searching for a suitable gong.

It is also advisable to check the pendulum. So often they have become separated from the clock in a saleroom and a substitute found. On many of the French movements corresponding numbers are stamped upon the backplate of the movement and the pendulum. The other numbers on the pendulum indicate the length of the required pendulum, i.e. 5 6 meaning 5.6 French inches. One French inch (*pouce*) equals 1.0657 English inches and so for all practical purposes they are the same. There is usually an adjusting nut on the pendulum for regulating the clock, although many examples also have an adjusting device attributed to Achille Brocot of Paris. Born on the 20th July, 1817, the son of a clockmaker, Louis-Gabriel Brocot, Achille Brocot is reputed to have been an above average pupil. He left school when he was fifteen. His parents were pressed by his tutors to allow him to continue studying at the école polytechnique, but his father refused as he felt that Achille was already sufficiently well versed in mathematics to join him at the workbench. There is a maze of diverse evidence as to whether it was the father or the son who deserves the credit for the various mechanisms bearing their name. The Brocot suspension in Figure 2a appears in the addition to the patent taken out in 1840 by Louis-Gabriel Brocot for various improvements mainly related to temperature compensation. However, it is claimed by Redier in his obituary of Achille Brocot that Achille devised the suspension one day while conversing with his father's colleagues upon the merits of a good suspension. He goes on to say that his father was delighted with the new mechanism and quickly set about manufacturing them commercially somewhere outside Paris. In fact this was their sole occupation for some time. There is no doubt that they quickly became popular with the trade, and with only minor changes continued to be used on most French clocks until the end of the century. Apart from their effectiveness as a method of regulating the clock, they had the added advantage of removing the need to turn the clock round to adjust the timing, as the adjustment was by means of the small arbor above the number 12 on the dial.

The escapements illustrated in Figures 3a and 3b raise further speculation. The father did patent in 1826 an escapement which is in all probability that shown in Figure 3a, but it was most likely to have been the son who perfected it and provided the decorative visible escapement seen on so many French clocks, as illustrated in Figure 3b. Variations do occur, but the principle remains that of Brocot. He was awarded a Medal for this at the 1851 Exhibition and eleven years later at the International Exhibition in 1862 another Medal for his "general manufacture and inventive genius". Another type of visible escapement is shown on the clock in Figure 33c

The perpetual calendar as illustrated in Figure 4a is attributable to Achille Brocot and it received much praise when shown by him at the 1849 Exposition. Again according to Redier, Achille was responsible for designing simple, less ornate marble cases and four-glass cases for mantel clocks. It is known that he came to England in 1845 to promote sales and examples have been seen of clocks made by this family. One being a marble-cased clock with all parts signed and numbered, with a fourteen-day movement, striking on a bell, with a centre seconds hand and Brocot perpetual calendar. Carriage clocks bearing his trade mark have also been found — this according to Charles Allix in *Carriage Clocks* being the letters A and B within a five-pointed star. Eventually after his father's death, Achille sold the business in Paris and opened one in St. Petersburg, Russia where his work had always been favourably received. This failed to be the anticipated success and he returned chastened but not discouraged to Paris where he returned to work with the successor to his old business until he retired through ill health. He died in January, 1878. Obviously some original research is necessary to clarify firmly the role of each member of the family, but sufficient is known to realise that any clock with a movement (not just utilising one of the Brocot inventions) made by either of them would be a desirable piece.

Any difference in quality between one French maker and another is barely discernible, which is hardly surprising when it is remembered that all the *blancs roulants* originated from a mere handful of factories all using the same methods and were often only finished and cased by the 'makers' — the Japy Frères organisation being by far the largest. Not content with manufacturing the movements they also processed their own brass, iron and steel. Other equally commendable names that appear on the backplates of the movements — those on dials are inevitably those of retailers — are 'S. Marti and Co.', 'F. Marti' and 'A. Mougin'. From the *Dictionnaire des Horlogers Français,* by Tardy, it is possible to obtain the approximate dates between which any manufacturer or retailer was operating. Details will frequently be stamped on the backplate of Awards received by the manufacturer at the various Exhibitions held in Paris, London, etc. This does not always refer to the clock upon which the information is recorded, but frequently is intended to indicate that a particular maker has received an Award for one of his products. However they do assist in documenting and dating a particular example as the specific dates when some of the award stamps were commonly in use are known:

Japy Frères	Medaille d'Or	1850-1858
Japy Frères	Grande Medaille d'Honneur	1855
Japy Frères et Cie	Medaille d'Honneur	c.1888
Vincenti et Cie	Medaille d'Argent	1855
Production ceased about 1870		
Marti et Cie	Medaille de Bronze	1860-1869
	Medaille d'Argent	1889-1900
S. Marti	Medaille d'Or	1900-1931
S. Marti	Grand Prix, Paris	Post 1931

(Another source states Marti ceased movement making in 1912.)

Some of the movements are German and it is possible to see a marked difference in quality between the French and German movements without having to refer to names on the backplates. Comparison of the French movement in Figure 5 with that of the German in Figure 6 demonstrates some of the major differences. The French movements have a higher machining standard which gives sharper angles which are usually only associated with the appearance of the best hand-finished clocks. The pallets are finished

Figure 5

A typical French movement found in a great variety of cases including those made of black Belgian marble. Note the thick plates with sharp, well finished edges, solid pinions and well crossed out wheels. This example has no provision for a Brocot suspension.

Figure 6

A typical German movement made by the Hamburg American Clock Company (the trade mark of crossed arrows is stamped on the backplate). Note the thin plates, lantern pinions and heavy arbor in contrast to those of the French movement. Further more subtle differences can be seen by studying the replacement parts available for the two types of movements in the clock material section appearing in the Appendix. Any of the clocks in this Chapter with a German movement would command a much lower price than one with a French movement.

dead hard and brightly polished and have worn well through their years of wear. The German movements have thinner plates, ill-defined angles and edges, soft pallets which have frequently worn with use and most important of all, they have lantern pinions. Their French counterparts have solid pinions. Further subtle differences can also be evaluated by studying the appropriate entries in the clock material section in the Appendix.

The movements are held in place by straps extending from the back bezel to the front as shown in Figure 6. One of the commonest reasons for these clocks stopping is that, as either the front or back has been opened, the whole movement has been inadvertently slightly rotated. It is easily done and equally simply rectified.

Examples of styles of hands, dials, etc., found on these clocks can be noted by studying the illustrations in the pages of the clock material section in the Appendix. It is also informative to note the difference in prices depending upon either the quality of finish or material used for each part as the same principles apply today when assessing a purchase.

Most collectors ask for some guide lines regarding pitfalls to avoid when contemplating a purchase. If the item in question is only a few pounds, possibly there is no better and cheaper teacher in the long run than personal experience. However, for more expensive ventures it is best to be cautious, especially if "the movement only wants cleaning"! Cleaning a French movement is no task for an inexperienced amateur and many professionals are wary of the task. The movement is extremely delicate and it is only too easy to damage the fine pivots. Further complications arise if the movement has a striking train. For those who insist upon proceeding themselves, *Watch and Clockmaking and Repairing* by W.J. Gazeley is one useful source of information.

As well as assessing the timekeeping qualities of the clock, one other point to check is whether the clock is authentic in all details, i.e. original pendulum, dial, hands, etc. Some of these are obviously not vital but, if the price is high, should be correct. At the other end of the scale it is possible to be fobbed off with a 'marriage' of a movement in a different case from that in which it started life. It is sometimes argued that as originally the movements and cases were manufactured separately such an interchange nowadays would be quite legitimate restoration. This opinion is endorsed so long as the replacement is commensurate with the original design and neither the case nor the movement have had to be adapted for this marriage. An example of what should not be done is ably demonstrated by looking at the illustrations in Figures 2a, 2b and 5. The bezel and dial shown in Figure 2b should have a movement with a Brocot suspension adjusting device — the aperture is made for the arbor through the bezel and top of the porcelain dial. However, upon dismantling the movement it was discovered that it was the basic movement as shown in Figure 5 without a Brocot suspension and never had one. Once the movement is cased there is no way of checking for this discrepancy apart from confirming that there actually is an arbor and not just an aperture in the dial.

Further examples of clocks with more exotic marble and ormolu cases will be found in Chapter XI.

The movements depicted in Figure 7 are curious. These illustrations show movements advertised in the 1906/8 catalogue of Anthony Mayer & Sons Ltd. of Aldersgate Street, London. Mayers were one of the major wholesale importers of both movements and complete clocks in London. From the dates taken from the London Directories it appears that this firm was founded during the 1870s and by their continued advertising in trade periodicals were active to pre-World War I. They had extensive showrooms, packing departments, casemaking workshops, stores etc., in Aldersgate as well as just off the Goswell Road. These trade catalogues were intended to be used by the retailer (with his own name superimposed) and only too often the original source of the clocks being offered is deliberately disguised. Some useful clues appear however in these two pages. The 'F' and 'P' movements are self explanatory and are obviously varying qualities of French roulants. The 4 inch 'M' movement has "solid Hardened Steel Pinions" so although it is logical to assume this is of French origin it is not unknown for American movements to have solid pinions. The 'Rex' movement has lantern pinions so is almost certainly German or American in origin. As a further page of this catalogue (Figure 8) states "French Marble Cases fitted with 8-day American Cathedral Gong Movements" this leaves little doubt that these movements are of American manufacture. A feature of both movements is that the pendulum suspension allows for the pendulum to be hung outside the gongs and is thus theoretically easier to attach without entanglement with the gong. There is also a spring on the end of the rack tail to prevent the rack from stopping the clock should the striking train run down.

FRENCH MARBLE CASES.

FITTED WITH 8-DAY AMERICAN CATHEDRAL GONG MOVEMENTS.

Height 10½in.   No. **6525/239**.   Price £2 0 0.

Height 11in.   No. **6684 240**.   Price £2 11 0.

Height 13½in.   No. **6619 241**.   Price £2 19 0.

Height 14½in.   No. **6567 242**.   Price £3 14 0.

ABOVE PRICES INCLUDE VISIBLE ESCAPEMENT.

A. M. & S. L.

Figure 8
Page of French marble clocks fitted with 8 day American Cathedral gong movements ranging in price from £2. 0s. 0d. to £3. 14s. 0d. taken from the catalogue of Anthony Mayer & Sons Ltd.

As American 'marble' clocks are unusual in this country £80 — £200
Similar examples with French movements £100 — £250

Figure 9

An impressive clock, 17½ins. high, the case is of black Belgian marble decorated with cast bronze mounts and mouldings. The porcelain chapter ring has a recessed gilt centre with a visible Brocot escapement; an escapement invented by Achille Brocot (1817-78) of Paris. This particular example has steel pins but it is not unusual to see agate used for this purpose. As well as the adjusting nut on the pendulum, the length of the suspension spring may be altered by means of the small arbor protruding through the dial above the number 12. This particular form of adjustment is also attributed to Brocot. The movement is numbered with matching pendulum and with rack striking on a gong. The size of this example is a possible disadvantage.

Similar examples without visible escapement and poor quality bronze mounts would be about £200. This clock with visible escapement and good quality mounts would be in excess of £300.

Figure 10

The case of this clock, 17ins. high, is of black Belgian marble with inlay of green marble and brass pillars and finials. The porcelain chapter ring has a recessed gilt centre inscribed with the name of the Bournemouth retailer. The exceptional feature of this clock is the bevelled glass front that provides a sight of the two-jar mercury pendulum. Access to this is through the back of the case by means of a neatly hinged marble door. Similar clocks have been seen with an Ellicott pendulum.

A clock of this quality would realise between £200 and £250. A visible escapement would increase the basic price by £50.

Figure 11

An unusual black Belgian marble case with brass mounts and feet. The porcelain chapter ring has a recessed gilt centre decorated with a brass rosette. The movement is numbered with a matching pendulum and the maker's name — Jules Rolez Limited, Paris — is stamped on the backplate. Jules Rolez is known to have made movements during the nineteenth century at St. Aubin-sur-Scie, Paris. This is a good quality movement with rack striking on a gong. Features that would raise the price of this clock would be the practical size, the unusual shape of the case and an uncommon maker's name recorded on the movement.

£100+

Figure 12

A similar clock to this appears in the 1900 catalogue of S. Smith & Son Limited, 9 Strand, London retailing at £1. 10s. with an eight-day movement, and £2. 2s. with a fifteen-day movement striking on a bell. The height of the case is 12½ins. high and is made of black Belgian marble with a green marble contrasting inlay. The bezel is highly decorative and the chapter ring is of polished black marble with engraved and gilded numerals and a recessed gilt centre. This was obviously a presentation clock and as the small brass plaque is only glued in position it provides interesting historical information while not damaging the case. The fifteen-day movement is numbered with a matching pendulum with the maker's name — Japy Frères et Cie — stamped on the backplate. It is a matter of personal opinion as to whether the additional documentation from the catalogue enhances the price.

£120+

Figure 13

The case of this clock, 13½ins. high, is again made of black Belgian marble, with cast bronze feet, ornamental pillars and drop piece. The frieze in the tympanum is in bas relief. The dial has a polished black marble chapter ring in which the numerals have been engraved and then gilded. The recessed gilt centre is engraved with stylised foliage. The movement is numbered with a matching pendulum and bears the maker's name — S. Marti et Cie. Striking is on a rack and gong. This example has all the virtues of being a reasonable size, of having a highly decorative and exceptionally good quality marble and bronze case as well as having a named movement.

£120+

Figure 14

A black Belgian marble-cased clock, with bronze relief and thin brass columns. The dial is of porcelain with a gilt recessed centre. At some stage the pillars have been lacquered black — a common phenomenon which some attribute to a deep sense of mourning upon the death of Prince Albert, but one suspects is more likely to be due to a desire to avoid the necessary cleaning and polishing. Movement by Japy Frères et Cie.

£50 in this state
£75+ if case restored to former glory

Figure 15

Although a similar architectural style case to that in Figure 14, the use of contrasting coloured marble for the pillars instead of brass makes the overall appearance rather heavy. When inspecting this style of case care must always be taken to note that the pillars have not been cracked and are only being held in place by the metal rods through their centres.

£90+

Figure 16

An extremely well finished case in black Belgian marble with a green marble top and inlay. The dial has a black marble chapter ring with green marble centre, and skeltonished brass numerals. The door at the back is glass. This is not an unusual feature in the better quality cases although it is more common to find a metal door with ornamental piercing covered with gauze. The backplate has the stamp of the maker (Vincent et Cie) and the information that he had received a Silver Medal for his clock movements in 1834.

£100+

Figure 17

An interesting and attractively designed case of black Belgian marble with contrasting insets of green marble. The quality of the case is demonstrated by the use of solid marble for the scrolls, etc., and the deeply incised decorative markings. The dial is plain white enamel without protective bushes to the winding apertures. In common with most of these clocks there is a Brocot suspension. Striking is on a locking plate and bell. The movement is numbered with corresponding pendulum and the name 'Japy Frères et Cie' is stamped on the backplate.

£120+

Figures 18, 19, 20, 21

The similarity of these four clocks is misleading. Figures 18 and 20 have fifteen-day movements, striking on bells, while the other two have eight-day movements and are timepieces only. Although the clock in Figure 18 has a well shaped solid case the dial is white enamel and very puritanical. It is uncommon to see the letters 'S' and 'F' indicating adjustment for 'Slow' and 'Fast' either side of the adjusting arbor above the numeral 12 in this type of clock. The other striking movement in Figure 20 would have a slightly higher value as the dial is of porcelain and a few pieces of decorative brown veined marble have been added to the case. Although simple both of the timepieces have a good feature. The clock in Figure 19 has a fancy bezel and an undamaged dial. That in Figure 21 while having a plain dial, slightly damaged due to careless use of the key when winding, does have some shaped pieces of veined red marble decorating the case.

£40 — £60
A damaged porcelain is difficult to restore satisfactorily so commands a low price.

Figure 22

Although solid this case with its classical columns and pediment is well proportioned. The black Belgian marble surfaces are broken up by the red marble columns and plinths, while the incised traceries are gilded. In most instances the original gilt has faded or worn away, although this was by no means a standard feature. Many of the cases relied purely on the effect of plain grooves or markings for decoration. The chapter ring and recessed centre are both of porcelain with protective brass bushes to the winding apertures. The movement in this clock and that shown in Figure 23 were made by F. Marti et Cie. Fritz Marti is known to have been making movements at Vieux-Charmont in 1876 and won medals in 1908 for his work.

£85+

Figure 23

A well constructed case of black Belgian marble with green marble pillars. The movement is by F. Marti et Cie.

£75+
If just a timepiece £40+

Figure 24

There is no doubt that the value of this clock lies largely in the aesthetic appeal and quality of the bronze figure surmounting the case. By far the greater proportion of the total value would depend upon this feature rather than the movement or the fact that it still retained its matching side vases. Unfortunately the clock in many of these elegant sets has been cannibalised in order to obtain a desirable bronze. Price of similar examples would depend on whether the figure was spelter or bronze.

£800+

Figure 25

This timepiece, 15½ins. high, has been included as an example of a marble-cased clock with an English movement. The architectural styled case was made of solid marble with the figure of a sphinx on top. The engine-turned dial is gilt with the name 'Barraud & Lund, Cornhill, London' engraved between the hour markings. Lund was in partnership with Barraud between 1838 and 1869 but the style 'Barraud & Lund, Cornhill, London' was only in use between 1844 and 1864. It is known that the firm made at least two black marble Egyptian-style clocks c.1858 and shortly after 1864. In his book Paul Philip Barraud, the author, Cedric Jagger, feels that the work on the Suez Canal between these dates may have had some influence on the choice of case decoration. Both documented clocks had a circular fusee movement with recoil escapement. These additional facts concerning the makers and the probable circumstances in which the clocks were made should greatly enhance their value.

The English movement greatly escalates the price
£450+
Similar with French movement £300+

Courtesy Sotheby's

26a

Figures 26a, 26b, 26c

This is an extremely high quality well made case with brass gallery, pillars and feet as well as a brightly coloured frieze and dial centre depicting birds, butterflies and flowers.

The movement is unusual and interesting as it embodies the keyless wind mechanism patented by Etienne Maxant (U.K. Patent No. 933, 10th March, 1879 — American Patent No. 220,401, 7th October, 1879). Maxant is recorded as having worked at Rue de Santonge, Paris between 1880-1905. Winding of both strike and going train is by means of the two 'handles' seen clearly in Figure 26b in the extended position at the bottom edge of the dial bezel. When retracted they fit snugly into the inner bezel and are therefore not normally visible.

The regulation is controlled by a further 'handle' in the upper section of the inner bezel. As a few other examples have been noted this was obviously a commercial production. A worthwhile acquisition for any collector.

Unusual and attractive £500+

26b

26c

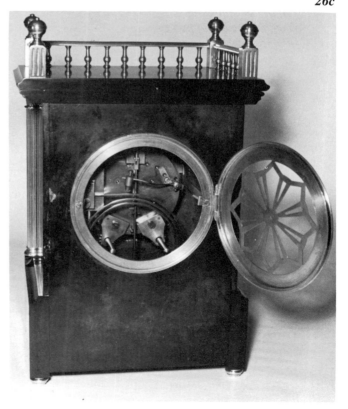

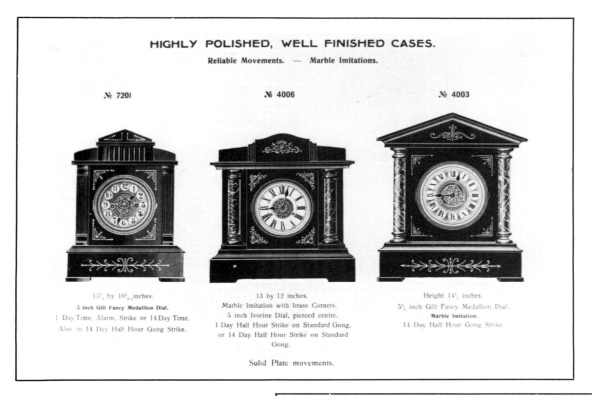

HIGHLY POLISHED, WELL FINISHED CASES.

Reliable Movements. — Marble Imitations.

№ 7201

№ 4006

№ 4003

13¾ by 10½ inches.

5 inch Gilt Fancy Medallion Dial.

1 Day Time, Alarm, Strike or 14 Day Time.

Also in 14 Day Half Hour Gong Strike.

13 by 12 inches.

Marble Imitation with brass Corners.

5 inch Ivorine Dial, pierced centre.

1 Day Half Hour Strike on Standard Gong,

or 14 Day Half Hour Strike on Standard Gong.

Solid Plate movements.

Height 14½ inches.

5½ inch Gilt Fancy Medallion Dial.

Marble Imitation.

14 Day Half Hour Gong Strike.

Figure 27

Page taken from catalogue of Hamburg American Clock Company showing the marble imitation cases supplied by them at this date (1912).

£40 — £70

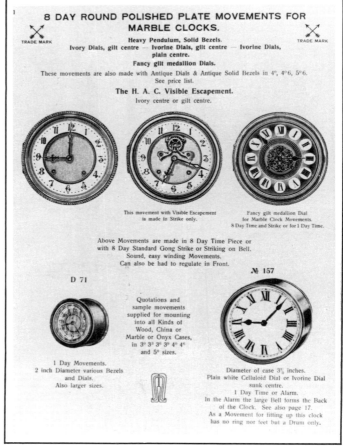

8 DAY ROUND POLISHED PLATE MOVEMENTS FOR MARBLE CLOCKS.

Heavy Pendulum, Solid Bezels.

Ivory Dials, gilt centre — Ivorine Dials, gilt centre — Ivorine Dials, plain centre.

Fancy gilt medallion Dials.

These movements are also made with Antique Dials & Antique Solid Bezels in 4″, 4″6, 5″6.

See price list.

The H. A. C. Visible Escapement.

Ivory centre or gilt centre.

This movement with Visible Escapement is made in Strike only.

Fancy gilt medallion Dial for Marble Clock Movements.
8 Day Time and Strike or for 1 Day Time.

Above Movements are made in 8 Day Time Piece or with 8 Day Standard Gong Strike or Striking on Bell.
Sound, easy winding Movements.
Can also be had to regulate in Front.

D 71

№ 157

Quotations and sample movements supplied for mounting into all kinds of Wood, China or Marble or Onyx Cases, in 3″ 3½ 3½ 3½ 4″ 4″ and 5″ sizes.

1 Day Movements.

2 inch Diameter various Bezels and Dials.

Also larger sizes.

Diameter of case 3½ inches.

Plain white Celluloid Dial or Ivorine Dial sunk centre.

1 Day Time or Alarm.

In the Alarm the large Bell forms the Back of the Clock. See also page 17.

As a Movement for fitting up this clock has no ring nor feet but a Drum only.

Figure 28

Further page from Hamburg American Clock Company (note crossed arrows trade mark) advertising the movements sold by them for marble clocks and smaller wood, china etc., cases.

Figure 29

Black wooden imitation marble clock with white enamel dial signed

> *J. Collbran*
> *295 Regent Street*
> *London.*

Note the presence of the small alarm dial.

£300+

Figure 30

View of 8 day movement with bell removed. Note outside count wheel. Although this usually indicates an early movement it is frequently noted that cases that would be most definitely dated as late 1920s or 1930s still house movements with the earlier method of controlling the strike. This can only be interpreted as the ultimate example of the cases being of more importance than the movement. The maker's mark BR, Movement No. 2456l (also visible on pendulum bob although badly stamped), 5.5 (denoting pendulum length) and a repairer's scratched date 6.9.83 are visible on the back plate. It has not been possible to trace the maker. This is a most unusual manifestation of an alarm clock and therefore can only be categorised as desirable!

Figure 31

A rather magnificent late 19th century black marble clock signed Potanie Leon, Paris, with perpetual calendar flanked by thermometers, dead beat 'open Brocot' escapement and standing some 18ins. high. The centre seconds enhance interest and value. Estimated at £800 — £1,200 when offered in 1983 it remained unsold at £580. A straight visible escapement is considered less desirable than the curvaceous example seen in the following illustration.

£950+

Courtesy Sotheby's

Figure 32

2ft. 1in. high

A truly magnificent example of a mantel clock in brown and black marble case with gilt metal mounts. Note the fine detail on these mounts — good sharp detail to the mask and stylised foliage to bezel, feet etc.

The movement is signed S. Marti et Cie. A perpetual calendar and barometer can be seen in the lower section of the case. The quality is excellent and the provenance important — this clock came from Camden Place, Chislehurst formerly the property of the Empress Eugénie, mother of the Prince Imperial — but of course the size is somewhat overpowering for most modern homes. This clock was estimated in 1982 at £900 to £1,200 but it actually realised £780 (including buyer's premium and VAT).

Well over £1,200

Courtesy Bonhams

Figure 33a

This fantastic clock stands nearly four feet tall, is twelve inches wide and weighs nearly two hundred-weights. It is included to demonstrate that not all clocks were small shelf clocks, nor were they plain and dowdy. The identical clock was offered in the 1901 catalogue of Smith & Sons Ltd., of 9 The Strand, London for £30:

"Beautifully designed in Green Mexican and Algerian Marble, Solid Ormolu Gilt Mounts, inlaid with Cloisonne Enamel of various colours, 15-day Movement of the Finest quality, Striking Hours and Half Hours on Deep Toned Cathedral Gong".

This description adds considerably to the interest of this piece. How many clocks were made to this design is not known, although the existence of several has been noted. One appeared in a house sale at Efford Cottage, Everton near Lymington, Hants by order of Mr. Mrs. H.G. Smith in June, 1972 and sold for £1,160. A further example appeared in a London saleroom in 1975 and realised £1,300, while yet another offered at Sotheby's Belgravia in December, 1977 reached £2,300. Twice the price within five years.

£4,000+

A French green onyx and champ-levé enamelled grandmother clock. Sold with non-matching side plinths at Christie's South Kensington for £5,500.

Figure 33b

Figure 33b shows the movement which is of good quality and French. The word 'Brevete' and initials 'S.G.D.G.' around the letters 'GLT' are stamped on the backplate. According to the Dictionnaire des Horlogers Français by Tardy this was the mark of Thieble of Paris. Thieble took out a patent for an improved pendulum in 1865.

Similar with non matching side plinths £5,500 in 1984

Coupes accomp

Nº 6884. Pendule marbre et bronze
verni ou argenté

Haut. 23 cent. — Larg. 40 cent.

Haut. 15 cent.

Modèle déposé

Figure 34

The illustrations show a selection of marble, bronze and enamel decorative clock cases offered in the catalogue of Hour, Lavigne & Cie of 7, rue Saint-Anastase, Paris, earlier this century.

The clock above would be priced at about £300 and more if with side ornaments, those below have an art deco appeal and would probably fetch £250 at auction.

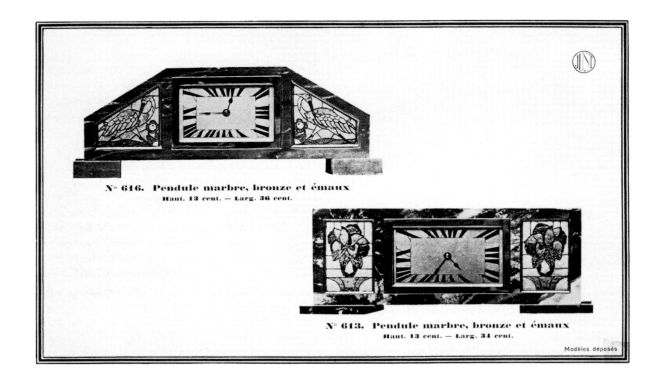

Nº 616. Pendule marbre, bronze et émaux
Haut. 13 cent. — Larg. 36 cent.

Nº 613. Pendule marbre, bronze et émaux
Haut. 13 cent. — Larg. 34 cent.

Modèles déposés

CHAPTER III

Skeleton Clocks

It was not surprising that in an age which was rapidly becoming industrialised and mechanically minded that there should develop a demand for clocks with visible movements. To the retailer it was an 'eye catcher'; to the family, especially the children, it satisfied the desire 'to see the wheels go round'. Although never acknowledged by the horologists of their day as having any special merit, these clocks admirably blended with the décor of that period and joined the wax fruit, imitation flowers and stuffed birds — also under glass domes — in front of the large overmantel mirrors on the chimney-pieces. They were not cheap clocks as can be ascertained from the prices given in the 1865 catalogue of Smith & Sons of Clerkenwell. The prices quoted — £2 10s. for a simple timepiece and £10 to £12 for 'York Minster, striking half hours on bell and hours on gong' — would not have suited the pocket of the average working man.

While iron-framed clocks were common in the sixteenth century, the brass-framed skeleton clock first appeared in France during the 1750s. The names of Le Roy, Lepaute, Berthoud and Lepine being some of the French makers associated with their design and manufacture. The most noticeable difference between the French skeleton clocks and those made later in England was that the French rarely, if ever, used a fusee and the general design was far more ornate with delicate crossing out of the wheels. Even the bases, as can be seen from the example in Figure 35, were far more ornate with many added embellishments.

Although skeleton clocks were conceived on the Continent, the English clockmakers soon followed suit and from 1820 until 1914, when commercial production finally ceased, there was a steady market for these clocks. Naturally examples were on display at the Great Exhibition of 1851, including one superb chiming skeleton clock by Moore and Son of Clerkenwell. The location of this clock is not generally known. It is at present in the possession of the Norwich Union Insurance Group, who have generously supplied the following interesting information concerning the events leading up to their acquiring this desirable piece. The clock remained unsold in the showrooms of Moore and Son from the time of the Exhibition in 1851 until 1863 when it was sold to Mr. Joseph Langhorn for the sum of £215. Mr. Langhorn having been prosperous in business, decided at an advanced age to study medicine. Having qualified as a doctor he eventually took up residence at Ashburton in Devonshire. In 1878 he returned the clock to its makers with instructions to repolish and enamel it, as it had been "somewhat damaged by the humid Devonshire air".

In the meantime he decided to donate it to the Directors of the London and Westminster Bank while still alive sooner than their having to await his death. He had bequeathed the clock to them in his will. The Bank declined to accept this gift as they were unaware at this stage in the developments of its value and felt it was rather a large item to accommodate in their offices. After receiving their refusal he approached the Norwich Union Fire Office, as it was then known, who accepted the gift "with pleasure and gratification". However, Dr. Langhorn died before the clock reached the Norwich Union and, as he had not altered the bequest in his will to the directors of the Bank, legal problems arose. Eventually the Bank directors waived any claim and the clock was set up in 1878 at the entrance to the Boardroom of the offices of the Norwich Union in Norwich where it has remained in spendour to this day (Figure 35). It is thought that two comparable models exist — one in Russia and the other in the United States.

Smith & Sons, J. Moore and Son and other Clerkenwell makers provided fifty per cent of the total output of skeleton clocks. Some provincial towns such as Birmingham, Liverpool, Ashbourne (Derbyshire) also had skeleton clock manufactories. The name on the clock is rarely that of the maker but usually that of the retailer. In fact many of the London business houses objected to the maker's name or trade mark appearing anywhere on the clock — no doubt wishing to pass them off as one of their own products. Exceptions to this rule are the clocks bearing the name of 'James Condliff of Liverpool' (see Figures 51 and 52). Although later members of the family carried on manufacturing skeleton clocks they all retained the charateristic unique to James Condliff — a horizontal balance and helical hair-spring. His name is prominently displayed — usually on a small name plate fixed to the base of the clock. The example in the Liverpool City Museum has the additional information engraved on the rear of the base that 'This clock was made entirely by Thomas Condliff for the late James Condliff'. Not all these clocks are in museums and examples do appear on the market from time to time, at a high price!

As can be seen by the examples illustrated the designs for the plates fall into two categories: architectural or ornate scrollwork. The architectural designs were intended to portray in some instances specific buildings, cathedrals or monuments. Lichfield Cathedral with its spires, York Minster, Westminster Abbey, Brighton Pavilion and the Scott Memorial in Edinburgh being a few (see Figure

Courtesy of Norwich Union Insurance Group

Figure 35

A description of the Moore Clock is taken from the house magazine of the Norwich Union Insurance Group. "This clock was made for the Great Exhibition of 1851, and is one of the finest specimens of workmanship in existence. It chimes the quarters upon eight bells and strikes the hours on a ten inch hemispherical bell concealed in the base; it also discharges each hour a very finely constructed machine playing twelve operatic selections; all the pinions are finished in the same manner as the best astronomical regulators, and the greatest accuracy is displayed in the fashion of every other portion. The frames are very massive, and having been polished and lacquered, were afterwards enamelled by a process then only known to the makers of the clock. The escapement is a finely constructed dead beat, and the pendulum has a steel rod and polished steel plates under enamelled and gilt rosettes. The clock is supported upon four pedestals standing upon a polished steel plate, which rests upon a massive carved walnut stand. It was one of the clocks which gained for Messrs John Moore and Sons, a Medal for excellence, and by the most competent Judges was considered the best specimen in the Exhibition." Details of how this company came to acquire the clock will be found on pages 40-41.

46). The Scott Memorial design was exclusive to W.F. Evans of the Soho Factory, Birmingham. It is readily identifiable, even to those not familiar with the monument, by the gilt figures of Sir Walter Scott and his reclining dog at the base of the building. The story that only one was ever manufactured is completely fallacious. Admittedly only one was made originally and shown at the Crystal Palace Exhibition in 1851 and this is now in the City Museum and Art Gallery, Birmingham. However, many more were made in the ensuing years as it proved an extremely popular design especially in Scotland. Being identifiable with an interesting history these clocks tend to fetch a good price.

Normally the ornate frames holding the movements were cast and then hand finished, lacquered or frost gilded. The latter being a softer less brilliant finish. Designs to customers' personal specifications would be more likely to be hand pierced from the solid brass. It is impossible to describe how to differentiate between cast and pierced brass as it is a question of varying textures. It is necessary to be shown the two types. By the nineteenth century the brass manufacturers were expert at casting and so it is only rarely that a blow-hole can be seen in a piece of cast brass. It is equally unlikely, although not impossible, to be able to detect sawmarks on the scrollwork of an example that had been hand pierced, as these should have been removed in the finishing.

The bases were of marble (white marble usually), walnut, mahogany or pinewood painted black with a velvet covered centre. The marble bases being stepped to take the glass shade while many of the wooden bases had an incised groove into which the shade neatly fitted. The final touch to exclude dust and pollution was a ring of chenille encircling the base of the dome. Chenille is difficult to obtain now and the more prosaic cording for cushion covers is an admirable modern substitute. To achieve the best results measure the circumference of the base of dome and deduct an inch. Measure off the length of cord required but before cutting encircle the sections of cord where the cuts will be made with a piece

Courtesy of Keith Banham

Figure 36

An elegant French 'boudoir' skeleton timepiece of the first half of the nineteenth century. The large wheel is typically French, but it is unusual to find a French example with a fusee. This is a highly decorative piece and would demand an exceptionally high price for this reason alone.

£1,750+

Figure 37

The clock shown in this illustration is a later example with a very ornate frame and fretted dial; it has, however, full striking on a gong. It is an attractive example — having well crossed out wheels, domed collets to the two ratchet arbors, fusee and chain and stands on a pleasing base of white marble and mahogany inlaid with brass.

Two train movement commands a price of £600+

Figure 38

17ins. including base and dome. This example is an eight-day timepiece standing on a white slightly veined marble base with marble bun feet. The design of the frames indicates that it was made by Smith & Sons of Clerkenwell. It is a well proportioned clock with an exuberantly fretted dial, recoil escapement and conventional lenticular bob to the pendulum. This would have possibly been the style of the clock that Smith & Sons were selling in 1865 for £2. 10s. This figure would need to be greatly multiplied to reach today's market price!

£450+

of sticky tape. Then cut through the cord and tape at the appropriate places. If the tape has been bound sufficiently firmly this will ensure that the ends of the cord do not unravel. Next butt-joint the two cut ends with epoxy adhesive. The sticky tape can be removed when it is certain that the join is holding and the result should be a virtually invisible join. Crimson is inevitably the best choice of colour for the cord.

The problems of damaged or missing domes are not so easily solved. Repairs can be achieved but they are far from sightly. When it is recorded that W.E. Chance & Co. of Oldbury, offered in their 1856 catalogue some two hundred different sizes the unlikelihood of finding a replacement of the correct dimensions is fully realised. The shades from stuffed birds and dried flowers are usually too narrow from back to front and in any event too tall. It is possible to have domes cut down, but few firms will attempt the task. This is always undertaken at your own risk, as old glass is notoriously unpredictable and it is not unknown for it to shatter into a pile of fragments. Added problems are created by the fact that many of the domes are not oval, but shaped at the 'corners'. A price in the region of £80 to over £100 could be expected for a dome for a medium to large skeleton clock. Some collectors have resorted to having rectangular glass cases made with brass corner strips but these do not have the same appeal. This case style was fleetingly introduced by Smith & Sons of Clerkenwell which supposedly gives an excuse for the substitution.

The simple timepieces with anchor or deadbeat escapement were by far the most common, but many incorporating a striking train were also made. This ranged from an uncomplicated single blow on the hour with possibly another at the half-hour to those striking on two or more bells or even musical chimes. In English examples striking was always on a rack. Wire gongs appeared on later examples. The duration of time for which the clock ran ranged from one year, three months, one month to the more usual eight days. The dials were either solid chapter rings silvered or gilt with engraved or painted numerals. The more ornate fretted examples appeared later. Although these are attractive in their own right they do not make for easy reading of the time as can be seen from the examples in Figures 37 and 38.

Figure 39

An English year-going skeleton clock with two subsidiary dials. The left hand dial records the weeks of the year, while the right hand dial shows the date. The points to note are the exceptionally thick plates, large barrels and extra sturdy chain all of which are essential to withstand the tremendous torque of the very strong main spring used due to the long duration of the clock's running time. In contrast the wheel work is exceptionally fine and delicate. The rectangular brass base is engraved with the name 'B. Parker, Bury St. Edmunds'. These year clocks are extremely rare and fetch several thousands of pounds on the open market.

This example without subsidiary dials £2,000 depending upon quality of workmanship.

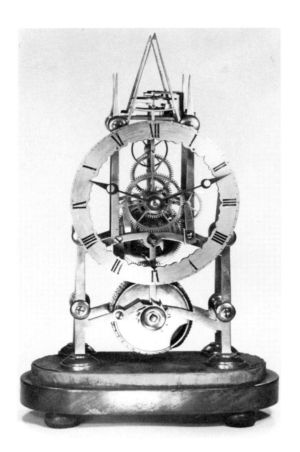

Figure 40a

This is an early example of a skeleton clock as is borne out both by the plain silvered dial and the dates of the retailer whose name appears engraved on the dial. Joseph Watson and Son, The Market Place, Cambridge, can be traced in Directories as having been in business between 1823 and 1853. It is an exceptional example in so far as it has a horizontal platform lever escapement. These more unusual escapements greatly enhance the price of these clocks and most certainly would not be found on any of the modern reproductions. The base is mahogany. The unusual escapement is the important feature of this clock.

£500+

Figure 40b

Close up of platform lever escapement.

Figure 40c

View of movement of skeleton clock by Joseph Watson and Son. Note fusee and fusee chain. In some instances these clocks originally had a gut or steel line but many that had chains have had it replaced by alternatives as chains are often difficult to obtain nowadays. It should be possible to detect a later substitution by noticing the shape of the groove in the fusee. For a gut or steel line it would be rounded whereas if intended for a chain it would be flat. The chains for these clocks as well as those for chronometers and verge watches were made in Christchurch, Hampshire partly as a cottage industry and partly in small factories by women and young girls.

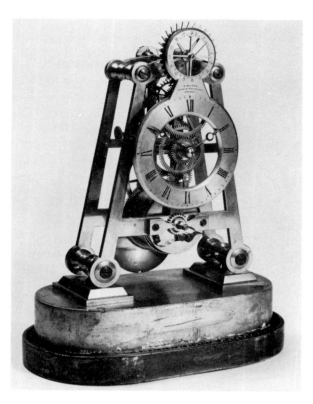

Figure 41

This small compact skeleton clock is on a gilt gesso and wood base, and has a gilt chapter ring and mount dial engraved with the legend 'C. Elisha, Duke Street, Piccadilly, London.' The sunburst effect around the month dial is due to the position chosen for the calendar wheel. There are two unusual features concerning this clock. The first being that it is an English skeleton clock with a going barrel. The omission of a fusee would imply a French origin, but upon studying the movement it can be seen that this had been omitted in order to plant an extra wheel and pinion to obtain a duration of a month. The style of the frame and general finish make it indisputably English. A further feature is the pendulum bob. In order to add to the effective length of the pendulum the bob has been half filled with lead shot. The alternative would have been a longer pendulum with a conventional fully-filled bob and the movement mounted on pillars to gain the required height.

The name of 'Calib Elisha' is that of a maker of some consequence. It has been possible, with the assistance of Mr. Clive Osborne, to ascertain that he was in business between 1823 and 1850 in the vicinity of Piccadilly, London. The fact that he was engaged in watchmaking prior to this is known from the watch that was hallmarked in 1821 bearing his name and the information that he was Watchmaker to the Duke of York. An Astronomical Regulator shown by him at the Great Exhibition, 1851, had a compensating pendulum ball (spherical bob) of his own invention. Other items displayed by him were a watch 'with a radii compensating apparatus' and a device using extremely sturdy chains and Bramah locks to constitute 'a safety door'. By now his advertisements carried the accolade "Watchmaker to His late Royal Highness, the Duke of York, Her late Majesty the Queen of Hanover and His Royal Highness, the Crown Prince of Hanover etc. etc." Additional details of the clock can be found in an article published in Antiquarian Horology *for December, 1972, and written by C.K. Aked and Dr. F.G.A. Shenton.*

If placed on the market the price of this clock would be greatly enhanced by the unusual technical features and the data that research has uncovered regarding the maker. Extremely difficult to evaluate as similar clocks do not appear on the market.

Possibly unique — £1,200+

A few years ago it was jokingly stated that the domes were worth more than the clocks, but this is no longer the case. Apart from the decorative appeal of skeleton clocks, they also embody a large number of interesting escapements and deserve a longer description than space allows here. One of the finest specialist books ever published was written by F.B. Royer-Collard and entitled *Skeleton Clocks*. A great deal of intensive research was carried out by this author and the book deals exhaustively with the many small variations of these clocks. The only information that can be added here are a few comments pertinent to a prospective purchaser of one of these clocks.

The problems surrounding replacement domes have already been discussed. As the ornate frames serve as the back and front plates of the movement, cleaning this is only made possible by completely dismantling the clock. Whether this can be successfully accomplished by an amateur depends to some extent upon the complexity of the example in question. Unless the frames are badly tarnished it is a matter of opinion whether there is any great merit in undertaking the task. If the dome is not tightly fitting and the brass not lacquered it is going to acquire a fresh patina in a relatively short period. As

Figure 42

18ins. high including base and dome. This is an eight-day timepiece with 'one at the hour' strike on a brass bell. Closer examination shows several features that make this clock a desirable piece to a collector. Firstly it has a dead beat escapement. Secondly the shape of the brass-covered bob is cylindrical as opposed to the more usual lenticular shape. Thirdly the seconds dial is not a very common feature on a skeleton clock. Lastly it has the curiosity of having the number VI incorrectly engraved (this has been reversed). A nicely proportioned example with some interesting technical aspects.

£500+

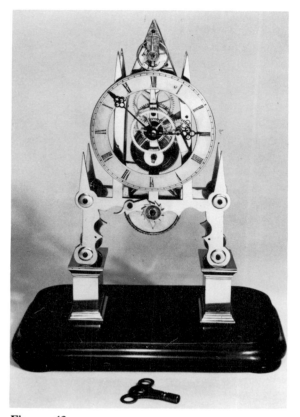

Figure 43a

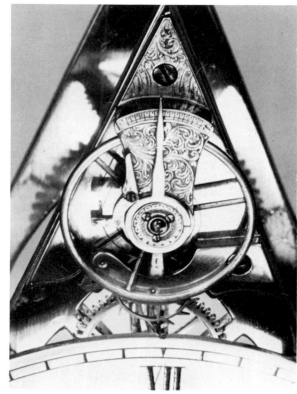

Figure 43b   *Close up of escapement.*

17ins. including base and dome. The highly commendable feature of this rather severe looking eight-day timepiece is the vertical English lever escapement. This is an extremely unusual escapement (see Figure 43b). Other worthy points are the fusee with chain, maintaining power, nicely finished wheels and rounded collets on the screws holding the pillars. The plinths are cast brass.

Unusual escapement would compensate for severity of design £850+

lacquering needs to be carried out under professional conditions to look well, it is not advisable for a home handyman to attempt this. However, a gentle rub with a coat of any silicone polish does delay discoloration.

As with all items that increase in value the point is reached where it is commercially viable to start 'reproducing' examples. This has occurred with the skeleton clock. The motion work of a basic skeleton clock is identical to that in a fusee movement in many a good quality English dial or three train board room clock, so it is not surprising that when the latter were not fetching a good price many felt it made more sense commercially to cannibalise them, have plates cast and reassemble them as skeleton clocks. After all, this is virtually what was happening at the cheaper end of the market at the turn of the century. A small jobbing clockmaker would purchase plates — possibly from Smiths of Clerkenwell — and a fusee movement and finish them to the best of his ability. To detect these 'conversions' can be difficult for the newcomer to the field of horology as it is only by having seen and studied the genuine pieces that the subtle differences in finish are noted. The finish of an old piece would have been achieved ardously by hand methods. Only in this way can the plates be nicely finished off with sharp edges polished along the greater dimension. Modern short cuts using buffing machines tend to blur and round the angles. If is often stated that the number of crossings out of the wheels given some indication of the authenticity of a clock, but this is an over-simplification. While it is true that there is more likelihood of a good quality three train movement having had many hours of work put into it which would include a high number of crossings out, a lower number do appear quite genuinely on the early simple examples.

It is the quality of the work not the quantity that is important. As can be seen in Figure 47 a good example has them slightly taper from the centre with all the angles sharp and the corners in no way rounded.

In 1973 a series of articles appeared in the *Horological Journal* entitled 'How to make a Skeleton Clock' by John Wilding. These were later published in book form. Alternatively a small publication entitled *Working Drawings for a Skeleton Clock* has been produced by Chronos Designs Ltd., in 1982. These publications led to a great deal of activity among the workshop enthusiasts as can still be seen by the fine displays each year at the Model Engineer Exhibitions. It is doubtful if any of these ever came on to the open market and in any event the total output from this source was minimal. However, several firms have produced castings of plates for these clocks (see Figures 55, 56 and 57) and if sufficient additional hand finishing is carried out the end result could be deceptive. It must be emphasised that the motivation behind the manufacture of these plates was above reproach. The trouble lies in the motives of some of the people into whose hands they have fallen. Apart from the English sources, some reproduction skeleton clocks have been imported from Spain. Those on the market several years ago had plastic bases, but those manufactured by Marton and Gain, S.A. of Spain were on wooden or onyx bases.

Two further examples of these clocks are to be found in Figures 381 and 430a. One is a French miniature skeleton alarm and the other an electrically rewound skeleton clock.

Figure 44

9½ins. high including base and dome. This is a small neat timepiece in a simple Gothic style, with a recoil escapement and fusee with steel line. As can be seen through the centre of the dial one wheel has not been crossed out. As the provenance of this clock is known it is possible to state that this is not due to any conversion but merely typical of a number of the early examples. Although a pleasing clock it was one of the more basic types without a great deal of embellishment, as can be seen by the fact that there are only four crossings out in the wheels and the lack of collets to the screws holding the pillars.

£350+

Figure 45

English striking skeleton clock in the form of Westminster Abbey. Note the unusual chapter ring; high number of fine crossings out on the wheels, and the shape of the bell and pull repeat cord. A similar ornate example with a three train movement plus an unusual escapement made by an important maker was realising about five years ago an excess of £4,000, now the price has stabilised to nearer £3,000.

Figure 46

Scott Memorial skeleton timepiece on grey veined marble base with the shaped circular chapter ring pierced with quatrefoils. The movement has a chain fusee. The most interesting feature is the vertical pointed tooth lever escapement.

Interesting escapement should give this a value of £750+

Courtesy of Sotheby's

Figure 47

This example is by Liverpool maker Edward Evans, the design of which is in the French style. It has a going barrel (not fusee), a great wheel and the gong contained in the oval mahogany base. In March 1979 it realised £650 in the auction rooms. A remarkably similar clock has been recorded with the name plate engraved "Mudge and Dutton, London". A very doubtful origin.

£750+

Figure 48

An elaborate skeleton timepiece designed to take full advantage of the most important characteristic of skeleton clocks i.e. full visibility of the complicated mechanical detail. Several variations of this design have been seen. Maker is usually anonymous but the design is felt to be of French origin. This example has anchor escapement with micromotion beatsetting. Surmounted by a gilt eagle the semi-circular bar carries on its extremities two small pulleys over which run the chains bearing the weights. One weight drives the movement whilst the other acts as counterpoise. The smaller bullet shape weight provides maintaining power while the clock is being wound. A highly decorative and interesting piece. Realised £1,900 when sold in 1981.

£2,200+

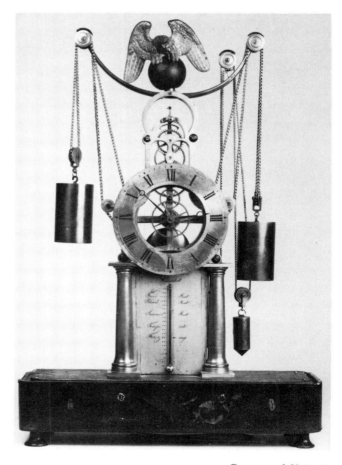

Courtesy of Christie's

Courtesy of Christie's

Courtesy of Sotheby's, Chester

Figure 49

William Strutt, FRS was a brilliant mathematician and engineer. Some time during the early 1800s he designed a skeleton clock movement with epicyclic gearing based on the sun and planet wheels used in steam engines. Unfortunately their production proved to be an uncommercial proposition and only twenty clocks were made. The first examples were signed W. Wigston, Derby (the manufacturer) and underneath Wm Strutt Esqr Invt while others made later carried just the signature Wm Strutt. Strutt died in 1830 and Wigston five years later. About 1850 further examples appeared on the market. They were manufactured by D. Bagshaw of London. The design was virtually the same apart from a degree of opening out of the frame, a skeletonised minute ring, the use of Roman numerals and a marble base. A few years ago a limited number of reproductions made to the Strutt design appeared on the market.

The example shown here is signed Wm Strutt, Derby. Example sold in 1992 for £4,250

Figure 50

A two train chain fusee skeleton clock striking on a gong, with half hour striking on an overhead bell dating from the second quarter of the 19th century. The pierced fretted silvered chapter ring is highly decorative but not the easiest to read. Only too often the elaborate design of these clocks tends to overlook the fact that they were intended as 'time tellers'.

Estimated at £600 to £900 this clock actually sold for £1,045 in 1982.

£900+

Figure 51

A particularly fine example of a skeleton clock by Jas Condliff, Liverpool. Although several generations of this well known family were responsible for the manufacture of clocks of a individualistic character they nearly all retained the helical hair spring and horizontal balance at first used by James Condliff.

These are both decorative and technically interesting clocks and it is not surprising that this clock realised £3,800 when offered at auction in February, 1980. Certainly would not have fallen in value

Figure 52

A further example of the work of James Condliff, Liverpool dated c.1860. It is described in detail in Skeleton Clocks *by Royer-Collard. As the author states this version is a two train quarter striking skeleton clock striking hours and quarters on three gongs. The gongs are typically concealed in the wooden base. There is a remarkable similarity between the frames of this clock and that shown in Figure 51 . Likewise the base, feet style, etc. The dial however is of glass within a brass bezel. The outer chapter ring has Roman gilt numerals on a black background. The inner ring is white with the Arabic figures continuing to complete the 24 hours. The centre is clear to allow the motion work, etc., to be fully visible. The clock has the added features of a seconds hand and maintaining power with a duration of 8 days. A further example of a Condliff skeleton clock was on view at the City of Liverpool Museum.*

Well over £2,500

Figures 53a and 53b

19ins. high including dome and base. The name 'F. Butt — Chester' appears on the small plate screwed below the barrel. This retailer has been traced as 'Frans Butt,' a retail jeweller and watchmaker who was in business between 1880 and 1887 at 32 Eastgate, Chester. The design of the frames and the fretted, silvered dial is compatible with these dates. The movement is an eight-day timepiece with recoil escapement and 'one at the hour' striking on a brass bell. Note the shape of the hammer — these hammers were often in the image of a halberd, flower, bell or even human hand.

As can be seen in Figure 53b, the lower half of the pendulum rod is wood; this would have been intended to overcome the expansion and contraction of the more conventional brass rod. The cylindrical zinc bob is unusual — they are usually lenticular.

Complete with original dome, striking movement, and a traceable retailer it would be difficult to find a similar example under £350.   £450+

Figure 54

The castings, dials hands, etc., shown in this illustration are a selection of those manufactured by a firm in Gloucestershire. In the early 1970s they supplied the entire kits including a single train fusee movement, base and dome.

Courtesy of Stow Antiques

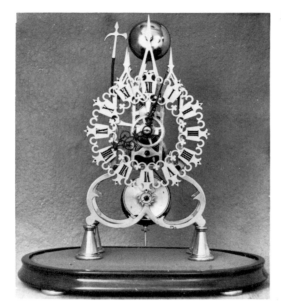

Courtesy of Stow Antiques

56

57

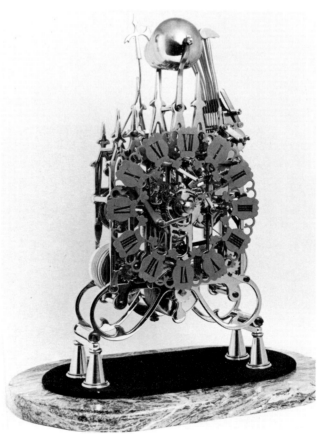

Courtesy of Classic Clocks

Figures 55, 56 and 57

These three clocks are examples of modern reproduction skeleton clocks. The first two have eight-day movements, with striking 'one at the hour' on a brass bell. However, comparison between the crossed out wheels on the two clocks exemplifies the desirability of this being light and fine. The example in Figure 55 is of an average standard, whilst that in Figure 56 is rather heavy with wide crossings. Both examples have a wheel that is not crossed out.

The third example Figure 57 is a quarter chime triple fusee movement built from a Classic Clocks kit. Two tunes are played on 8 bells.

Although all of these clocks are attractive and have merit it is worth remembering that such reproductions are available on the market.

If original £200+

CHAPTER IV

Torsion Clocks

These clocks can be referred to as 'Torsion Clocks', '400 day Clocks' or 'Anniversary Clocks'. The first name alludes to the twisting motion of the pendulum bob, while the second indicates the duration for which most models run, with the last drawing attention to the use to which they are often put — namely presentations upon birthdays or other anniversaries. The last name originated in America but is becoming widespread in the jewellers' advertisements for modern examples.

Mention is made by a watchmaker Robert Leslie 'of Merlin Place in the Parish of Saint James, Clerkenwell in the County of Middlesex' in his patent of 1793 which, among some twelve improvements in 'Clocks, Watches and other Time-keepers used either at sea or on land', mentions a pendulum with 'a horizontal circular...motion'. This is possibly the first mention of a torsion pendulum, but it is doubtful if the idea ever left the drawing board. The first torsion pendulum clock produced commercially was that patented in 1841 by Aaron Dodd Crane (1804-60) of New Jersey, U.S.A. His first patent of 1829 is basically for a clock with only two wheels and the phrasing of the reference to the pendulum, although suggestive of a torsion pendulum, is somewhat ambiguously worded. The 1841 patent, however, definitely was for a torsion pendulum clock — one that was intended to run for a year before needing to be rewound. This was manufactured and marketed by The Year Clock Company of New York, with later models being introduced which ran for eight days or a month. Examples are extant today, but it is doubtful if they would be found outside the United States. The cases were those of typical ogee shelf clocks of the period, and the pendulum 'bob' comprised one or more balls depending upon the model. A very few striking models were also manufactured.

Another type it would be unlikely to find outside the United States but which should be mentioned is the torsion pendulum marine clock patented by Samuel B. Terry of Plymouth, Connecticut in 1852. This appears in a number of cases typical of that period. The term 'marine' is used in America to mean a clock with a balance rather than our definition which infers a chronometer for use at sea. It could be said that the electric clock described and illustrated in Figure 414, the mystery clock described and illustrated in Figure 266 as well as several others with similar mechanisms, could be referred to as 'torsion clocks'. While this is correct as a technical description, most horologists intend the term to apply to the clocks described in this Chapter.

Lorenz Jehlin of Sackingen, Germany took out the first German patent in 1877 in which the application of a torsion spring to a clock was described. Upon his death in 1879 Anton Harder who had been working with him for several years took over the patent rights and in 1880 applied for a patent (Austria/Hungary) for the torsion pendulum as we know it today – a movement with a rotating pendulum usually on a circular base under a glass dome. Cast styles, however, do vary and four-glass cases both simple and highly ornate have been seen, and there is one model of a lighthouse with a torsion pendulum movement. The clocks were initially produced by the Willman Company, Gustav Becker and the Fortuna Clock Company. It would appear that these first models had some minor technical or

Figure 58

The circular wooden base of this clock has a velvet covered centre, with only a brass trim round the centre and exterior of the base which suggests a date prior to 1900. The dome rests in the deep groove. The dial is white enamel with the words 'R.L. Patent 2182, U.S. Patent 269052, DR Patent 2437' encircling the centre. The U.S. Patent number is for the patent taken out by Aaron Harder in 1882. This was made by Jahresuhren-Fabrik. The other points to note are the fact that the bob is 'pinned' to the suspension spring and not hooked, and that the pendulum bob itself is of the earlier design without a crown. Although this model still appeared in the 1893 Jahresuhren-Fabrik catalogue as can be seen in Figure 59, this was one of their first products with the patent numbers prominently displayed to discourage competition. Later examples do not have these numbers.

In 1981 a limited number of replicas of this model were made to mark the centenary of the house of Schatz (Jahresuhren-Fabrik).

Early example £200+

production problems as production soon ceased. The early examples included models with verge or cylinder movements which further demonstrates that the experimental stages had not entirely been solved.

Jahresuhrenfabrik (Aug Schatz & Sohne) were the first successful manufacturers of the Harder torsion pendulum clocks, with production starting in mid-1881. August Schatz was born in 1854, trained in the Black Forest area of Germany and eventually in 1881 founded in conjunction with five other clockmakers the Wintermantel Company. This company changed its name in 1884 to Jahresuhren-Fabrik AG. After introducing mass production methods similar to those so successfully employed in the United States their range of clocks was extended to include wall regulators and alarms. By 1923 all of the original partners apart from August Schatz had dropped out of the firm and therefore when his two sons — August junior and Karl — entered the business the name of the company was changed once more — Jahresuhren-Fabrik GmbH Aug Schatz & Sohne, Triberg. In 1939 Remington Rand bought shares in the firm. During the Second World War their production of fuses kept the factory open and subsequently they were able to turn once more to the domestic market. Examples of early models can be seen in Figures 59 and 60. In 1981 Schatz, in order to mark their centenary, produced a limited edition of 3,000 pieces (£300) of an early model very similar to that seen in Figure 59. It is interesting to see from the reproduced catalogue pages on page 58 that striking examples were also introduced into the range. Whether these proved too costly or difficult to produce is not known but they disappeared from the range being offered at a later date, and few extant examples have been reported. It does appear,

Jahres-Uhr.

Höhe 300 mm = ⅓ natürlicher Grösse.

Fein poliertes, sichtbares Werk.

ı Jahr Gehwerk.

Miniatur.

Nr. 2.

400 day Clock with open movement polished brass. Height of Clock 8 inches Height af Shade and Stand 10½ inches Scale ¼ **Timepiece.**	Pendule à 400 jours mouvement visible laiton poli. Hauteur 300 mm. ¼ de la grandeur naturelle **sans sonnerie.**

Figure 59

A page from the 1893 catalogue of August Schatz showing the miniature timepiece manufactured by them at this date. Note height 8ins.

A missing or damaged dome would devalue the clock by £20+
£220+

Figure 60

A further page from the same catalogue of 1893 but showing a striking model. Striking examples are very rare and this must double the basic value of a similar non-striking example. Note this example is 12ins. high.

£200+ if with strike

Jahres-Uhr.

Höhe 400 mm - ¼ natürlicher Grösse.
Fein poliertes, sichtbares Werk.

ı Jahr Geh- und Schlagwerk.

Nr. 2a.

400 day clock with open movement polished brass Height of clock 12 inches Height of Shade and Stand 16 inches Scale ¼ **Striking.**	Pendule à 400 jours mouvement visible laiton poli Hauteur 400 mm ¼ de la grandeur naturelle **à sonnerie.**

however, that many more models were exported to the United States than this country so judgement on rarity could be inaccurate. The strike was either on a bell or gong.

Harder sold the patent rights to F.A.L. deGruyter in 1884. Also in 1884 deGruyter took out an English patent (Patent No. 3724 — 21 February, 1884) relating to a striking example that ran for 600 days and struck the hours and half hours. He allowed his patents to lapse in 1887 and other manufacturers quickly seized the opportunity to commence manufacturing these clocks. They were all German firms except for one French manufacturer Glaude Grivolas, who in 1907 took out patents in this country and his own. Whereas the weights for adjusting the timing of the clock are normally visible on the top of the pendulum disc, he envisaged containing them within the disc itself. The weights were adjusted by means of a key inserted through a small hole in the rim of the disc. Within the next few years he took our further patents replacing the adjustable weights with tubes of mercury. A further patent in 1907 made provision for the ready accessibility through the backplate to the escapement. Examples are illustrated in Figures 61a and 61b.

The list of German manufacturers prior to the First World War as listed by Charles Terwilliger were Lenzkircher, Friberger, Becker (they had reintroduced torsion clocks into their range after their first unsuccessful attempts), Kieninger & Obergfell, Wendes & Metzgar, Ideal, Hauck, Haas, Bauer, Schneckenburger, Wintermantel, Vossler, Junghans, Kienzle and a few others. See Figures 63 and 72 for examples emanating from the workshops of Hauck and Vossler.

The case style adopted by most manufacturers was that of a movement on a circular brass base covered with a glass dome, although some examples are recorded as having a four-glass case. Later catalogues of this century do depict examples in art nouveau and other cases, but it must be remembered that catalogues can also be listing styles they would be prepared to provide if requested, and not styles that were already in stock. During the 1920s and 1930s competition was extremely fierce and finances stretched and it would have been foolish to have wasted money and resources on making case styles that no one was going to buy. There was always a casemaker who would quickly make one up as and when required. This does not mean that the complete catalogues were an antiquated form of market research — the majority of the models were sitting on the shelves waiting to be purchased.

From the collector's point of view it is the clocks produced before the First World War that are the most desirable. It would be difficult to select any maker as having produced vastly superior clocks to his competitors as they were nearly all of a high standard. Those that were perhaps not quite so solidly constructed or well finished are so much in the minority that they are therefore rare and sought after for this very reason! The example illustrated in Figure 63 demonstrates this point. A further example would be the model with pin pallets and lantern pinions introduced by Kienzle around 1900 for a short period.

Few manufacturers stamped their names on the backplates. It is more likely to find serial numbers, patent numbers or possibly the name of the importer. One helpful book has been written by Charles Terwilliger entitled *The Horolovar Collection,* which is at present out of print but much of its historical content can be found in the 9th edition of the *400 day Clock Repair Guide* by the same author. By studying the patents taken out by the various manufacturers together with contemporary catalogues he has managed to identify many of the makers and their products. The vast majority of the patents taken out were concerned with either the problems arising from the effect on the pendulum spring by changes in temperature or methods of protecting it from damage during transit, etc. Several attempts were made to invent a temperature compensated pendulum (one example by Claude Grivolas has already been described) either mechanically or by altering the composition of the metal used for the spring. In 1904 a Frenchman, Eduard Guillaume, patented an alloy that was eventually marketed under the name of 'Elinvar', but although theoretically the solution some problems arose with its use in torsion clocks and so most of the springs manufactured prior to the First World War continued to be made of steel. By 1949 all springs were made of phosphor bronze which was later replaced by yet another alloy.

A guard in the form of a metal strip or a slotted tube running the length of the backplate to protect the spring was another innovation. That patented by Gustav Becker can be seen illustrated in Figure 65b. Other attempts to avoid kinking the spring when moving the clock involved changes in the design of the top suspension. Eventually a device using gimbals was introduced.

In the interests of accuracy it is possibly best to confine suggestions as to points to note when attempting to date one of these pre-1914 clocks to the following general observations. The earliest examples had a flat disc, with two adjusting discs mounted on the top surface, which was attached to

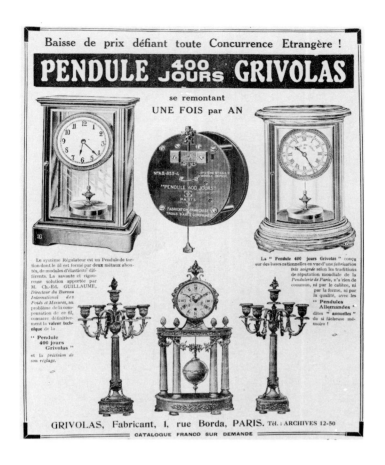

Figure 61a

This is an extremely interesting advertisement which appeared in 1924 to promote the torsion clocks manufactured in France by Claude Grivolas. The points to note from the three examples shown are:

1. The use of his patented pendulum bob with the adjusting discs mounted internally.
2. The four-glass and oval glass case styles although unusual for a German-made torsion clock were common in French examples.
3. The totally unconventional and distinctive style of the third example with matching candelabra.
4. The fact that the movement is 'hung' in the two top cases and not supported in the normal manner by at least two pillars. See Figure 61b for more comments on the movement.

It is unusual to find one of these clocks with a French movement and this, together with the fact that they are so well documented, would raise the price. The final price would be influenced by the case style.

£350+

Grivolas had taken out the following patents in this country:

Patent No. 26452	29th November 1907
Patent No. 26453	29th November 1907
Patent No. 17212	19th February 1909
Patent No. 299	5th January 1910

Figure 61b

This enlarged view of the movement patented and manufactured by Claude Grivolas of Paris could not be more informative! It confirms that these clocks were entirely of French manufacture, provides the trademark of Grivolas (a linked C and G) at this date, and clearly shows his patented method of easy access to the escapement to avoid having to dismantle the movement completely.

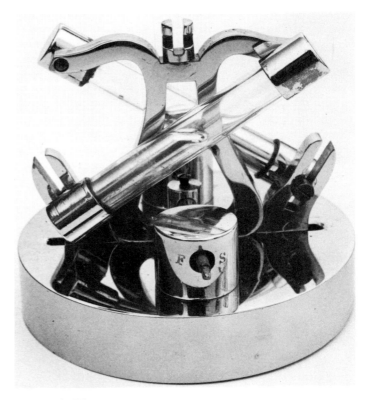

Figures 62a and 62b

This is an extremely rare and unusual example of a torsion pendulum clock with a unique method of temperature compensation i.e. crossed tubes containing mercury. It is not the design patented by Grivolas — the tubes in that particular model were contained within the pendulum bob. This model was illustrated in Deutsche Uhrmacher Zeitung for 1926 but gave no patentee or other details as to the inventor or manufacturer of clocks utilising this device. One other example has been recorded and this carried the Gustav Becker trademark and was dated as c.1908. Another reference attributes it to Andreas Huber who patented a twin loop temperature compensating pendulum in 1902.

**In view of its rarity and excellent working condition this particular example would
be of a value in excess of £400**

the lower end of the spring by a device with a pin. The base would be of soft wood with a separate spun brass outer cover and inner rim to define the margins of the glass dome. The centre would be covered with velvet. An example of one of these early models is illustrated in Figure 58. Within a few years, but still before the turn of the century the flat disc of the pendulum acquired further ornamentation in the form of a 'crown', and was attached to the suspension spring by means of a hook arrangement. The base now became the conventional circular spun brass base. An example of this type of clock is illustrated in Figures 64a, b and c. Within the first ten years of this century further innovations were adopted, some decorative and others to overcome inherent faults in the suspension spring. The four-ball pendulum of varying designs was introduced, as were various spring guards and it became common to find two circular apertures either side of the spring high up in the backplate in order to facilitate adjusting the pallets. This last feature can be seen in Figure 65b.

Naturally the First World War halted production and the exportation of these clocks from Germany, with only a relatively small number being manufactured between the Wars. The changes during this period were nearly all concentrated upon additional decorative features rather than any technical advancements. The illustrations in Figures 66 to 69 are from the 1936 catalogue of A. Schatz & Sohne of Triberg, Germany. The plates and bases of many of these later examples were thinner than their predecessors. Not all of the movements ran for a year — some of the cheaper examples only ran for a month. The term 'cheap' here refers to their original cost as they are now rare and sought after as far more of the true 400-day variety were manufactured.

After the Second World War several of the German factories retooled and recommenced production. After a period of tremendous sales, supply and demand settled into the steady pattern held to this day. The features found on these post-1953 models, that would not be seen on earlier more collectable examples, would be levelling screws to the base, a small spirit level on the base and a device to lock the pendulum to prevent it swinging during transit. Today only four manufacturers still make key wind versions although there are many models with quartz movements.

Although true of all repair work it is even more important not to commence 'adjusting' these particular clocks before fully understanding their mechanism otherwise a single fault can be rapidly turned into a multiplicity of shifting variables! Again Charles Terwilliger's book *The Horolovar 400-day Clock Repair Guide* is of tremendous assistance. One soon learns that it is vital that they are kept level. The next most common reason for their malfunction is that the spring has been damaged in some way. It is possible to obtain replacement springs — the best source being Southern Watch and Clock Supplies Ltd. of Orpington, Kent. Replacing domes is again a problem, but not such an overwhelming one as in the case of the skeleton clocks. A few of the standard sizes manufactured for the modern torsion clocks also adequately fit the older models. Sadly more and more of these are of plastic rather than glass, but it should still be possible to obtain the latter from firms in Clerkenwell, for instance.

Figure 63

Although the quality of the finish of the brasswork of this clock is thin and poor, it is a rare example made c.1905/6 by Franz Vossler. This manufacturer had ceased production by the beginning of the First World War. The points of interest about the clock are the fact that it has a front wind, only runs for thirty days, and has a silvered dial, in contrast to the more usual white enamel or porcelain dials.

The price in this instance is governed by the name of the manufacturer.
£200+

Figures 64a, 64b and 64c

A sturdily constructed and yet decorative example of a torsion clock with a porcelain dial and pie-crust bezel on a circular spun brass base. The date of this example would be early 1900s. The points to note in order to reach this conclusion are:

1. *Lack of spring guard.*
2. *Lack of viewing aperture.*
3. *'Crown' to the pendulum bob.*
4. *Hook attachment between the suspension spring and the bob.*

Good quality example £200+

64a

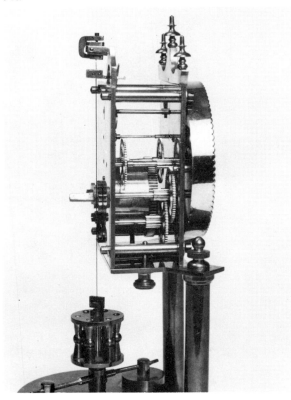

64b

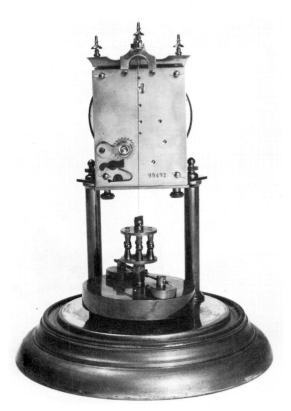

64c

65a

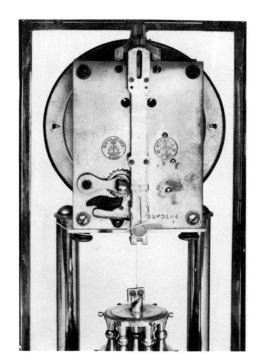

65b

Figures 65a and 65b

This clock was manufactured about 1905 by Gustav Becker of Silesia, Germany. Although he did produce a few clocks between the two Wars, his main period of productivity was prior to the First World War. Some further details concerning this maker will be found on page 211. His trademark of an anchor with GB can be seen on the backplate, together with the stamp referring to the fact that he was awarded a Medaille d'Or at the Schlesien Industrial Exhibition of 1852, which had no connection with the torsion clock, but was given for his work on regulators.

The suspension guard clearly seen on the rear view of the movement was introduced by this maker about 1905, as were the two small circular apertures either side of the suspension guard that were intended to facilitate adjusting the pallets.

A pleasant example by a documented maker.
£200+

Figure 66

16ins. Illustration taken from the 1936 trade catalogue of August Schatz of Triberg, Germany. Note the garland decoration to the dial and decorative rim to base. This model was offered with two finishes — either gilt with enamel dial or chrome with a silvered dial. To the modern eye a chrome finish is not as acceptable as gilt but this was a novelty at this date.

£150+

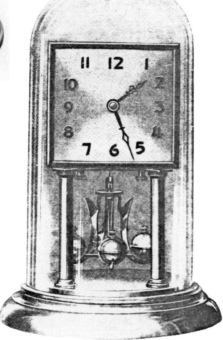

Figure 68

Height including shade 12½ins. The base and pendulum were chrome plated and the dial silvered with raised figures.

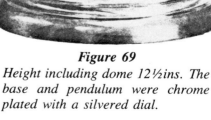

Figure 67

Height including dome 12½ins. In a gilt or chrome finish. Note hexagonal dial.

Courtesy of Jahresuhren Fabrik Aug. Schatz & Sohne

Figure 69

Height including dome 12½ins. The base and pendulum were chrome plated with a silvered dial.

70a

70b

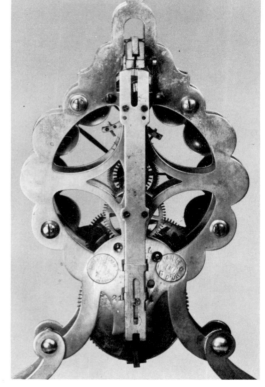

70c

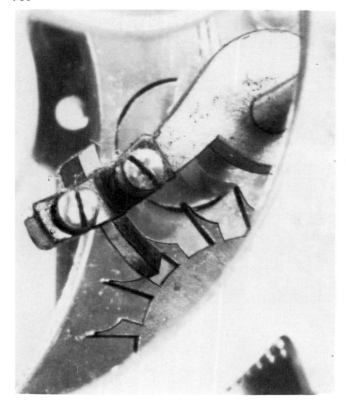

Figures 70a, 70b and 70c

13½ins. A rare 400 day torsion clock by Gustav Becker with skeletonised dial on oval mahogany stained base with original thick glass shade c.1905/6. Two models have been noted, the variations occuring in the chapter ring and pendulum bob design. An identical clock was being marketed by Hirst Bros, in 1910 at 56/- each. The manufacturer is given as USC. Note the Becker suspension spring guard and view of steel adjustable pallets. When a similar clock was offered for sale in the United States in 1981 it carried an asking price of £700.

£800+

71a

Figures 71a, 71b and 71c

Torsion clock by Gustav Becker again with suspension guard, pallet adjusting apertures and trademark of crown, anchor and initials GB.

£175+

71c

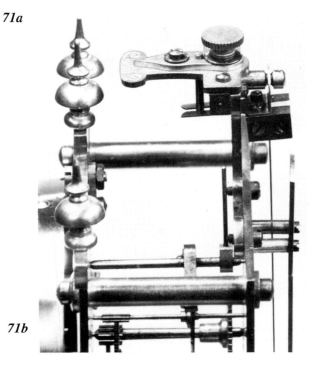

71b

Figure 71b

Side view of Becker overhead suspension saddle with special attachment for adjusting the pendulum.

Figure 72

Advertisement appearing in Watchmaker, Jeweler, Silversmith and Optician *for 1st October 1906 showing a Half Year Clock (200 days) by Ph. Hauck. They claimed that being of shorter duration more power was possible, irregularities caused by hardening of oil were overcome by more frequent winding and that the compensated pendulum overcame temperature variations. In other words a very superior clock according to their British Agent S. Moser & Co.*

£300+

CHAPTER V

Carriage Clocks

It would be presumptuous to attempt to add any information to that already copiously recorded in the book *Carriage Clocks* by Charles Allix, but for the more general reader a few interesting facts can be recounted here.

The imagination of the general public, rather than that of the horologist, has been caught by these clocks and the prices in many instances have risen out of all proportion to their true worth. Fine and complicated examples by French makers such as Bréguet and Garnier (see Figure 74), or the English makers McCabe or Dent, naturally command and deserve a high price but many of the late 'bread and butter' timepieces made in the early years of this century have no great horological merit. In many instances the intrinsic value lies in the ornate case only.

The origin of these clocks stems from the *pendule de voyage* made by Abraham-Louis Bréguet (1747-1823) in the first decade of the nineteenth century. They were masterpieces of design and craftsmanship with many complicated mechanisms. Later Paul Garnier (1801-69), with the introduction of a simple basic design using his escapement, made it possible for these clocks to be produced more cheaply. He is generally regarded as 'the creator of the Parisian carriage clock industry' — an industry that existed into the 1900s. Strangely enough few of these clocks were sold in France but thousands were exported to England.

Between approximately 1826 and 1845 the framework of the cases was pinned and brazed, with a sliding rear door, but after this the parts were screwed together thus allowing for a greater degree of variation in the case designs. To facilitate the mass production of these clocks the manufacturers tended to keep to a few standard sizes: 'Mignonnettes' below 4¼ins. with the handle raised; 'Full size' between 5½ins. and 9ins. with the handle raised, and 'Giant' over 9ins. with the handle raised. Decoration of the cases embodied many arts, with ornate pillars of cast caryatids, or brass with cloisonné or enamel work, etc., while the side panels were frequently finely pierced or given enamelled designs on porcelain. The presence of a pair of fine Limoges panels naturally escalates the price of the clock dramatically. Designs were obviously influenced by prevailing fashions in other fields, and these can be used as a very rough yardstick in dating a particular example. The oriental influence is noticed in the 1905 to 1910 era while designs characteristic of the art nouveau style soon followed.

Although the clocks were finished and cased in Paris, virtually all the movements (*blancs roulants*) came from either Saint Nicolas d'Alierment near Dieppe or the Jura region of Franche Comté, the main source being the factory run by the Japy Fréres. The escapements were obtained from specialist manufacturers working along the French-Swiss frontiers, with the mainsprings coming from yet another source. Thus, while using standard parts, a wide diversity of clocks could be produced with variation in type of escapement, with or without strike or complicated calendar work, etc. Several alternative strikes were employed which could be on a bell or gong, the use of a bell predating that of a gong with

Lever escapement as found on early carriage clocks. Note polished steel two armed brass and steel half cut balance wheel — it is extremely rare to find a fully cut balance wheel in a carriage clock e.g. one that provides effective temperature compensation. The ratchet tooth escape wheel is of brass. General points denoting quality are the hand engraved Fast/Slow scale, and overall finish to shaping of lower edge of platform and pallet.

Figure 73b

Original cylinder platform escapement with brass balance wheel.

Figure 73c

Modern typical Swiss platform escapement for use in carriage clocks. Note general austere appearance with scape wheel, pallets and balance in straight line, simplification of crossings to balance, indicator etc. The jewels would be synthetic and pushed in.

Figure 73d

Later lever platform escapement as found on carriage clocks with club foot steel escape wheel, and plain balance i.e. uncut and non functional. The jewels — possibly garnets — are set in.

the changeover period occurring in the mid-nineteenth century. The variations are as follows:

Plain Strike:	On the hours and half-hours, with or without the repetition of the hour by single blows.
Petite Sonnerie:	This sounds on two bells or gongs usually of differing tones with the result that this is referred to as ting-tang. At the quarter past the hour one ting-tang (two blows) is struck. At the half-hour two ting-tangs (four blows) are struck. At the three-quarters past the hour three ting-tangs (six blows) are struck. The hour is struck normally on single blows.
Grande Sonnerie:	This time the preceding hour is struck at each quarter by single blows as well as the appropriate number of ting-tangs.

Minute Repeater:	The striking is similar to a *Grande Sonnerie* but with the additional feature of having the number of minutes past the last quarter sounded as well. The time of 3.50 would sound thus:
	   Three blows (three o'clock)
	   Three ting-tangs (three quarters past)
	   Five single blows (five minutes past the three-quarters)

The best makers signed their names and it is needless to point out that these pieces fetch hundreds of pounds in the salesrooms. Some of the earliest makers were Garnier, Lépine, Bolviller and some of the later Jacot, Drocourt and Margaine.

In the event of the erroneous impression having been given that all carriage clocks were made in France, mention must be made of the English carriage clocks. They are, horologically speaking, important clocks made by important makers and often those examples appearing on the market are well documented from workshop records which is always an added advantage. They tended to be larger and heavier than their French counterparts and made use of a fusee and chain. Early examples, by such makers as J.F. Cole or Vulliamy, had lever escapements but later makers of the second half of the nineteenth century, such as Frodsham and Dent, tended to use a chronometer escapement. Although still referred to as 'carriage clocks' this escapement did not readily lend itself to safe transportation. These clocks may have been used more often than not as mantel clocks which perhaps explains the mahogany or rosewood cases that are often found in place of the more usual brass. The English carriage clocks were never intended for other than the extremely wealthy, and although many manufacturers shared a common source for the rough movements, escapements, etc., as did the French makers, each evolved his own style of finishing which was superbly carried out.

The manufacture of French carriage clocks declined after the turn of the century and those that were produced in the first years of the 1900s, before production finally ceased, were of rather poor quality — thin plates, poor crossing out of wheels, very plain mass produced cases, etc.

English carriage clocks of a continuing high standard are produced to this day. One firm so doing being Thomas Mercer Ltd. of St. Albans now at Cheltenham, Glos. Early in the 1970s there began to appear reproductions of carriage clocks in the French style (without fusee) but made in England. The only point in mentioning them here is to give a word of caution. They are extremely well-made and desirable clocks, just so long as the purchaser realises their lack of years! Unfortunately a few, a very few, are being 'aged' rapidly. It is obviously financially worth the time and trouble to work on one of the small miniature reproductions now available to make it look a great deal older and then sell it for two or three times the intended price. It is difficult enough for an experienced eye, let along an inexperienced one, to detect any tell-tale signs. One of the only safeguards is to purchase from a reputable dealer who is jealous of his good name.

Another point to note is whether the escapement has been changed. The majority of the mass produced French timepiece carriage clocks utilised a cylinder escapement. Although this escapement appears more complex, it is in fact easier to mass produce than a good quality lever escapement. It does not, however, rival the good timekeeping qualities possessed by a standard lever escapement. It was not surprising, therefore, that as the cylinder became worn or required servicing, many were replaced by a new platform lever escapement. In assessing any purchase it is important to differentiate between this replacement

74a

74b

Figures 74a and 74b

6ins. high with handle raised. This is a beautiful clock made by an important French maker — Paul Garnier, 1801-69. Note the one-piece case, the white enamel dial panel with chapter ring, alarm dial and inscription 'Paul Garnier H^{er} Du Roi'. Points to note from the rear view are the shutter on the door to exclude dust from entering the movement through the winding holes, the bell and the matching numbers on the case, movement and door (No. 2562). Striking is on the hour and half hour. The escapement is the original (chaff-cutter escapement) as invented by Paul Garnier and patented by him in 1830.

£1,300+

lever escapement and an original quality lever escapement that may have been original. An original lever escapement would command a higher price than an original cylinder escapement, whereas a replacement lever escapement would be lower. It is usually quite easy to note a replacement as the quality of the conversion work is not up to the standard of the rest of the movement. Frequently the words 'Swiss' or 'Swiss Made' appear on a replacement. (See Figures 73a, b, c and d).

By virtue of their purchase price carriage clocks are not suitable clocks for an amateur to contemplate repairing. The only exception could be the replacement of a cracked or broken glass panel in a standard case that can be easily dismantled (i.e. screwed, not brazed at the joints). As the thickness of the glass varies it is as well to take the damaged panel when ordering the replacement.

It is hardly necessary to indicate what is a desirable carriage clock — but to reiterate some of the points already covered and to add others — ornamentation of case whether it be engraved, pierced or enamelled panels or just an attractive well finished example obviously influences value. Miniature highly decorative cased examples realise a very high price.

Originality of escapement whether it be a lever or cylinder is important while an unusual escapement such as a Garnier chaff cutter would be highly desirable. Complexity of movement, i.e. strike, petite or grande sonnerie, minute repeater alarm or calendar is again a plus point. It is always preferable to have an identifiable maker or even better one that is known to have been a reputable maker of high class carriage clocks!

English examples with fusee and chain are sought after and consequently realise a high price. It is, however, essential to be selective as simple timepieces in standard plain cases or damaged examples are not so highly valued and indeed can be difficult to sell. A cracked enamel or porcelain dial or panel is a great disadvantage as neither a modern replacement nor a repair are truly acceptable substitutes to a connoisseur.

Figure 76a

There is no record of Richard Watson, King Street, Cheapside, London in the standard reference works but this is a good example of an English carriage timepiece with lever escapement.

£900+

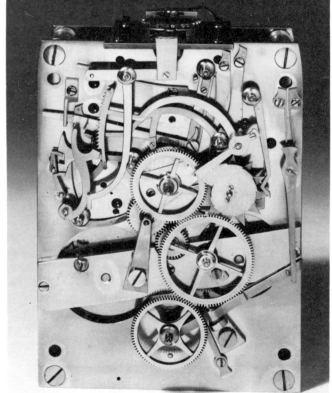

Figure 76b

View of movement.

Figure 77

Edward White, 20 Cockspur Street, Pall Mall, London, exhibited at the International Exhibition, 1862 a display of chronometers, watches, clocks and gold chains. It is highly likely that this carriage clock came from his premises. Sold in 1981 for £3,800 is is an extremely acceptable example of an English repeating carriage clock. The white enamel dial was signed E. White, London, 635. The date given was c.1850

£4,000+

Figures 78a and 78b

This is a good example of a typical English carriage clock. The points to note are the enamel dial, and engraved dial surround. It is doubtful if the carrying handle is original. The door is opened by means of a small lever in the base of the case. The side view of the movement in Figure 78b gives a clear view of the fusee and chain.

As shown here £800+

Courtesy of Bonhams

Figure 79

This example has its original carrying case and key. Note the removable slide to the front of the case which may be stored at the rear of the case when the clock is in temporary use. Provision for storing the key has been made in the framework of the case below the hinge to the lid. The double ended key is sized one end to fit the winding arbor and the other for the hand set. The case is exceptionally elegant with decorative pillars, finials and filigree work depicting two dragons and foliage. The dial is of ivory porcelain with a gilt centre trim. The movement is that of a plain timepiece with its original cylinder escapement. An almost identical clock was selling in 1900 for £2.2s.

£300+ if with strike further £100 — £150

Figures 80a, 80b, 80c, 80d and 80e

An excellent set of illustrations of a grande sonnerie carriage clock with repeat and alarm made by Drocourt. Figure 80a is a general view.

Figure 80b shows the backplate with winding arbors, going train, strike and alarm, and the setting for the hands and alarm. The two silvered bells are one within the other but it is possible to see the two hammers which strike the ting-tang on the right and the hour striking hammer on the left. The trade mark of Drocourt (the letters D and C with a carriage clock between them) is above the left winding arbor.

Figure 80c shows the base of the clock with the strike, silent and grande sonnerie levers. The serial number of the movement (6615) can be seen. The name Darley is that of the retailer.

Figure 80d shows the barrel for the going train with the smaller barrel for the alarm appearing above.

Figure 80e shows the barrel for the strike work — this is nearly always larger than that for the going train. It is just possible to see the fine wheelwork and pinions above.

£1,950+

Figure 80a

Figure 80b

Figure 80d

Figure 80c

Figure 80e

Figure 81

8ins. high with handle raised. Although somewhat similar in outward appearance to the clock in Figure 79, this example has a repeater movement (strikes the hour and half-hours and repeats the hour and five minute intervals at will), and an original lever escapement. The pierced panel dial has a motif of birds and strapwork with an enamel hour circle and recessed gilt centre. The movement is numbered 3509 and was made by Gay, Lamaille & Co. who are mentioned as working in Paris and London in 1880.

With mentioned complications £900+

Figure 82

5ins. with handle raised. This is an extremely unusual example and provides a fine debatable point. The carriage clock is frequently defined as a small portable clock with a horizontal platform escapement. In this instance, however, there is in a typical carriage clock case a movement with a vertical balance wheel escapement mounted on the backplate. There is no doubt that it is an identical movement to that of the VAP drum clock depicted in Figure 198b.

While this example cannot, by any stretch of the imagination, be regarded as a high quality timepiece it is substantial and well made. Moreover it is of great interest to a collector by virtue of its unusual escapement. Very few other examples have been seen.

£150+

Figure 83

This appears to be yet another simple 'bread and butter' carriage clock but upon closer examination some interesting details appear, which enhance its value. Firstly, it is complete with its maroon carrying case and key and the movement has retained its original escapement. Secondly, and most important, it is possible to identify the 'Lion' trade mark stamped on the backplate. Albert Villon opened a factory in Saint Nicolas d'Aliermont in 1867 which later became Duverdrey & Bloquel. From 1910 they produced a wide range of finished carriage clocks which are distinguishable by this trade mark. Today the factory manufactures alarm clocks but does produce a few carriage clocks signing these on the backplate and back door 'Duverdrey & Bloquel, France'. Without this documentation and original case and key this type of late clock would hardly qualify as a collector's piece.

£250+ with case

Figure 84

An English carriage clock made by Jump of London in a 'humpbacked' case style that originated with Bréguet. The silver case is hallmarked and so it is possible to date the clock. Jump of Mount Street was in business from the mid-nineteenth century to early twentieth, the life of the business spanning two generations. This particular example has an aperture showing phases of the moon and subsidiary seconds dial. Other examples have been noted with perpetual calendar mechanism. A choice collector's item.

£2,000

Courtesy of Keith Banham

85b

85a

Figures 85a and 85b

6ins. with handles raised. The carriage clock on the left hand side of the illustration (Figure 85a) is in a one-piece case which indicates a date prior to 1850. This is commensurate with the fact that striking is on a bell and not a gong. There are two marks stamped on the backplate; one of the maker (Japy Fréres) and the other the retailer (Henri Marc of Paris). Apparently Japy Fréres not only supplied blancs roulants *to the rest of the trade but also sold complete carriage clocks. The movement is numbered 37703.*

The example on the right (Figure 85b) is included to provide an example striking on a gong instead of a bell.

£450+

85c

Figures 85c and 85d

Rear views of the two carriage clocks in Figures 85a and 85b.

85d

Figure 86

7½ins. high with handle raised. It is hardly surprising to note from the inscription at the base of the clock that it was a presentation clock. It is a handsome example. W.J. Walsham, (1847-1904), to whom it was presented, was a surgeon of some note at St. Bartholomew's Hospital — a fact which adds interest to the clock. The gilt brass case is solidly constructed and has two fine hand-painted side panels — both signed. One has a lady stooping to remove her overshoes before entering the house, while the other shows her, still in her outdoor apparel, playing with a pet bird. The backplate carries the stamp GL with the words 'Patent Surety-Roller'. GL indicates that the clock was made by E.G. Lamaille of London and Paris. Mention is made of a Gay, Lamaille & Co. of London and Paris in 1880. The movement is French, has its original lever escapement and striking is on the hours and half-hours with a repeater mechanism.

As illustrated £1,400+

Figure 87

Not surprisingly this is a particularly valuable carriage clock (estimated at £4,000—£4,500) with an alarm of the 'giant' size. 9ins. high. The central dial shows hours and minutes with the three upper dials showing months, seconds and day of the month. The dial between the mercury thermometer and flyback weekday indicator is for the alarm. The three train movement has a lever platform with the striking work planted on the back plate. Strike is on a gong and three further gongs in the base.

£6,000+

Courtesy of Christie's

Figures 88a and 88b

A most unusual carriage clock with mechanically separate musical box in the base connected to the dancing tight-rope walker contained in the glazed top tier. Three tune titles are engraved underneath with operational lever. The movement is stamped Japy Frères with lever platform, strike, repeat and alarm on bell. Total height 11¾ins. Estimated value in 1980 was between £3,000 to £4,000.

Similar example sold at Christie's for £10,846

Figure 89

9¼ins. high. This carriage clock is signed Sewill, Maker to the Royal Navy, Liverpool, Glasgow and London. It has a substantial movement with Earnshaw's spring detent platform escapement with split balance, blued helical spring with terminal curves, high count train, fusee and chain, thick mottled plates and double screwed pillars. Not surprisingly its estimated value in 1980 was £3,500 to £4,500.

£5,500+

Courtesy of Christie's

CHAPTER VI

Longcase Clocks

The popularity of the longcase clock was at its peak during the eighteenth century, but the demand began to decline from the beginning of the nineteenth century. By the mid-1800s the dominance of the London maker was waning but provincial makers continued to produce these clocks for another twenty to thirty years. It is among these later provincial clocks that the collector will find sensibly priced examples of interest. The cases were made locally and reflected their regional styles, and were therefore of a more individual style than many of their London counterparts. Quality of case varied according to the skill of the craftsman. It would appear that he could be either the local carpenter or cabinet maker as chance had it. The cases of provincial clocks tended to be larger — both taller and wider — than those made in London and were later often of mixed woods or mahogany. Figure 90 illustrates the more individual approach often found in these cases.

Examples with brass dials are exceedingly uncommon. By now the white painted dial — often with gaily painted scenes and corners — had made its appearance. Various reasons are given for this change. Some say that it was enforced by the high price of the brass and further skill needed to work it, whereas others say that the white painted dial was easier to read. Opinion varies according to whether the answer is coming from a northern or southern source! Most of these dials were supplied as were the movements from the clockmaking centres that had evolved, e.g. Birmingham. The names on the dial are rarely those of the makers and more likely to be those of the retailers. It is possible to find stamped on the dial plates, and possibly the movements, the initials or name of the true maker. However, it is necessary to dismantle the clock to find these and this is not always possible before purchasing the clock. Mention is made of these Birmingham makers, together with a great deal of additional information on these clocks in the book *The White Dial Clock* by Brian Loomes. The names on the dial can be checked in *Clock and Watchmakers of the World,* Volumes I and II, but it has not been possible for the compilers to confirm in all instances whether the names are those of makers or retailers.

At the date in question few clockmakers made their own movements. The majority as already indicated came from Birmingham, Coventry, etc., where they were produced on semi-factory lines with a certain amount of standardisation of parts. This is one reason why these clocks have, until recently when the supply of older longcase clocks outstripped the demand, found little favour with the collector. Often the movements were thirty-hour only, although obviously more expensive eight-day movements were necessary for the more wealthy customers. Calendar work is found on a fair number, but far more show the phases of the moon either in the dial arch or through a small aperture in the dial. To the countryman it was far more important to be able to foretell how dark the night was going to be before he ventured forth than which day of the week or month. In seafaring areas times of high tide are commonly found.

Not all the movements were mass produced; as late as 1880 or 1890 there were a few, a very few, genuine clockmakers scattered through the countryside, and in the preceding years there had been several well known clockmaking families whose work is now much sought after. They had frequently

Figure 90

8ft. 1in. high. The case of this clock is typical of one of the styles to be found on North Country clocks. The case is oak inlaid with fruitwood and ebony and whether or not the overall result pleases it must be admitted that a high standard of craftsmanship went into the making of these cases. The dial is signed 'R. Snow Padside' with a seconds dial and a revolving day/night dial in the arch. A Snow family of Padside have been mentioned in Country Clocks and their London Origins *and* The White Dial Clock *both written by Brian Loomes. Apparently they were a farming family (father and three sons) who carried on clockmaking as a secondary winter occupation. This clock style is rather late to have been made by one of these Snows, as is the design of the dial! The overall size of this clock would be a deterrent to most collectors in this country although the Dutch and German market would be more receptive. This clock realised £480 in the saleroom early in 1977.*

Doubtful if any significant price increase

Courtesy of Sotheby's Belgravia

originated from clockmakers who had left the big cities through religious or other personal reasons and set up working in areas where these did not apply. Many of these clockmakers had Quaker origins or were refugees from other countries, especially from the Continent following the unrest of the 1848 period. The best sources of information on these makers are to be found in the books devoted to a specific region and its clockmakers. A list of these will be found in the Bibliography.

When contemplating a purchase it is wise to bear in mind the possibility of the movement and case being a 'marriage'. As previously commented this is acceptable so long as the movement and case are contemporary and neither have had to be adapted to accommodate each other. Telltale signs could include:

1. A completely new seatboard.
2. Indications that the position of the seatboard has been altered.
3. Superfluous holes in the seatboard that had obviously accommodated another movement at some previous date.
4. An ill fitting dial, i.e. spandrels partly obscured by edges of door; too great a gap around the edge of the dial and door.
5. Winding squares that do not fit centrally in the winding holes, extraneous holes in the front plate of the movement, and filled holes on the dial would all indicate a marriage between movement and dial.

The presence of active woodworm is a further possibility, although there are successful modern methods of controlling this. Some renovation to the cases is to be expected, especially the repair or even replacement of parts of the plinth. Many of these clocks stood on damp stone floors that were frequently washed down and therefore the wood became rotted in parts.

It is possible to find dial restorers who will repaint a dial, and whether it is preferred to keep the old authentic worn dial or have it restored is a matter of individual opinion. Again dealers in clock materials can supply modern replacement weights, finials, keys, etc., that are quite acceptable. It should be possible to undertake the cleaning of the movement from one of these clocks and more than adequate instructions can be obtained from many of the old handbooks or new books stocked by most horological booksellers. Problems will arise if new pinions or wheels are needed to be cut, as few amateurs can undertake this task. The problem is not so much the cutting of the wheels but obtaining the correct cutters.

Figure 91

This is an attractive example of a provincial mahogany clock with a white painted dial with a subsidiary seconds dial and calendar aperture. The dial arch is painted with a scene depicting a castle, river and bridge, with floral corner pieces. The name John Thomson on the dial could be the maker, but is more likely to be that of the retailer. The price at auction in 1977 was £300. Today a more realistic price for a reasonable clean example would be in excess of double this figure even with 30 hour movement.

Any similar provincial clock would now be well over £700+

Figure 92

This is a good example of a well-proportioned mahogany longcase clock of Scottish origin. The white painted dial has Commerce as the subject for the colourful painting in the dial arch. The maker's name — George Lumsden, Pittenweem — on the dial is not without interest and would, therefore, add to the value of the clock. There were two George Lumsdens — the father known to have been active between 1818 and 1849, and the son known to have been active between 1849 and 1899. The former was apprenticed to John Smith a noted clockmaker, also of Pittenweem, who excelled in complicated movements including an elaborate musical clock with several dials and playing eight tunes. The swan neck pediment was a popular style that continued on throughout the 19th century.

In view of maker and gain in popularity of Scottish clocks £850+

Courtesy of Kingston Antiques

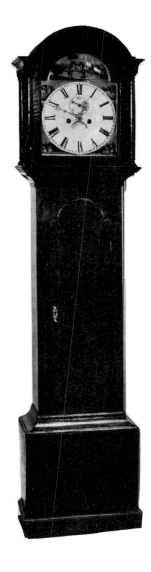

Figure 93

A further example of a provincial white dial clock, this time in an oak case. The case is well made — a point to note as the quality of the case varies considerably depending upon whether the maker was a carpenter or joiner. The dial has, apart from the gaily painted picture in the dial arch and spandrels, the added features of a seconds dial and an aperture showing the date.

**In today's financial climate £600+
More if in mahogany case**

Although there are London made longcases of this period, the style that continued to be made irrespective of current fads and fashions was the regulator. Here the floor standing longcase style of regulator is being referred to and does not include the Vienna regulator which is quite a different matter. As the primary purpose of a regulator is precision timekeeping all unnecessary motion work or complications such as strike, etc., were avoided. There was little point in losing, through the friction introduced by the operation of the striking train, the advantage gained by the high standard of workmanship put into these clocks. Many of them were made as one-off pieces by skilled clock repairers for their own use, while others were made for the trade by such makers as Dent, Barraud and Lund, and J. Smith and Sons of Clerkenwell. Depending upon the source the quality will vary. The Victorian case style for these clocks was a mahogany round topped case, with a fully glazed door at the front and little embellishment, as in the example shown in Figure 98a. Most jewellers and clock retailers kept one on the premises by which to regulate and adjust clocks and watches in for repair and many of them have been retained as part of the shop furnishings. The dials are plain and simple, usually silvered with the minute markings around the circumference and the two smaller subsidiary dials showing seconds and hours. This unusual arrangement is due to the movement being designed to avoid friction caused by the more usual central arbor and concentric hands. Various forms of pendulum compensation were used, with a compromise being made between the theoretical ideal and the aesthetic appeal. As a consequence

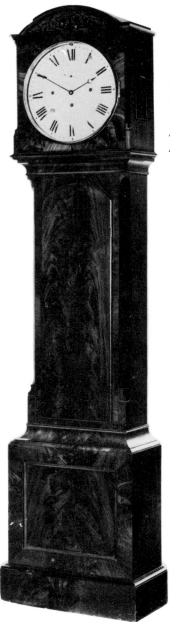

Figure 94a

Mahogany quarter chiming longcase, with 14 ins. diameter enamel dial, striking the quarters on two bells and striking the hours on a subsidiary bell. The date given was the second quarter of the 19th century. It was not possible to trace details of the maker — Westwood, London. The estimated price in 1981 was £1,500 to £2,000.

£2,500+

Figure 94b

This illustration of the movement of the clock shown in Figure 94a clearly shows the exceptionally thick shouldered plates, four massive baluster pillars and adjustment through the dial at the top. The escapement is deadbeat, the pendulum rod was wooden. Obviously a quality clock with features commensurate with a high rate of timekeeping.

Courtesy of Sotheby Beresford Adams

Figure 95

A mahogany longcase of Scottish origin by C. Winterhalder, Glasgow. The subsidiary dials are for seconds and calendar. The movement with anchor escapement has rack striking on a bell. This was also dated as being from the second quarter of the 19th century and although a pleasant clock with an attractively styled case could in no way be compared favourably with the clock shown in Figure 92. The circular dial is typical of Scottish clocks after about 1820.

Many German refugees took up employment in Scotland after fleeing from persecution, thus accounting for the name on the dial.

This clock was estimated in 1981 at £300 — £400 and although there is general sales resistance to the rounded case style £500+ would be more realistic now.

Courtesy of Sotheby Beresford Adams

the mercurial pendulums were with glass jars as opposed to metal; the Harrison gridiron pendulum with its alternating rods of brass and steel proved another favourite. For the purely functional regulator the invar pendulum tended to be chosen. Invar is an alloy of nickel and steel discovered by Dr. Guillaume in 1904. Its expansion at different temperatures is negligible — the name comes from *'invar*iable'.

Whether there is any credence in the story that the sonorous Westminster chime of Big Ben greatly influenced the domestic taste for deep-toned chiming clocks is not certain, but it is certain that there was a great upsurge of popularity for this type of clock during the late Victorian and Edwardian period. Harrington patented his tubular chimes in 1884 (Patent No. 14270) and there is little doubt as to their immediate success. Their tone was particularly pure. They won a gold medal at the Paris Exhibition in 1885, a bronze medal at the Inventors' Fair held the same year and a further gold medal at Liverpool 1886.

Fewer longcase clocks were manufactured with bells at this time. The alternative to the tubular chimes being 'cathedral' gongs — gongs with a deep resonant note. Depending upon the chime there were either eight or four tubes with an additional one for striking the hour which was in some instances replaced by a gong. This was possible to obviate the need for an even wider case to house the ninth tube, as well as providing a contrasting note. Cheaper models only had tape gongs, which in turn were replaced in the 1920s by rods. This last change was both for reasons of economy and the demand for smaller clocks. There is no doubt that the size of the weight-driven clock with tubular chimes became and still is a problem. Although now considered choice collectors' pieces, especially in America and Germany, few modern homes will readily accommodate a clock of this size. It must be admitted, however, that even though it may appear ostentatious to the modern eye, the large heavily carved case with its glass panelled door enabling a view of the polished tubular chimes and pendulum is certainly most arresting!

The most commonly found chimes used were the Westminster (on four tubes), or the Whittington (on eight tubes). The Westminster chime being by far the oldest, as it originated from the fifth bar of Handel's 'I know that my Redeemer liveth' and was expanded into the present chime by Dr. Jowett and Mr. Crotch when this was needed in 1793 at St. Mary's Church, Cambridge. It was then known as the Cambridge chime and only became known by usage as the Westminster chime after being chosen for Big Ben at Westminster. The story regarding the Whittington chime is that it was so named after Dick Whittington who, on hearing the sound of this chime from the Bow Church, Cheapside, retraced his steps to London and ensuing fame and fortune. Sometimes the St. Michael's chime (on eight tubes) is

Figure 96

8ft. 5ins. high and not particularly well proportioned! The heavy hood and pediment, wide trunk and base tend to give this clock a short squat appearance. This is unfortunate as a high standard of work has gone into the corinthian columns, shaped door and satinwood boxwood crossbanding and inlay. The 14ins. enamel dial with roman numerals contains seconds subsidiary dial and is inscribed 'Smales Rochdale'. A Richard Smailes is recorded as working in Rochdale in the mid-19th century. The painted spandrels depict hunting scenes and an automaton sailing ship is painted in the arch. The quarter striking movement has an anchor escapement. The estimated value in 1982 was £1,200 to £1,600. The automaton obviously affects valuation.

£2,000+ now

Figure 97

The maker of this rather 'sturdy' mahogany longcase is given as Elias E. Jones, Bethesda. Again the spandrels are painted (floral) and with lunar work in the arch of the dial. The second edition of Brian Loomes' book on the white dial clock goes into some interesting detail concerning the designs for these lunar paintings. The movement has an anchor escapement and rack striking on a bell.

In excess of £500 although not the most attractive of proportions.

Courtesy of Sotheby's, Chester

Courtesy of Sotheby's, Chester

used as a third alternative. There would appear to be some divergence as to its origin. It is usually attributed to St. Michael's Church in Hamburg, but the following interesting comment was seen recently in an issue of the *American Horologist and Jeweler* magazine. A set of bells was cast in London and installed in the steeple of St. Michael's Church in Charleston, South Carolina, in 1764. During the War between England and America the bells were captured and shipped to England, where they were later found by a Charleston merchant and returned to Charleston. However, cracks were found in some of the bells and in 1823 they were sent back to London to be recast. Unfortunately, during the Civil War in America they were destroyed but the fragments were returned to the original London bellfounders and the bells once again recast. Finally, in 1867, the eight bells were yet again installed in the steeple and the tune they rang out was 'Home again, home again, from a foreign land'. Since then they have been left undisturbed.

Many of these clocks were exported to America and the east. Most manufacturers recognised the need for, and adopted, alternative construction techniques to allow these clocks to function well in hot climates with the minimum of skilled attention. It is only necessary to read the extensive list of countries to which one firm, Gillett and Johnston of Croydon, sent their turret clocks as well as a large range of domestic clocks, to realise how sought after were our quality clocks of that period.

This firm also manufactured these weight-driven clocks with tubular chimes and made the clock presented by the Borough of Croydon to H.R.H. the Prince of Wales. This had a handsome carved rosewood and ivory case. William Gillett began work as a small clockmaker in Hadlow, Kent, under the patronage of Lord Sackville of Knole. He later worked in Clerkenwell before establishing his business at Croydon in 1844. He was subsequently joined by Charles Bland and then in 1877 by Mr. A. Johnston. The name of the firm became Gillett and Johnston which it has retained until this day. They are still world famed for their turret clocks, but have ceased to make clocks for the domestic market. This part of their business was transferred to F.W. Elliott Ltd. in 1923. This firm was founded by James Jones Elliott in 1886, who had served his apprenticeship with Bateman of Smithfield in London. There appears in the Bulletin of the National Association of Watch and Clock Collectors Inc. for December, 1981 an interesting account of one Walter H. Durfee who is generally acknowledged to be the 'father of the modern grandfather clock' (? in the United States). In 1883 Durfee commissioned two longcase clocks to be made by an English manufacturer, J.C. and J. Jennens of London. These clocks were an instant success and his business expanded rapidly. On a subsequent visit to England in 1886 he met J. Harrington who had just developed and patented his tubular chimes. Eventually Durfee purchased the rights to be the sole agent for these chimes in America. Elliott's American agents at this time were Harris and Harrington so what could be more natural than Elliott should be approached to make the movements for the proposed longcase clocks with tubular chimes. A business association that continued well into the 1900s. A recent reprint of the 1915 catalogue of Harris and Harrington (Fine Imported Clocks) of New York lists them as "American Agents for the Celebrated Elliott English Chime Clocks". Examples with either 9 or 5 tubes are illustrated.

Prior to 1901 Elliott's carried on their business from Percival Street, Clerkenwell, London where they manufactured fusee and chain quarter chime bracket clocks as well as longcase movements. Their weight driven tubular chime clocks were such a success that larger premises were needed and the firm moved to Rosebery Avenue, Clerkenwell, London.

In 1909 J.J. Elliott Ltd. amalgamated with Grimshaw and Baxter and the factory moved in 1911 to Gray's Inn Lane, and in 1917 to St. Anne's Road, Tottenham. The association did not last and the two firms parted company in 1921, when Frank Elliott, James' son, sold the name of J.J. Elliott to Grimshaw and Baxter. Frank joined the firm of turret clock and domestic clock manufacturers, Gillett and Johnston Ltd. in Union Road, Croydon. In 1923 Frank Elliott formed the present company, F.W. Elliott Ltd., and took over the production of the Gillett and Johnston range of domestic clocks. Unlike a number of other firms, which made their own movements but sent out to casemakers for cases, both of these firms manufactured movements and cases. As the production methods and case designs of both firms were somewhat similar, identifying differences are too subtle to be readily noticeable. Neither firm is aware of any numbering sequence being stamped on movements, nor of varying trade marks, although from the advertisement shown in Figure 100 it would appear that Elliott's did at one time possibly use one. Examples of both firms' clocks appear in the illustrations.

6ft. 6ins. This is a good example of a Victorian mahogany longcase regulator with a mercurial pendulum. The points to note are the minute markings around the perimeter of the dial, the seconds dial below the 12 position and the twenty four hour dial above the 6 position. It is more usual to find the conventional twelve hour dial.

Courtesy of Derek Roberts Antiques

Figure 98b

Movement of the regulator shown in Figure 98a. The points that denote a high quality movement in this instance are: the substantial plates and pillars, the fine crossing out of the wheels, a high count train and jewelled pallets. Notice too the sharp angular appearance of the dead beat escapement in contrast to the rounded form of the more commonly found anchor escapement and the consequent upright teeth of the escape wheel.

A good regulator inevitably reaches £4,000+

From the catalogue published by Smith & Sons Ltd. of 9 Strand, London, c.1900, it is interesting to note that they offered among the range of clocks manufactured by them some twenty-one registered longcase designs housing movements striking either on bells, tubular chimes or gongs. One case is in the form of a 'miniature' Big Ben — Westminster Tower and all — which stood some 10ft. 6ins. tall (see Figure 104). The examples striking on bells could be had for a further £10 striking on 'Nine tubular Bells', or there were the alternatives of a repeating action for £1 extra, a mercurial pendulum for £5 extra or a moon dial for £1 extra. Most of the models illustrated had moon dials, similar to that in Figure 105.

Other manufacturers advertising the fact that they made longcase clocks in the years immediately after the First World War were A. & H. Rowley, 4 Theobalds Road, London, W.C.1., who were established in 1808 and W.H. Evans & Sons, Soho Clock Factory, Handsworth, Birmingham. Further details of the former will be found on page 131.

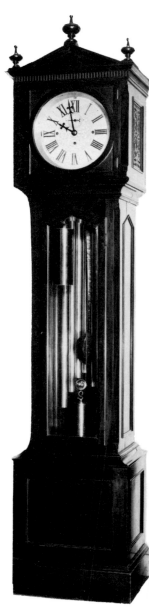

99a

99b

99c

Courtesy of F.W. Elliott Ltd.

Figures 99a, 99b and 99c

An early weight-driven tubular chime clock by J.J. Elliott Ltd., of Croydon. Although the clock is large the proportion and classical design of the case makes this an elegant example. The dial is silvered and has the typical flowery engraving where the eighteenth century example would have had a simple matted surface. It has raised skeletonised Roman numerals, engraved centre with name 'Elliott, London' engraved below the winding aperture and a sunk seconds dial. The huge movement is three train striking on a tape gong, with the alternative chimes of Westminster or Whittington on eight tubes. The points to note from the interior view of the trunk in Figure 99b are the three brass-cased weights for chime, time and strike, together with the eight tubular chimes suspended in parallel across the back of the case. The small cord on the left-hand side of the striking weight is to activate the pull repeat mechanism. As with most chiming/striking clocks of this period provision has sensibly been made for silencing them at will. In this particular example this is by means of small levers through the dial above the number XII and to the left of the number IX. The third lever to the right of III is for selecting the chime. As can be seen from Figure 99c this activates a series of levers which in turn move the pin-barrel horizontally for the appropriate series of pins to be set in position. The other points to be noted from the illustration of the movement are the method of suspending the tubular chimes, and the thick, good quality plates of the movement with a spotted finish (this finish is generally associated with marine chronometers).

£1,000 — late tubular chime but not in the most attractive of case styles

The British Empire Exhibition which opened at Wembley on the 23rd April, 1924, was intended to promote home industries — 'Buy British' was a common slogan at this time. Among the horological exhibitors the following are mentioned as manufacturing longcase clocks:

F.W. Elliott Ltd., Union Rd, Croydon
John Smith and Sons, 42 St. John's Square, London, E.C.1.
Pleasance and Harper Ltd., 4 Wine Street, Bristol.

The model displayed by Pleasance and Harper Ltd. being an extremely rococo production of a Louis XVI case but housing a movement with tubular chimes, was similar to that seen in Figure 106.

Interest in these Edwardian chiming clocks has never completely disappeared. Although F.W. Eliott Ltd. discontinued the model with nine tubular chimes and only retained that with five tubes, they have now reintroduced the former due to the increasing demand for these clocks. One firm in Germany — Joseph Keininger — also recommended production. Neither of the firms are contemplating reintroducing the truly massive examples but rather those standing 6ft. to 6ft. 6ins. tall.

Figure 100

This advertisement shows a weight-driven tubular chime clock manufactured by J.J. Elliott Ltd. sometime between 1911 and 1917. The trade mark and statement related to it are not without significance. Although present members of the company (F.W. Elliott Ltd.) have no recollection of a trade mark it appears that one was used "unless otherwise requested". As already mentioned, when the two firms of Grimshaw, Baxter and J.J. Elliott parted in 1921, the name of J.J. Elliott was sold to Grimshaw and Baxter and hence continued to be used by them. As this trade mark appears on clocks sold by them as late as 1927 it would appear that care has to be taken when interpreting who actually made the clock and when. It is not known with certainty who manufactured clocks for Grimshaw and Baxter in the 1920s although it was possibly Williamsons of Coventry.

£2,000+

ENGLISH (LONDON MADE) GRANDFATHER CLOCKS.

No. 1.
Height 8ft. 11ins.
Handsomely Carved Mahogany or Oak Case.

No. 1090.
Height 7ft. 9ins.
Mahogany or Oak Case.

Nos. 1 and 1090 —8-day Best Carved Mahogany or Oak Case. Dead Beat Escapement, Maintaining Power, Brass Cased Weights, Compensated Pendulum. The Chimes can be changed at pleasure by indicator hand on dial, also hand to shut off Chimes, making Clock strike the hours only. There is a further action to shut off Hour Strike, making Clock silent. Dial Corners and Centre finely Hand Pierced and Engraved. Silvered Hour Circle with Raised Gilt Figures. The Tubes are well tuned and heavily nickelled.

	No. 1.			No. 1090.		
	£	s.	d.	£	s.	d.
Chime on 9 Tubes, giving at will St. Michael, Whittington, and Westminster Chimes	170	0	0	72	0	0
„ 9 „ „ „ Whittington and Westminster Chimes	168	10	0	70	10	0
„ 8 „ „ Westminster Chime or 9 Gongs	160	0	0	62	0	0
„ „ 8 Bells and 5 Gongs giving at will Whittington and Westminster Chimes	158	10	0	60	10	0
„ „ 5 Gongs Westminster Chime	155	0	0	57	0	0
„ „ 8 and 4 Bells giving at will Whittington and Westminster Chimes	154	0	0	56	0	0

Mercurial Pendulum £8 0 0 extra. Plain Brass Mounted Dial with Cast Corners, £3 0 0 less.
Movements chiming on Tubes can be made, if desired, to strike the hours on a Gong at same price.
Cases specially manufactured for India and other tropical climates at a slightly increased cost.

G. B. & J. J. E. Ltd.

Figure 101

Two further examples of clocks made by J.J. Elliott sometime after 1909. These illustrations are taken from the trade catalogue of Grimshaw Baxter and J.J. Elliott Ltd. and show clocks made by J.J. Elliott. The variety of choice of styles, woods and movements together with prices is of great interest. They were far from cheap clocks starting at £89. 10s. when a working man's wages at this time were 30s. per week, and it is understood that a number were exported to America at the time of manufacture. Since then they have become choice collectors' pieces in that country.

An extract from an Elliott manual below gives details of fitting and setting up instructions for Elliott clocks.

"The tubes should next be suspended from the tube rail. These are hung according to size, starting with the shortest tube on the right-hand side looking at the clock. Loop the cord at the top of the tube from slot to slot in the tube rail as shown in this diagram:

"The longest tube of all is the hour tube, and this is suspended from a separate bracket on the left hand side of the movement. Make sure that the tubes do not touch each other when hanging, as this will result in a "jangle" of the chimes. If they do touch, adjust each tube in turn by holding the bottom and moving slowly from side to side so that each swings clear of the other. Adjust the chime hammers by bending the stems so that the heads are $^1/_{16}$ in. away from the tube when at rest. The hammer heads must not lie on the tube, as this will result in a double note being struck and a dull tone. It is important that the hammers strike the tubes squarely and centrally.

"These clocks are fitted with a chime sequence corrector, so that if the chimes get out of sequence as a result of the clock running down and stopping, the chimes being repeated, or the hands being moved without waiting for each quarter to chime, etc., etc., the chimes will automatically correct themselves."

Example on left £3,000+, example on right £1,000+

Figure 102

It is interesting to note that Grimshaw, Baxter and J.J. Elliott Ltd. were still advertising and marketing this style of clock as late as 1937.

Figure 103

This advertisement for clocks manufactured by Gillett and Johnston of Croydon that appeared in 1921, does not really do justice to their products. In common with other manufacturers of this date they also manufactured "the more expensive and massive designs with quarter chiming movements". Their 1906 catalogue stresses "with correct design, sound and perfectly seasoned wood, skilled carvers and wood workers, and no piece work, we are able to supply the finest procurable specimens of English Grandfather Clocks, while the workmanship of the Movements is in every way in accord with the high finish of the case". At this date their eight-day grandfather, in sold oak or walnut, hand carved, brass engraved and silvered arch dial, striking the hours and quarters on deep toned gongs, was priced from £80.

£1,400+

This firm was, and still is, a noted manufacturer of turret clocks with many a church or public building in England or abroad housing one of their products. This side of their business remains extremely active to present times. In the past they were also bell founders and again many famous bells came from their shops including the Bourdon Bell of the carillon of the University of Chicago (weight seventeen tons).

Courtesy of Christie's

Figure 104

The large carved oak longcase shown here is made doubly interesting in so far as the identical clock appears in the 1906 catalogue of S. Smith & Sons Ltd., Strand, London entitled Smith's Guide to the Purchase of a Clock, as well as a later edition under the description of:

"No. 10583 — 'Big Ben' Registered design. Height 10ft. 6ins; width, 2ft.; depth, 1ft. 5ins. No expense or trouble has been spared in order to make this chef d'oeuvre of English workmanship. The Carving is as fine as can be executed; every detail of the original is faithfully carried out, even to the Dial and Hands. The Movement is a very powerful one, and Chimes the Quarters and Hours on a set of Beautifully-toned Tubular Bells, faithfully representing 'Big Ben'. This fine piece of work, however, requires to be seen to be properly appreciated. The case alone occupies four months in manufacture, and the clock complete nearly twelve months; the weight is about 5 cwt. As regards its Timekeeping properties, we can safely recommend it as one of the finest that could be made, its variation being about one second per week; this wonderful accuracy is only attained by very high class work".

Although the inscription 'Made by Jonne Salvam in the reign of Queen Victoria I' has been carved below the dial the case, dial hands etc., are undoubtedly identical to those marketed by Smith and it is highly likely that this also applies to the movement. This was a massive three train movement with deadbeat escapement and the Westminster chimes on eight tubular chimes and a further one for the hours. The size makes it unacceptable for most domestic situations and although regarded with horror by many wives it is a magnificent piece in the correct setting. This particular piece sold at Christie's in February, 1982 for £1,700. Size was obviously a deterrent to buyers.

Novelty value makes price unpredictable
£2,000+

Figure 105

A similar clock to that shown here appears in the 1900 catalogue of S. Smith & Son, Limited, 9 Strand, London, with the following description:
" 'The Savoy' Grandfather Clock, in Splendidly Carved Oak Case, with Bevelled-Edge Plate-Glass Door, Best Quality Movement, showing Phases of the Moon and the Seconds on the Dial; There is also a Small Dial at the Right-Hand Corner, with a Hand, which, by Turning, effects a Change of Chimes either on to the Westminster Set of Four Gongs or the Cambridge Set of Eight Bells. The Hour is struck on a Very Fine-Toned Gong. The Small Dial on the Left is for the purpose of making the Clock Silent.

£2,000+
Fitted with Nine Tubular Bells, £10 extra;
or with Five Ditto, £5 extra
Repeating Action £1 extra
Height 8ft. 3in; Width 2ft. 0in; Depth 1ft.

This particular example has a month movement, chiming on five tubular bells, with maintaining power and a dead-beat escapement. According to the catalogue the dial was signed 'Diamont Merchants Jewellers and Silversmith'. Although not the de luxe model with nine tubular chimes and apparently needing some restoration carried out, this clock realised £1,450 in the sale-room in 1977. There is no doubt that these Edwardian chiming clocks are gaining in price and in view of their size it is surprising that they command a higher figure than older, smaller traditional longcase clocks, even if they do have a more complicated chime and strike.

Courtesy of Messrs. King and Chasemore

Courtesy of Sotheby's, Belgravia

Figure 106

9ft. 1in. The catalogue description for this incredibly ornate longcase clock was that it was "A good mahogany Long-case clock, the brass dial with rococo spandrels and silvered chapter ring, with subsidiary seconds, Whittington/Westminster and Chime/Silent dials, the movement striking on gongs, with a carved cresting inlaid with foliage and corbel supports, the glazed waist door with carved astragals, the plinth inlaid with a hanging basket of flowers and moulded block feet." There are no maker's marks on the backplate, but it is remarkably similar to the case of the clock made by Pleasance and Harper Ltd. of Bristol and shown in Figure 107. The price reached in the saleroom in 1977 was £2,200 and shows clearly the effect of overseas demand at that date for highly ornate pieces regardless of age. In 1980 a similar clock was being offered under £2,000. Now with a good exchange rate for overseas buyers prices are rising again.

£3,000+

Figure 107

This is an illustration of the reproduction Louis XVI Grandfather Clock that appeared on the Pleasance & Harper Ltd. stand at the British Empire Exhibition at Wembley in 1924. It is described as follows:

"The case is of solid carved mahogany with inlaid rosewood panels of correct period design. It has glass panels in front and sides, mounted with ornamental lattice carvings, and the dial is brass, with raised mounts and gilt figures. The height is 8ft. 6in. The movement is of best quality and finish, with dead-beat escapement, and the weights are gilt. An important feature is a specially adjusted seconds pendulum, built on scientific principles to ensure accurate timekeeping. The clock can be fitted with Whittington and Westminster Chimes and chimes every quarter of the hour on 8 and 4 Harrington tubes. It also strikes every hour on an extra large and deep-toned tube. The clock can be supplied with any set of three chimes; Whittington, Westminster, Canterbury, Magdalen, Guildford, etc. The value of this Louis XVI Clock is about one hundred and twenty guineas. This firm makes a speciality of Grandfather Clocks in all styles. For safe transit, care is taken to pack in special cases for despatch to any part of the world. An actual photograph, in natural colours, will be sent free, on request to this firm at 4, Wine Street, Bristol."

£2,200+

Figure 108

This clock dates from the late 1800s and the case is typical of the Victorian school of thought that delighted in over indulgence and extravagance of design. The 13 ins. brass dial has a silvered chapter ring enclosing a subsidiary seconds dial and calendar aperture. The moon dial in the arch has automaton rolling eyes. Striking is on a series of seven bells. The waist door is carved with a figure entitled Tempus Fugit. The estimated value in 1981 was £500 to £700. Certainly worth far more if judged on quality and quantity!

See Figure 116 for a similar case manufactured by H. Williamson Ltd.

In the right setting worth £2,000+

Courtesy of Sotheby Beresford Adams

The majority of the movements found in these large chiming and striking clocks are English, but there are a few that were manufactured between 1890 and 1950 by Winterhalder and Hofmeier of Germany. These are of such good quality that it is often only by the trade mark on the backplate (W. & H. Sch.) that the country of origin is realised.

Inevitably the American factories produced similar clocks but it is doubtful if any found their way on to the English market except through private channels. The only movements that were imported from abroad in any quantity were those used in cheaper chiming clocks as shown in Figure 121. These were from Germany and chimed and struck on rod gongs, which although surprisingly melodious are considered inferior to either the tubular chimes or tape gongs. Further details concerning the movements with rods will be found on page 119 where their use with regard to shelf clocks of the same period is discussed.

If purchasing an example in need of restoration it is wise to locate a source of supply of suitable replacement tubular chimes. To replace a whole set is simpler than restoring one or two of the initial chimes. If replacing a full set it is possible to choose tubes of the correct material and calibre to provide both visual and aural symmetry. When replacing one tube, however, it is difficult to trace a tube that matches the remaining parts, both in note and size. Care also has to be taken when hanging them to ensure that the tubes are freely suspended by cords passing through the holes in their sides and that they neither touch each other nor their support. The hammers originally had their faces softened by thin pieces of chamois leather. A certain amount of trial and error is needed to adjust the hammers to strike the tubes at the correct point, and at the same time to achieve some equality of tone. If possible it is as well to hear the chimes through to verify that the pin-barrel is not damaged. This can be restored by any of the restorers that undertake work for music boxes, but this again adds to the initial cost of the clock and so needs taking into account when assessing the total outlay.

Henry B. Fried has written a definitive article in the Bulletin of the National Association of Watch and Clock Collectors (April 1982 — Whole Number 217 XXIV No. 2) which although relating to American tubular chime 'Hall Clocks' there is much of relevance to English and German made movements. As Henry Fried indicates even the correct arrangement and hanging sequence of the tubes for the various chimes can cause problems. An example with nine tubes can have 362,880 variations! Many of the models made for the American market by Elliotts and some German manufacturers had either nine or eleven tubes, the hour chime being hung on the far right and the longest chime tube on the left and graduated down to the shortest on the right. It must also be remembered that the Whittington Chime also has an American version as well as a German rendition and the English standard version.

Figure 109

A 8ft. 9ins. high walnut 'Renaissance' longcase heavily carved with strapwork, scrolls and caryatid figures flanking the dial. Almost as an afterthought there is a 12ins. lacquered dial with silvered chapter ring and plaque inscribed James Jay, 142 & 144 Oxford Street, W. The movement has an anchor escapement and rack striking on a bell. When this appeared in the saleroom in 1982 the estimated price was £500 to £800. It actually realised £935 so somebody must have loved it! It would be interesting to know if the buyer was from this country.

£1,500+

Courtesy of Sotheby's, Chester

Figures 110a and 110b

A handsome late Victorian rosewood longcase clock inlaid with marquetry of engraved ivory representing foliate strap-work, trellised panels, C scrolls and neoclassic ornament. The applied silver numerals on the silvered repousee 19ins. dial were hallmarked 1890 (see Figure 110b). The month going three train movement struck on eight tubular chimes. This clock realised £5,000 in 1979. As is only too often the case it was not possible to establish the maker of this clock. It would be reasonable to assume that the case was made to special order as it is exceptionally fine.

£5,000+

Courtesy of Philip's, Son & Neale

Courtesy of Phillips

Colour Plate 3

2¾ins. high. A late nineteenth century miniature French gilt brass and enamel carriage timepiece. Cylinder movement. Circular white enamel dial, Anglaise rich style with canted corners, the sides, back, door and dial surround with polychrome enamel plaquettes depicting figures with various landscapes. This realised £900 when sold in 1982.

£1,750+

Colour Plate 4

7½ins. high. A nineteenth century Swiss (P. Girard, Chaux de Fonds) gilt brass petite sonnerie striking carriage clock. The movement had a chronometer escapement, with the striking and repeating work visible on the backplate, and alarm train. The white enamel dial with subsidiary seconds dial at the XII position and two further subsidiary enamel dials for alarm set and calendar below. Estimated to reach £1,200—£1,600 in 1982 it actually realised £2,300. The maker, who is recorded as having exhibited at the Great Exhibition of 1851, together with the unusual escapement and external striking work, would all have contributed to the value of this piece.

£2,500+

Courtesy of Phillips

101

111

Figures 111 and 112

These two clocks are 'in the Gothic taste' and date from the late Victorian period. It is highly likely that Figure 111 with crenellated decoration and trefoil glazed arched waist panel door, with eight day three train movement with dead beat escapement and mercurial seconds pendulum chiming the quarters on eight tubular chimes would have come from Elliotts of Croydon. Figure 112 also had a three train movement but with striking on a series of eight bells and a gong. Estimated at £700 to £1,000 in 1981, and standing at nearly 9 ft. tall this could have been considered a bargain. Although they possibly stimulate feelings of disbelief, awe, horror and outright hatred there is little doubt that in the right setting these clocks can make imposing focus points. Quality of workmanship in both case and movement is immediately apparent.

£1,200+

112

Courtesy of Sotheby's, Chester

Figure 113

7ft. 2ins. This mahogany cased longcase clock with a silvered dial and subsidiary dials for seconds, chime/silent and choice of Whittington or Westminster chimes on nine rod gongs, has no identifying maker's marks but was manufactured at the beginning of the twentieth century. It fetched £900 in the saleroom in 1977. Possibly would not realise so high a price now in view of plain case and late date of manufacture. A traditional old longcase could be had for that price although admittedly without chimes.

£1,000+

Courtesy of Sotheby's Belgravia

Figure 114

6ft. 2ins. This caricature of a longcase clock with a two train movement was made sometime during the 1920s or '30s and could, at a generous stretch of the imagination, be considered a collector's piece by virtue of the exceptionally curious assortment of styles and designs. The combination of Oriental scenes on the painted panels, the exaggeratedly narrow trunk and the dial predictably engraved 'Tempus Fugit' culminating in putti riding on eagles must be unique! The price fetched in the saleroom in 1977 was a generous £320 (the pre-sale estimate was £80—£120!). Only the spurious 'Tho⁵ Tompion London Fecit' is missing to make it a fine example of an horological joke.

£300+

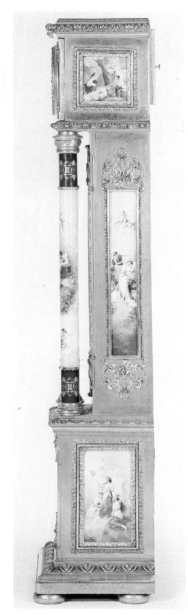

Courtesy of Sotheby's

Colour Plate 5

6ft. 4½ins. A fantastic example of a porcelain mounted pedestal clock thought to have originated from Vienna c.1900. The movement was signed Gustave Becker and numbered 397916. Estimated to realise £8,000 — £10,000, it reached £12,100 when sold in 1983.

Colour Plate 6

7ft. 9ins. A Boulle longcase clock in the French style. The movement has outside count wheel and is stamped Vincenti, which would date it around the late nineteenth century. A highly decorative piece. The realised price of £3,500 in 1983 was higher than the estimate.

£4,500

Courtesy of Sotheby's

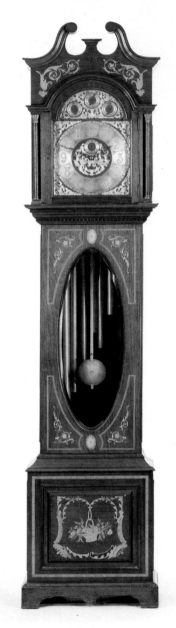

Colour Plate 7

A painted mahogany longcase clock c.1900 but in the late eighteenth century style. The silvered dial has subsidiary date and seconds dial with an oscillating figure of Father Time in the dial arch. This figure also appeared on the arch waist door together with a putto eating grapes. Realised £1,870 in 1983. **£2,000+**

Colour Plate 8

A quarter repeating mahogany longcase clock with quarter striking on eight tubular bells. The case is inlaid with foliage and crossbanded in satinwood. The given date was c.1910. Realised price in 1983 was £1,815. **£2,500+**

Colour Plate 9

A further example of a mahogany longcase clock of the early twentieth century with quarter striking on nine tubular bells. The similarity in style of this clock and Colour Plate 8 is quite noticeable. It realised £2,750 in 1983.

£3,500+

H. WILLIAMSON L^{TD.}

SPECIALITIES:

GRANDFATHER CLOCKS.

BRACKET - CLOCKS.

LANTHORN - CLOCKS.

OUR
New Showrooms
AT
87, Gt. Saffron Hill
(Adjoining our London Warehouse)

Are now replete with
a Splendid Selection
of
MARBLE, ONYX, WOOD,
CARRIAGE & NICKEL
CLOCKS.

SPECIALITIES:

TURRET - - CLOCKS.

JEWELLERS' DRUM CLOCKS

CHIME - - CLOCKS.

ESTIMATES
and
DRAWINGS
of every description
of above Clocks
sent on application.

CATALOGUES
on application.

English Clock Factory, SALISBURY

xviii

Figure 115

This advertisement by H. Williamson Ltd. appeared in the very early 1900s and was sold under the name of 'The Victoria'. It shows a three train striking longcase in outline very similar to a mahogany example of about 1780 but decorated with Sheraton type designs of inlay especially on the base.

A longcase clock with an identical case appeared in a Sotheby's sale in the mid-1980s and was estimated to realise £1,200 to £1,800.
A German example (Winterhalder) with plainer case was also given the same estimated prices.
£2,000+

The "Salisbury."

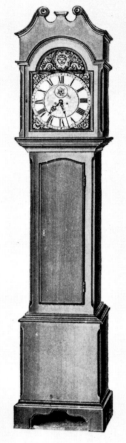

FUMED OAK CASE, Height 7ft. No. 192, 12 in. by 17 in. Arch Dial, Raised Hour and Seconds Circles, Cast Brass Corners.

	Fitted with Machine-made Movement.				Fitted with Hand-finished Movement.		
	£	s.	d.		£	s.	d.
Bell Strike	13	0	0	..	15	15	0
Gong	14	0	0	..	16	5	0
8 Bell Chime	21	10	0	..	—		
8 and 4 Bell Chime ..	—			..	29	5	0
4 Gong Chime	22	4	0	..			
8 Bell and 4 Gong Chime	24	9	0	..			
4 Tube Chime	30	10	0	..			

The above case can be supplied in Mahogany at **40/-** extra.

For description of movements, dials and price of extras see pages 22 and 24.

The "Lincoln."

SOLID OAK CASE, very richly carved. Height 8 ft. No. 192, 12 in. by 17 in. Arch Dial, Raised Hour and Seconds Circles, Cast Brass Corners.

	Fitted with Machine-made Movement.				Fitted with Hand-finished Movement.		
	£	s.	d.		£	s.	d.
Bell Strike	23	10	0	..	26	5	0
Gong	24	10	0	..	27	5	0
8 Bell Chime	32	0	0	..	—		
8 and 4 Bell Chime ..	—			..	39	15	0
4 Gong Chime.. ..	32	14	0	..	47	0	0
8 Bell and 4 Gong Chime	35	0	0	..	50	10	0
4 Tube Chime.. ..	41	0	0	..			
8 and 4 Tube Chime .	—				No. 146 Dial. 73	0	0

For description of movements, dials and price of extras see pages 22 and 24.

Figure 116

Williamson's made a wide range of case styles as is demonstrated by these two illustrations taken from their 1908 catalogue. The variation of movements included Bell or Gong Strike, 8 Bell Chime, 8 Bell and 4 Bell Chime, 4 Gong Chime, 8 Bell and 4 Gong Chime or 4 Tube Chime. Movements could be machine or hand finished. Hand finishing in a simple Bell Strike only commanded a further £2. 15s.

£700+   £850+

Colour Plate 10

2ft. 2¼ins. This is an extremely attractive gilt bronze and porcelain mantel clock. The movement is housed in a small drum case flanked by putti and surmounted by two birds. The central rectangular panel is painted with two young couples in a woodland setting and signed L. Simonnet, Sèvres. The factors affecting the price would have been the general high quality of materials and finish in the case, i.e. bronze figures, sharp detail to castings and inset porcelain panel.

£2,000+

Colour Plate 11

1ft. 10¼ins. Decorative mantel clock with heavy gilt bronze and champlevé case. Not a horologist's first choice but highly pleasing aesthetically.

£2,000

Courtesy of Sotheby's

Colour Plate 12

23ins. high. An extremely handsome garniture with gilt and enamel lyre case inset with brilliants. This style was introduced in the reign of Louis XVI and used by such makers as Kinable, Paris (1780-1825). An example can be seen in the Victoria and Albert Museum, London. The bob of the gridiron pendulum suspended from the top of the case takes the form of a circle of brilliants that effectively catches the light as it swings. This style of case became extremely popular and was continued for several decades. This particular example was dated as c.1890 and when sold in 1982 realised £3,000.

£3,000+

No. 192.—Brass Arch Dial, Raised Silvered
Hour and Seconds Circles, Brass
Ornamental Corners, Polished and
Lacquered.

12 in. by 17 in.   13 in. by 13½ in.   14 in. by 20 in.
£3 0 0   £3 10 0   £4 0 0

Square Dial as above, but without Arch,
12 in. by 12 in., **£2 10s.**

No. 67.—Brass Arch Dial, Raised Silvered
Hour and Seconds Circles, Brass
Ornamental Corners, Polished and
Lacquered, with Engraved and Silvered
Dome in Arch.

12 in. by 17 in.   13 in. by 18½ in.   14 in. by 20 in.
£3 4 0   £3. 15 0   £4 6 0

No. 217.—Brass Arch Dial, Raised Silvered
Hour and Second Circles, Brass
Ornamental Corners, Polished and
Lacquered, with Decorated Moon Plate
in Arch which revolves and indicates
the Phases of the Moon.

12 in. by 17 in.   13 in. by 18½ in.   14 in. by 20 in.
£4 4 0   £4 15 0   £5 6 0

Shakespeare Landscape, **10s.** extra.

No. 146.—Brass Arch Dial, Raised Silvered
Hour and Seconds Circles, Brass
Ornamental Corners, Polished and
Lacquered, with Raised Arabic Gilt
Figures, Engraved and Silvered Circle
in Arch for Changing the Chimes.

12 in. by 17 in.   13 in. by 18½ in.   14 in. by 20 in.
£5 1 0   £5 7 0   £5 17 0

Figure 117

These illustrations are also taken from the 1908 catalogue of H. Williamson
Ltd. and clearly show the variation of dials offered by them at this date.

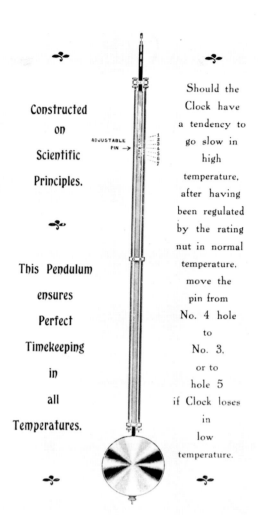

Constructed
on
Scientific
Principles.

This Pendulum
ensures
Perfect
Timekeeping
in
all
Temperatures.

ADJUSTABLE
PIN →

1
2
3
4
5
6
7

Should the
Clock have
a tendency to
go slow in
high
temperature,
after having
been regulated
by the rating
nut in normal
temperature,
move the
pin from
No. 4 hole
to
No. 3,
or to
hole 5
if Clock loses
in
low
temperature.

Figure 118

Williamson pendulum with provision for adjustment in order to regulate the clock.

Figure 119

Just one of the longcase designs produced by William Potts (Leeds) between the two World Wars. This was a style typical of that period including the cartouche inscribed 'Tempus Fugit'! Although the furnishings will inevitably come into the interior decorators' repertoire this is not a well proportioned or exceptionally attractive piece. The overall impression is that of a mantel clock sitting on a plinth. As the alternative of pendulum or lever movement is offered by the manufacturers this impression is proved correct!

£200+

GRANNIE ———— JACOBEAN

Made in several sizes, from
4 feet in height, upwards.

TIMEPIECE, HOUR STRIKE
OR QUARTER CHIME
MOVEMENT.

Pendulum or Lever
Escapement.

Can have Plain Brass Dial
(as shewn) or Enamel,
Silvered, or any other kind
desired.

Any type of Figures.

No. 12

Medium Arch Dial   £13 13 0

Colour Plate 13

8ins. high. The case of this clock is in the form of a gaily painted pottery dwarf. Introduced in 1929 when Walt Disney films were making their début and there is a label pasted to the vase stating that the design was 'By permission of Walt Disney. Mickey Mouse Ltd.' The clock was, however, made in England and has a basic uncomplicated movement – the eyes rotate as the clock beats the seconds, but the dial below is the true time teller (see also Figure 269).

Colour Plate 14

25ins. This particularly attractive weight-driven clock was made by Seth Thomas of Connecticut and has a well preserved veneered softwood case with fully glazed door and colourful tablet in the lower portion. See Figures 237b and c for details of the movement.

Figure 120

This longcase is characteristic of the 1930s and although for many years totally ignored by collectors it will now be coming into favour with those decorating their homes in period pieces of that date. It has the added advantage of being extremely pleasant to listen to — the chimes are most sonorous. The cases of these clocks are invariably oak finish and often of plywood bent and moulded and then veneered in oak. This would have been for reasons of economical production. This particular example was made by the Enfield Clock Co. Ltd. of Edmonton, London. Further details of this company appear on page 156.

£350+

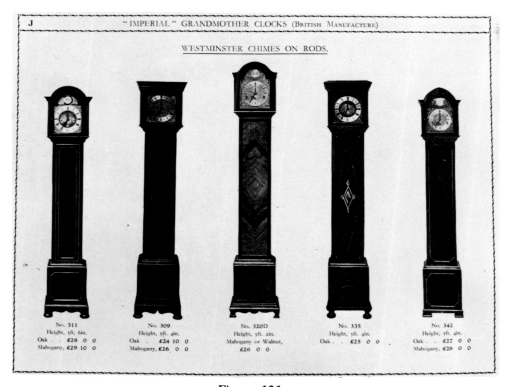

Figure 121

As homes became smaller the demand diminished for the larger longcase clocks striking on tubular chimes, and in an attempt to compete with the smaller, cheaper German clocks striking on rods, Elliotts introduced in 1932 the 'Imperial' range. This name continued to be used until around 1946 or 1947. The examples in this illustration have the Westminster Chimes on rods.

£150+

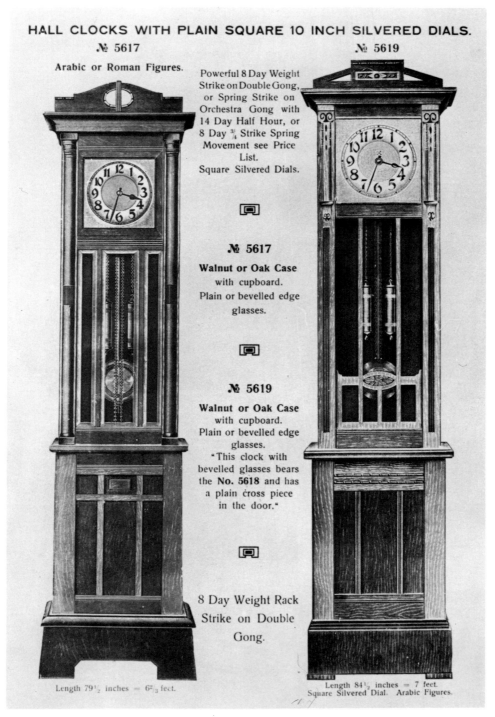

HALL CLOCKS WITH PLAIN SQUARE 10 INCH SILVERED DIALS.

№ 5617

Arabic or Roman Figures.

Powerful 8 Day Weight Strike on Double Gong, or Spring Strike on Orchestra Gong with 14 Day Half Hour, or 8 Day ¾ Strike Spring Movement see Price List.
Square Silvered Dials.

№ 5617

Walnut or Oak Case with cupboard.
Plain or bevelled edge glasses.

№ 5619

Walnut or Oak Case with cupboard.
Plain or bevelled edge glasses.
"This clock with bevelled glasses bears the **No. 5618** and has a plain cross piece in the door."

8 Day Weight Rack Strike on Double Gong.

Length 79½ inches = 6⅔ feet.

№ 5619

Length 84½ inches = 7 feet.
Square Silvered Dial. Arabic Figures.

Figure 122

Two of the Hall clocks marketed by the Hamburg American Clock Company in 1912.
Note the variation of movements offered and also with or without cupboard in the
plinth! This was very much a cheap version of a longcase clock.

£450+

Colour Plate 15

15ins. high. French striking movement (bell) in heavy brass and porcelain case with matching side ornaments. This particular set was entirely complete with both ebonised wooden base and gilt gesso velvet covered secondary base and shades — all original — for clock and ornaments. In view of the general quality and total originality price would be in excess of £1,000.

£2,000+

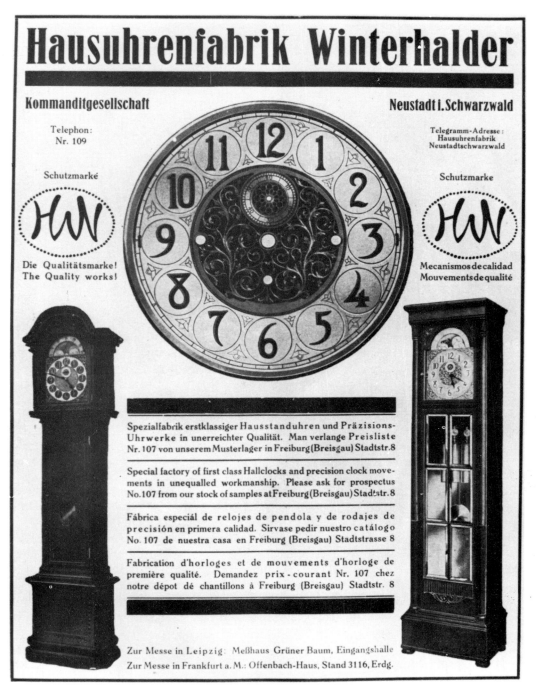

Figure 123

Examples of two of the styles of longcase clocks being manufactured in the 1918 period by the German manufacturers Winterhalder. The Winterhalder Clock factory was founded in 1810 and in 1850 joined by Anton Hofmeier to form the renowned Winterhalder & Hofmeier company. Other branches of the family established their own factories also in the Black Forest area of Germany and it is to be supposed that this was the work of Johannes Winterhalder from Neustadt.

£250+

CHAPTER VII

Bracket and Shelf Clocks

Bracket Clocks, Boardroom Clocks and English and European Shelf Clocks

The restrained style of the English Regency bracket clock continued into the early Victorian era and many examples from this period are highly desirable. By the mid-1800s, however, a complete polyglot of case styles had emerged. Some were faithful replicas of earlier styles, while others can only be described as over-embellished monstrosities. Unfortunately, it is usually the latter that are referred to as typical examples of Victorian clockmaking in England, with little regard for the fact that there were also on the market small, more elegant examples by makers such as Barraud and Lund, Vulliamy, Dent, etc. The only point to remember is that the demand for clocks by these makers far outstripped their output and, therefore, it was common practice for them to purchase some movements from makers in Clerkenwell — Thwaites and Reed or Moores being the two most quoted suppliers. Perusal of the Day Books of both of these makers confirms the practice, together with the fact that their marks can often be found on the dial plate or inside the barrel. Thwaites and Reed (previously Ainsworth Thwaites) have been in business since 1808 and John Moore and Sons (previously Handley and Moore) since 1824 until the end of the century. In a small leaflet of theirs dated 1886 (reproduced in the December, 1968, issue of *Antiquarian Horology*) they state that up to 1877 they had made 15,180 house clocks.

So long as their size is not a deterrent, the large so-called Boardroom or Directors' two or three train clocks of the late 1800s have much to commend them. In common with the longcase clocks of this period the chimes are usually those of Westminster, Whittington or St. Michael, and on bells and/or gongs. Circular tape gongs were used to strike the hour as this note was somewhat more reminiscent of the deep tone of Big Ben than that produced on a small bell. As can be seen from the examples illustrated these were far from cheap clocks. The solid cases being ebonised, mahogany, rosewood or oak either plain, carved or inlaid with brass, ivory, etc. The dials were silvered or brass with gilt brass overlay and small subsidiary dials for Strike/Silent, Chime/Silent, Fast/Slow rise and fall adjustment or choice of chime. The names on the dial being with few exceptions those of the retailers. Most of the movements were English with fusee and chain on both going train and striking train. Although, as can be seen from the examples in Figures 139 and 145 some German clocks of this type were imported. It must be stressed, however, that the movements made by Winterhalder and Hofmeier were excellent movements often incorporating a fusee and chain, and are frequently mistaken today for English movements by those who do not realise the significance of the letters 'W. & H. Sch.' stamped on the backplate.

In 1901 there was a famous test case regarding the use of foreign parts by manufacturers and yet still citing their products as 'Made in England'. After this date a new Merchandise Mark Act came into force and "precluded any British watch or clock factory from using more than sixpennyworth of material" if they wished to continue claiming their product to be British. In efforts to protect the British clock industry the British Clock Manufacturers pressurised for legislation. To some extent they were successful and after April 1st, 1933 all clocks and clock cases manufactured in and consigned from any

Figure 124

Mid-19th century mahogany and brass inlaid bracket clock, the arched case with ebony veneer canted angles, brass grille panels with carrying handles and painted dial inscribed Frodsham London. Subsidiary silent/chime dial in arch. The two train movement is signed Frodsham, Gracechurch St., London and chimes the quarters on eight bells and strikes the hours on a ninth bell. When sold in 1983 it realised £620. Price would have been influenced by pleasing design and proportions of case and a recognisable name of maker.

£1,000+

Courtesy of Phillips, Son & Neale

part of the Empire had to contain a minimum of 50% Empire material and labour in order to qualify for entry into this country under Imperial Preference conditions. Prior to this date only 25% had been necessary. If it had not been for such legislation American manufacturers for example of synchronous clocks could have sent movements for assembly and subsequent dispatch to this country across the borders into Canada.

A further Merchandise Marks Order came into force on the 22nd June 1934. Under this order all imported clocks, clock movements, synchronous electric motors of the type used for clocks and clockwork movements whose primary function was to record time had to bear an indication of the country of origin. The words Foreign Made sufficed in most instances without the need to actually specify the country.

The demand grew for a cheaper chiming clock and about 1890 rod gongs were introduced. The rod gong slowly superseded the more expensive bells and circular tape gongs; although sneered at by most collectors their tone is not unattractive. Prior to the First World War and again between the Wars large numbers of these movements were imported from Germany and cased here. According to one account the chime rods and strike gongs for the clocks made in the Black Forest area of Germany, at the beginning of this century were made in two small factories (Wagner, Schwenningen and Johan Filtner, Villingen) "by crude hand methods". The rods were mainly of phosphor bronze with a few of German silver. The former was reputed to produce a better note. The rods were cut to the required length and tapered at one end. Brass bushes or nuts were then forced on to the tapered end which was then screwed into a block. The tuning was done by reducing the length of each rod until the required sound was achieved, with a good musical ear being the only guide. The spiral gongs were made from a flat section

Courtesy of Derek Roberts Antiques

Figure 125

This is a good example of a rosewood bracket clock of about 1850, striking on a bell with a repeating mechanism. The silvered dial is signed 'Brockbank & Atkins, London.' Similarity of case style of clock illustrated in Figure 127 is not without interest.

£2,000+

Figure 126

This is an extremely attractive maplewood bracket clock of the mid-nineteenth century, striking on a gong. The silvered dial is signed 'Marsh — Maker to HRH the Duchess of Gloucester — Dover Street, Piccadilly.' It is also signed on the backplate.

£2,000+

Courtesy of Derek Roberts Antiques

of wire coiled on a hand coiler and then turned by reducing the length until the required effect was achieved. Apparently it is extremely important that these rods and spiral gongs are kept free from rust otherwise they lose their tone. After gentle removal of the rust with wire wool or a glass pencil (obtainable from clock and watch material suppliers) it is necessary to reblue them. This is a simple but skilled process, involving the heating and cooling of the metal to specified temperatures. The coloured fluid sold to colour the exterior of parts that should be 'blued', i.e. clock hands, etc., is not a satisfactory substitute as it is not the colour that improves the tone but the changes that take place during the heating.

After the First World War many small clock factories opened in England determined to compete with the cheaper pre-war German clocks. Makers, such as Gillett and Johnston and J.J. Elliott, of high quality clocks having been largely unaffected by the competition as their products were aimed at a different market.

To some extent the lesson regarding the necessity for greater standardisation of parts and thereby facilitate interchangeability had been learnt. The all out effort among those making munitions in the First World War had brought the point home and demonstrated the need for designs and methods of production that readily lent themselves to these results. Factories that had been well equipped for the war effort both with good machinery and skilled men now turned their attention to clock making as a furtherance of these skills. The Garrard Clock Company was one such. They had been founded in 1915 by C.E. Newbegin and Major S.H. Garrard — The Garrard Engineering Company — with premises at Stonebridge Park, London. During the war they had made bomb sights, fire detectors, etc. With the cessation of hostilities the decision was made to enter the gramophone component market. They concentrated on motors and soon made a name in this field for excellence of product. As space was required for further expansion they found it necessary to move to Swindon in Wiltshire. It was at this time that they turned their attention to clockmaking. Made at the Swindon factory the movements were examined, cased and dispatched from a depot in Golden Lane, London. See Figure 166 for examples of their products.

Some of these newly formed companies survived and others did not. It has been possible to trace the history of a few of these and there should be a certain amount of interest in the products from factories known to have existed for a short period only, as well as those who continued to flourish.

The name of Grimshaw and Baxter has already been mentioned in association with J.J. Elliott Ltd., makers of longcase and bracket clocks, etc. After this partnership broke up in 1921, Mr. Baxter and H.W. Williamson Ltd. of Coventry amalgamated their clock factories. The firm of Messrs. Williamson was founded in 1871 by H. Williamson and by 1903 they had factories at Coventry, Salisbury and Buren, Chaux de Fonds, Switzerland. Their main offices and showrooms being at 77-79 (and later 81) Farringdon Road, London. Watches were made at Coventry and Buren while the clocks and speedometers were being produced at the Salisbury factory. In 1909 a major fire destroyed this factory with a financial loss to the company in the order of £50,000. It was possible, however, to expand the Coventry factory to provide new and larger clockmaking workshops. One reason why this firm were able to offer such a wide range of cheap longcase and bracket clocks was due to the fact that their founder had realised the need for the 'new' mass production methods and had actually persuaded one of the leading men to come from the Hamburg American Clock factory in Germany to advise. The Buren factory expanded to include the manufacture of clock movements (mainly those with lever platform escapements) but unfortunately as the importation of clocks into this country was heavily restricted at this date (1914) only a few could be imported under licence. The cases were English made. H. Williamson died in 1914 but the firm continued with varying degrees of success until the early 1920s when they decided to amalgamate with Grimshaw and Baxter. By 1925 they had evolved a manufacturing and marketing policy that concentrated on low priced striking movements and thereby were able to compete with other manufacturers. These were marketed under the name of ''Astral'' with a later range ''Empire'' appearing in 1928. By 1930, however, they were once more in difficulties and it became necessary to appoint Receivers. H.W. Williamson Ltd. became part of English Watch and Clock Manufacturers Ltd. which was taken over by Smiths Ltd. in 1934. The names ''Astral'' and ''Empire'' were continued by the successive purchasers of the original Williamson company. Both Astral and Empire pendulum and platform lever movements were still in production in 1942 but by 1955 only the Coventry Astral lever movements were still being produced by Smith's English Clocks Ltd.

A rosewood bracket timepiece with fusee movement, bimetallic balance with compensation, shaped shoulder plates and long arm lever escapement. The enamel dial was inscribed Philcox, London, with the word "Patent" below the winding hole. The latter piece of information instantly awakens interest and the possibility of furthering documentation of this piece. Patents are invaluable as reference to the Abridgments of Specifications — Class 139 — Watches, Clocks and Other Timekeepers 1855-1930 *(or if prior to that to Class 9) provides the full name, occupation and address of patentee as well as details of device patented. If the original design has been subsequently modified one has a further date established. In this particular instance it was possible to ascertain that George Children Philcox had taken out six patents between the dates of 1839 and 1870. These had been in connection with improved escapements, balances etc. Rupert Gould rather harshly refers to his balance with two opposed helical springs "as simply absurd", however, his thoughts on escapements seems to have been a little more viable. His first patent (No. 8145 — AD1939) dealt with The Patent Diamond Lever Escapement. Later patents included modifications. Figure 127b clearly shows some of the features thus patented.*

Courtesy of Sotheby's Chester

Figure 127b

Pigot's Directory *lists George Philcox — Watch and Clockmaker — as working at 24 Great Dover Street in 1836 and by 1842 he had moved to 8 Southwark Street in London. Gould described him as a chronometer maker.*

Case style would indicate a date of first half of the nineteenth century with the gadroon top, shaped door and so on all making for an elegant appearance with the added attraction of an interesting maker. The estimated price for this clock in 1982 was £300—£400. The realised price was £352 including 10% buyer's premium. It should have been more and would be now!

£750+

The 'Newbridge' movements were manufactured by Horstmann Clifford of Bath, Somerset. The introduction of their range of gas controllers in 1921 led them to extend their range of products to clock movements. All of their advertisements lay great emphasis on the fact that their clocks were the outcome of *precision* mass production and that all parts were fully interchangeable.

'Tasty Bracket Clocks' was the appellation given in 1921 to one style of clock manufactured by the Hirst Bros. & Co. of Roscoe Street, Oldham! Their Tameside factory had been built in 1919 at Dobcross, Nr. Oldham, for the manufacture of cases and movements built up of standardised parts. Their clocks bear the trade name of 'Tameside'.

Hirst Bros had for many years been major importers and wholesalers of clocks, tools and materials. A range of the clocks being imported by them for resale to retailers is shown in a recent reprint of a selection from their 1910 catalogue, *Catalogue of Clocks*. This shows mantel clocks, 'Vienna regulators', alarms, etc., mostly of German origin. Pages from their materials catalogue (fourth edition of the Wide-Awake catalogue c. 1915) can be found in Appendix I.

Many of these small companies were eventually absorbed by what is now known as Smith's Industries and most of their history and archives lost in the take-over. It is not generally known that the Enfield Clock Company of Tottenham, London, prior to becoming part of the Smith empire after the Second World War had been founded by August Schatz and a few associates. Immediately prior to the war they brought over machinery from Guttenbach in the Black Forest of Germany and commenced production. Their clocks were excellent and retailed well.

The variations in the case styles of these clocks are far from inspiring. It would only be makers such as Gilletts and Johnston, F.W. Elliotts Ltd. and a few others that would make their own cases. The remainder of manufacturers would purchase them in bulk from a case maker.

Figure 128

A fine example of an eight-day striking clock in an overtly neo-Gothic styled rosewood case and matching stand. The legend on the dial (E.J. Dent London 655) would indicate that the clock was made by Edward John Dent, c.1845.

£750+

Courtesy of Kingston Antiques

Figure 129

A walnut Victorian bracket clock again in the Gothic style with slow/fast and chimes subsidiary dials, a three train fusee movement, an anchor escapement and with quarter striking on eight bells. This has been included in order to make comparison with the similar example shown in Figure 130. The difference in quality is immediately apparent in so far as this example has a well figured walnut case, more ornate carved embellishments with shaped top to door — all of which would have incurred additional cost at time of manufacture as well as the movement having choice of strike, etc. The fact that the first example has a known maker's name — Dent — whereas the maker of this example is unknown would of course detract interest for some collectors. This ably demonstrates the collector's dilemma of having to choose between the 'named' item or one that on face value appears of better finish and quality. A difficult decision if potential resale is being considered as sadly a 'name' carries a great deal of weight unless the purchaser is discriminating and prefers in some instances to have an unnamed but perfect example in his possession. This particular example realised £820 when sold in 1979.

£1,000+

Figure 130

An early Victorian rosewood timepiece with circular silvered dial. The movement has chain and fusee, screwed pillars, latched dial feet, deadbeat escapement and ebony rod pendulum. Both backplate and dial signed Vulliamy, London, 853. The Vulliamys were a noted family of clockmakers of Swiss origin. Justin Vulliamy came to London, early in the eighteenth century. His son Benjamin (1747-1811) and grandson Benjamin Lewis (1780-1854) both held the Royal Warrant, in fact at the time of his death the family had received Royal patronage for 112 years, during which period they had served five succeeding monarchs. Benjamin Lewis has often been condemned for his bespoiling of the work of other frequently eminent makers. Many of the clocks in the Royal collections commenced life with movements made by earlier clockmakers which were replaced by those of Vulliamy's making. It must be remembered however that at that date there was little antiquarian interest and practicality i.e. timekeeping ability, was of prime importance.

Full details of the family's numbering system will no doubt be provided in the publication at present being written by Nicholas Goodison on the family and their work.

Clocks with names of important makers such as Vulliamy ensure a good price. £3,000+

It is known that in 1933 J. Hodgkinson & Sons of 110-116 Elmore Street, London N.1 were making in excess of 80,000 cases a year in the proportions of 22,000 small shelf clocks, 28,500 bracket clocks, 650 wall clocks and 5,600 longcases of varying sizes. Other case makers working in 1928 included Wilson Warden & Co. Ltd. of Peartree Street, Goswell Road, and West Woodworkers Ltd. of 76 White Lion Street, both addresses in the Clerkenwell/Islington vicinity of London. The latter advertised Jacobean style longcases and the small mantel clock cases made from the solid with provision for a circular drum cased movement. Figure 168 shows an advertisement for yet another casemaker.

It is possible to obtain a very rough approximation of the date of these late clocks by the designs of the cases. Most readers will be familiar with the surge of mock Jacobean oak or veneered oak furniture and fabrics that were popular during the 1920s. Coinciding with this fashion was that of Japanese lacquer work "executed by native craftsmen on our own premises". Examples of these various styles will be found in Figures 166 to 168, which include the faithful 'Napoleon Hat' design.

Figure 129

Figure 130

Courtesy of Christie's

Figure 131

James Moore French became a Freeman of the Clockmakers' Company in 1810 and is recorded as working in various parts of the Royal Exchange Buildings (Sweetings Alley, Cornhill) between 1811 and 1846. William French is listed as working between 1849 and 1875. The occupations of both are given as watch and chronometer makers. It would appear from the heavily carved style of case that this would be a later piece. Not to everyone's taste it is however a good quality walnut case well carved, with much detail. It has a three train fusee movement with quarter striking on eight bells with provision for Strike/Silent at the top of the silvered dial. The dial is signed within the cartouche — French — Royal Exchange — London.

£500+

Figure 132

A plain but highly desirable bracket clock in walnut case striking on a gong. Both the dial and backplate are signed 'Dent 33, Cockspur Street, Charing Cross, London'. Edward John Dent (1790-1853) was associated with Lord Grimthorpe in the designing of the Westminster Clock (better known as Big Ben), with his stepson Frederick finalising the work after his death, from premises in the Strand, London. The other brother, Richard, having been bequeathed the Cockspur Street premises. The two businesses merged in 1921, when they moved to their present address in Pall Mall, London. See Edward John Dent and his Successors by Dr. Vaudry Mercer for a full account of the Dent family business.

£1,000+

Courtesy of Derek Roberts Antiques

As most factories were endeavouring to achieve a standardisation of parts and production methods there is not always a great deal of difference in the movements. However, a few interesting variations can be found in the designs used by the different manufacturers of self-correcting chime devices — an innovation that had obviously became necessary in order to obviate unwarranted complaints after customers had attempted to 'adjust' them. Many movements now had provision for the removal of the barrel without having to dismantle the whole movement. Anyone wishing to study these finer points in detail should consult repair books written in the 1930s and 1940s, *The Modern Clocks — Their Design and Maintenance* by T.R. Robinson or the *Practical Watch and Clock Maker* magazines. These frequently described in detail the 'new' movements as they appeared on the market.

For those who cannot afford the earlier bracket clocks there is a case to be made out for these later pieces. It cannot be ignored that they sound well! At the moment the price of these post-1920 clocks is being influenced by furnishing trends, but if they are ever to be considered potential collectors' pieces they must be in excellent and original condition. The quality of the movements does vary and if possible those with solid pallets against those with strip pallets, etc., are to be more desired. It is only necessary to pause for a moment and reflect as to which type or quality of component would have cost more to produce, and therefore have been a more expensive item originally, to access what will also be prized as a collector's piece in the future.

Figure 133

A further example of a bracket clock by Dent, in a walnut case, with quarter striking and repeating on eight bells or four gongs.

£1,000

Courtesy of Derek Roberts Antiques

Figure 134   Figure 135

Figure 136   Figure 137

Figures 134, 135, 136 and 137

These clocks usually stood on small matching brackets in offices, boardrooms and possibly libraries or halls. Comparison of the two examples in Figures 134 and 135 provides some points of interest. The similarity in design of the two mahogany cases leads to speculation as to whether they were manufactured by the same casemaker. The similarity ends here. The example in Figure 134 has a brass dial signed 'Thompson and Vine, London'. These were retailers known to have been in business in 1890 at 85 Aldersgate Street, London, E.C. The two train movement strikes on a gong. The example in Figure 135 has a painted dial with a centre alarm disc while the movement has a fusee with a chain. It is extremely unusual to find such a movement with an alarm in this quality case. The clock in Figure 134 sold for £130 (1977), and it had been estimated that the other example would fetch a little more, however it remained unsold probably through not reaching the reserve price.

The remaining two examples were in mahogany and walnut cases respectively. Again the names on the dials are those of the retailers. Both were striking, that in Figure 136 on a gong, and both fetched the sum of £140. (1977).

£200+

There is little doubt that this solid oak cased clock with its motto of 'Tempus Rerum Imperator' carved on the plinth was intended for anything but a boardroom! In common with many of the quality cases of this period the fretted sides are held in place by turn buttons and can be readily removed. The chapter ring and rings on the three subsidiary dials are silvered but the rest of the dial and overlay are gilt brass. The dials are Chime/Silent, choice of Whittington or Westminster Chimes and Fast/Slow adjustment through the dial arch. By turning the pointer on the appropriate dial the whole bar upon which the pendulum is suspended is raised or lowered thereby adjusting its effective length. Although there is no maker's mark visible on the backplate (there may well be on the front plate or in the barrels) there is no doubt that this is an English movement with fusees and chains. For comparison see the example in Figure 139. The quarter chimes are on four or eight bells depending upon chime chosen, with the strike on a gong.

£1,000+

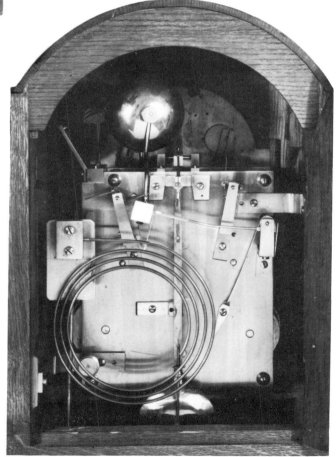

Figure 138b

Movement of clock shown in Figure 138a. Points to note are the general sturdiness of construction, nest of bells from back to front above the movement and coiled tape gong. The flat-topped knob at the bottom left-hand corner can be removed and reinserted in the threaded hole to the left of the pendulum rod — this is to hold the pendulum in place while the clock is in transit.

139a

Figures 139a and 139b

22½ins. high. The case of this clock is extremely ornate and with its ebonised wooden case and lacquered brass cast mounts and brass dial is reminiscent of the design of much earlier bracket clocks. It is, therefore, somewhat of a surprise to find that the movement is rather puny for the size of the case and is actually of Swiss manufacture, as can be seen in Figure 139b. Striking is at the hour and half hour on a single gong, and provision has been made for repeat at will. Note the maker's mark that can be clearly seen at the bottom of the backplate. This Swiss trade mark was registered in 1885 as being that of Paul Vuille Perret of Chaux de Fonds. The value of this clock would lie in its decorative quality rather than quality of movement.

£750+

139b

Courtesy of Christie's South Kensington

Figure 140

Figure 141

Figures 140 and 141

These illustrations show two clocks made by A. and H. Rowley of London, there having been a firm of this name since 1808. This name appears in many advertisements of clockmakers in the trade journals of the later half of the nineteenth century as well as among the list of exhibitors at the various Exhibitions. The International Inventors Exhibition of 1885 list A. & H. Rowley of 180 Grays Inn Road, London, W.C., as having the following items on display:

1. Large chiming clock with newly invented automatic figure work and perpetual calendar.
2. Clock with improved chiming on bells and gongs in newly designed Gothic oak case.
3. Lever clock, striking the ship's bells including the dog watch.
4. Parts of clocks showing the stages of manufacture.
5. Various chiming and other clocks with all the most modern improvements.

£750+
**an example in an oak case realising a lower price than a similar clock in a mahogany
or rosewood case.**

Figures 142a and 142b

A late Victorian chiming bracket clock with the case painted somewhat unusually, i.e. to simulate tortoiseshell, with ebonised mouldings and pofuse gilt metal mounts. The three train chain fusee movement with an anchor escapement, chimed on 8 bells or 4 gongs, and struck on a further gong. It was estimated when sold in 1983 at £800 to £1,200. A handsome although rather overpowering example, it actually realised £1,000.

£2,000+

Courtesy of Christie, Manson, Woods

142a

Figure 143

Large oak cased chiming bracket clock on matching plinth in the Gothic style. The name A. & H. Rowley, 180 Gray's Inn Road, London appears on the dial. The massive fusee movement chiming on a series of eight bells and five gongs has an anchor escapement with a steel pendulum rod and brass cased bob.

With matching pedestal

£2,000+

142b

Courtesy of Sotheby's, Chester

Courtesy of Kingston Antiques

Figures 144a and 144b

This is a handsome rosewood bracket clock, with good quality brass caryatids, pineapple finials and feet. To all outward appearances a pleasant late nineteenth century or early twentieth century English clock. However, upon looking at the movement it is quickly realised that it is German and actually manufactured by Winterhalder & Hofmeier (W. & H. Sch. on the backplate). It would appear that the case is of English origin made to house an imported movement, a common occurrence at this time. Although not one of their best movements it is an extremely attractive and well made clock. Again a recognised German manufacturer of repute.

£700+

Nº 7.

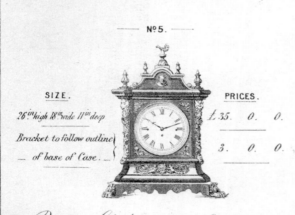

SIZE.		PRICES.
24ᵗʰ high 17ᵗʰ wide 11ᵗʰ deep		£30. 0. 0.
Superior Quality		34. 0. 0.
Bracket to suit		2. 10. 0.

*Quarter Clock to chime the Quarters on
8 Bells strike the hours on Gong with Ornamental Dial and Engraved and Silvered
circles for hours, Chime Silent, and Regulating,
in Ebonized Case with centre Pine on top of
Arch on plain square brass base, Pines at
the corners on bases, small Corinthian Pillars
at corners of body of Case ornamental feet.*

A. & H. ROWLEY, *180, Grays Inn Road.*

Nº 5.

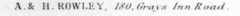

SIZE.		PRICES.
26ᵗʰ high 18ᵗʰ wide 11ᵗʰ deep		£35. 0. 0.
Bracket to follow outline		
of base of Case.		3. 0. 0.

*Quarter Clock with double Chimes, hours
on Musical Gong, with circular Dial, engraved
and silvered in Oak Case, with bronzed Ornaments consisting of Centre Vase, on top of Arch
4 Vases in front of top, below Arch, Lions Heads
with Festoons of Flowers at corners of body
of Case, Lions Claw Feet. Royal Head in front
of top Arch, Corner and front Ornaments
bright Bezel and Silvered bevel surrounding
Dial which is 8ᵗʰ diameter.*

Figure 145

*Four of the clocks shown in catalogue of A. & H.
Rowley while in business at 180 Gray's Inn Road,
London. It is known that by 1925 they were
working from 49 Farringdon Street, London. (Note
similarity of case in model No. 7 to that shown in
Figure 141.)*

— N.º 8. —

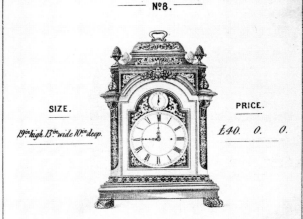

SIZE.

19.ᵢₙ high 13.ᵢₙ wide 10.ᵢₙ deep.

PRICE.

£40. 0. 0.

Quarter Clock to chime the Quarters on 8 or 4 Bells, strike the hours on Musical Gong with the Regulator in front, Ornamental Dial 6ᵢₙ across in very handsome black Case, with brass ornaments at the angles, Bell shaped top with brass mouldings and scroll work, brass top handle, Corners with scrolls, scroll feet, Top Vases Cone shape, same kind of ornaments at back and sides as in front.

A. & H. ROWLEY, 180, Grays Inn Road.

— N.º 17. —

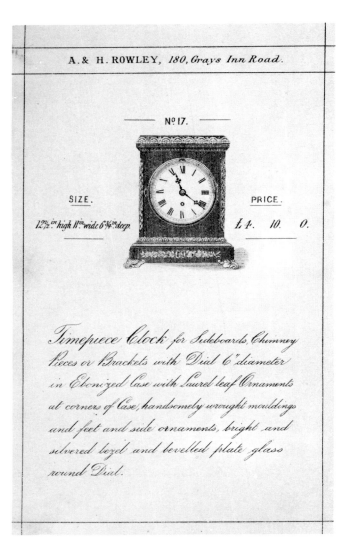

SIZE.

12¹/₂.ᵢₙ high 11.ᵢₙ wide 6¾.ᵢₙ deep.

PRICE.

£ 4. 10. 0.

Timepiece Clock for Sideboards, Chimney Pieces or Brackets with Dial 6ᵢₙ diameter in Ebonized Case with Laurel leaf Ornaments at corners of Case, handsomely wrought mouldings and feet and side ornaments, bright and silvered bezel and bevelled plate glass round Dial.

Figure 146

19¾ins. high. It is not uncommon to find lacquered cases to clocks manufactured in the 1920s — the oriental influence was, however, attributable to our association at this time with Japan and not China and showed itself in much of the furniture, and other designs of this period; the Japan-British Exhibition held in 1910 no doubt being an influencing factor. This is a particularly handsome example with a silvered dial and striking the hours on a gong. Small decorative bells hang from the pagoda style top of the case. There are no identifying marks on the backplate of the movement and this clock could have come from a number of manufacturers including B.T. Greening Ltd., of Hatton Garden, London, E.C.1., who are known to have exhibited similar cased clocks at the Empire Exhibition at Wembley in 1924. This clock was sold by auction for £380 in 1976.

£500+

Courtesy of Kingston Antiques

Figures 147a and 147b

A highly decorative and ornate three train bracket clock attributed to F.W. Elliott Ltd., with silvered dial and subsidiary dials. The chime is either the Westminster, Whittington or St. Michael on bells or gongs.

£2,000+ to an appreciative purchaser

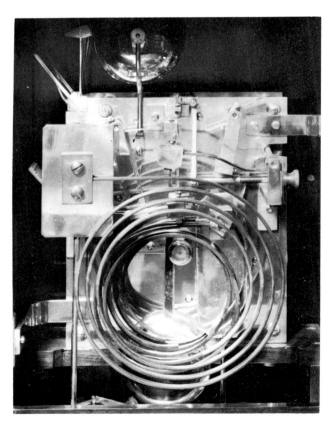

ENGLISH (LONDON MADE) BRACKET CHIMING CLOCK.

No. 35.—Finest Mahogany Case, with finely Chased and Gilt Brass Mountings well Lacquered, Fine Hand Pierced Dial, with Silvered Circles, Raised Gilt Bevelled Arabic Figures.

Height 30ins.   Width 15ins.   Without Bracket.

						£	s.	d.
8-day Fusee and Chain Movement, Chiming on 8 and 4 Nickelled Tubes	...	...	...	73	0	0
,,	,,	,,	,,	,,	,, 8 Bells and 4 Gongs	66	0	0
,,	,,	,,	,,	,,	,, 8 and 4 Gongs ...	69	0	0
,,	,,	,,	,,	,,	,, 4 Gongs ...	63	0	0
,,	,,	,,	,,	,,	,, 8 and 4 Bells ...	62	0	0

In each case the Hour is struck on a deep-toned Gong.

Engraved Silvered Dial, £2 0 0 less.   Bracket to match, £19 0 0

G. B. & J. J. E. Ltd.

Figure 148

The point to note in this full-page advertisement of an 'English (London Made) Bracket Chiming Clock', that appeared in the Grimshaw Baxter and J.J. Elliott Ltd. catalogue sometime after 1909, is that it was offered with "Chiming on 8 and 4 Nickelled Tubes" as well as the more usual gongs or bells. Although it is stated that the case was some 30ins. in height and 15ins. wide the inclusion of tubes would have been no mean feat. It is a magnificent example, and it may well be that this clock was intended for the overseas market. India, in particular, was an ardent supporter of the English clockmaker at this date, and many makers offered to adapt the construction of cases and movements to withstand the warmer climate.

Figure 149

Another example of a quality mahogany quarter chime bracket clock made about the turn of the century. At 19ins. high it most certainly would not have stood on many mantelpieces. The three train movement with strike on eight bells and three gongs with choice of Whittington/Westminster chimes at quarters. The dial was inscribed with the retailer's name, Russell's Ltd., 18 Church Street, Liverpool. The manufacturer could have been any of those previously mentioned in connection with similar clocks. When sold in 1982 it realised £620.

£1,000+

Courtesy of Sotheby's, Chester

Antique Brass Clocks

Copies of
OLD BRASS
ELIZABETHAN
TABLE CLOCKS
Eight Day
Lever Timepieces

By Bartholemew Newsam
Date about 1580
Height 8¼ins.
£6/0/0

By
← Emmanuel Bull
Date about 1600
Height 8¾ins.
£6/0/0
1 blow at the hour Strike
£6/10/0

An Antique Brass Reproduction
Elizabethan Table Clock
From the original by
Johan Scheirer   *Date about* 1590
Height 5½ins.   *Width* 5½ins.   *Depth* 4½ins.
8-day Lever Timepiece £7/0/0

Antique Brass Case
BRASS DIAL, SILVERED CIRCLE
Height 9½ins.   *Width* 5½ins.
From the original by
Richard Peckover
8-day Lever Timepiece £8/15/0

Antique Round Brass Table Clock
From the original by
L. Definod of Cien
Date about 1560
Height 4¾ins.   *Diameter* 5ins.
8-day Lever Timepiece £7/0/0

Figure 150

Reproductions of Antique clocks offered by Grimshaw, Baxter and J.J. Elliott Ltd., 29 to 37 Goswell Road, London, E.C.1., and 81 Mitchell Street, Glasgow — Clock Factory, Coventry. It would appear that this brochure appeared some time after the amalgamation of the Williamson and Grimshaw factories, i.e. after 1925. Extremely interesting to note that case styles were taken from originals although the thought of a John Knibb with Westminster Chime is rather mind shattering.

£100+

Antique Reproductions in Walnut

~[MADE ENTIRELY IN ENGLAND]~

Antique Walnut Case
BRASS DIAL, SILVERED CIRCLE

From the original by
THOS. TOMPION, *London*
Date about 1690
Height 13½*ins.   Width* 9*ins.*
ELLIOTT 4/4 *Westminster Chime*
£44/0/0
ELLIOTT *Strike* £28/0/0
EMPIRE *Strike* £11/10/0

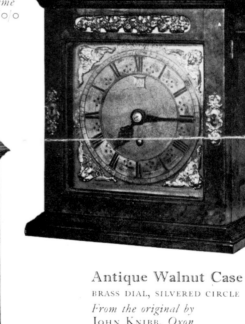

Antique Walnut Case
BRASS DIAL, SILVERED CIRCLE

From the original by
CHAS. GOODE, *London*
Date about 1690
Height 15*ins.   Width* 9*ins.*
ELLIOTT 4/4 *Westminster
Chime* £44/0/0
ELLIOTT *Strike* £28/0/0
EMPIRE *Strike* £10/15/0

Antique Walnut Case
BRASS DIAL, SILVERED CIRCLE

From the original by
JOHN KNIBB, *Oxon*
Date about 1700
Height 14*ins.   Width* 9*ins.*
ELLIOTT 4/4 *Westminster Chime*
£44/10/0
ELLIOTT *Strike* £28/10/0
EMPIRE *Strike* £10/15/0

Models offered with our EMPIRE *quality movement have
plain wooden sides to cases and solid wooden back doors*

Lever Escapements 25/- *extra*

£200+

Antique Reproductions in Walnut & Mahogany

·[MADE ENTIRELY IN ENGLAND]··

Antique Mahogany Brass Mounted Case

BRASS DIAL, SILVERED CIRCLES
From the original by
JOSIAH EMERY
Date about 1800
Height 19½*ins. Width* 11*ins.*
ELLIOTT 4/4 *Chime*
£58/0/0
ELLIOTT *Strike*
£42/0/0

Antique Walnut Case

Silvered, Oxydised, Chased & Engraved Top
BRASS SILVER OXYDISED DIAL,
SILVERED CIRCLE
From the original by
RICHARD JARRETT, *London*
Date about 1690
Height 15*ins. Width* 11*ins.*
ELLIOTT 4/4 *Westminster Chime*
£60/0/0
ELLIOTT *Strike*
£44/0/0

Lever Escapements
25/- extra

Antique Walnut Case

BRASS DIAL, SILVERED
CIRCLES
From the original by
WM. TOMLINSON, *London*
Date about 1700
Height 19*ins. Width* 11*ins*
ELLIOTT 4/4
Westminster Chime
£55/10/0
ELLIOTT *Strike*
£39/0/0

£200+

141

Figures 151a and 151b

The case of this clock is mahogany, quite unembellished but of pleasing proportions with small bun feet and silvered dial. Although no maker's marks are apparent, the words 'Made in England' together with a serial number do appear on the backplate. The movement is three train chiming the quarters and striking the hours on gongs. The mechanics of this can be readily seen in the back view. The quality is far superior both of case and movement to the example shown in Figure 162.

£100+

Figures 152a and 152b

What appears to be a simple kitchen clock has, on closer inspection, several interesting features. The softwood case is veneered with mahogany and the painted zinc dial has the name of the retailer, 'Camerer Kuss & Co. 56 New Oxford Street', inscribed on it. The firm was established in 1788 when their founder came from the Black Forest area of Germany in order to open a clockmaking and importing business in London. Initially they specialised in typical Black Forest clocks but soon expanded to include all types of timepieces. The only change made to the name of the firm is the substitution of the letter 'C' for that of 'K' in the name 'Cuss'.

This clock was an attempt to offer a reasonably priced but nevertheless better quality example than the average German and American kitchen clock of this date. Examination of the backplate (Figure 152b) reveals the initials 'W. & H. Sch.' These are significant and indicate that this movement was a product of the firm Winterhalder and Hofmeier of Neudstadt in the Black Forest area of Germany. The plates are well made and the movement generally lacks the machine stamped-out appearance of most of the German and American mass produced movements of this period. The manufacturer has long been recognised as having produced some very fine quality clocks – longcase, bracket clocks, dial clocks – which rival those of many English makers of the late nineteenth and early twentieth centuries. In many instances it is only upon observing the letters 'W. & H. Sch.' stamped on the bottom of the backplate or movement that positive identification is possible. If merely unidentified German make price would be £40+.

£75+

English Bracket Clocks.

No. 2812.
INLAID MAHOGANY CASE.

Size, 13½ in. high, 8½ in. wide. 5¾ in. Convex Silvered Dial. Solid Bezel. 8-day Polished English Movement.

Striking on Gong	.. ..	**£5 10 0**
If fitted with Unpolished Movement	..	**4 10 0**

English Bracket Clocks.

No. 2811.—INLAID MAHOGANY CASE.

Size, 13½ in. high, 8½ in. wide. 5¾ in. Convex Silvered Dial. 8-day Polished English Movement. Striking on Gong **£5 10 0**
If with Unpolished Movement **4 10 0**

No. 2713.
INLAID MAHOGANY CASE.

Size, 15½ in. high, 12½ in. wide. 7-in. Convex Silvered Dial. Solid Bezel. English Fuzee and Chain Movement.

Strike Hours and Half-Hours on Gong .. **£10 10 0**

26

No. 2723.—POLISHED MAHOGANY CASE.

Size, 15 in. high, 12½ in. wide. Brass and Silvered Dial. 8-day English Fuzee and Chain Movement, Strike on Gong **£10 0 0**
Fitted with Ting Tang Movement.. .. **10 10 0**

27

Figure 153

Two pages of typical clocks being manufactured by H. Williamson and Son in the 1930s. It is interesting to note the case styles also commonly used by other manufacturers.

£100+

Figures 154 and 155

Two examples of run of the mill pre-Second World War case designs. That in Figure 154 is intended to suggest a Jacobean style and would fit in well with the catalogues of Jacobean furniture, c.1924, which were particularly popular at that point, although the decorative dial centre and hands do not make for easy time telling. The example in Figure 155 is more suggestive of the art deco theme and has a plain silver (anodised) dial. Both have Chime/Silent levers on the outer edge of the dial. It has been suggested by one contemporary that the striking and chiming clocks proved to be a constant source of irritation to those wishing to listen on their wireless sets to the nine o'clock news, evidently the event of the day in many households! The solution was either to set the clock fast or slow or purchase one with a Chime/Silent lever. These clocks have been very much under-estimated in recent years.

£75+

Figure 156

The majority of these 'Napoleon Hat' shaped clocks are oak veneered, although this is in good quality mahogany with a silvered (anodised) dial with Silent/Chime lever at the three o'clock position. Other quality features are the clever use of multi-wood veneers.

£75+; the quality of the case playing an important role

Figure 157

Two of the small bracket clocks manufactured by Wm Potts of Leeds during the 1920s. See page 239 for further details of this manufacturer.

The Enfield Clock Company was founded in 1929 by Carl Schatz. Factory premises were purchased at Pretoria Road, Edmonton London N.18. Marketing came under the auspices of the newly formed partnership of Schatz Rombach Ltd. Workshop machinery and initially some clock parts were shipped from Germany as Schatz intended that production methods would be similar to those employed by the German manufacturers. The first movements were sold to the trade in February 1932. These were strike movements with lantern pinions or Westminster chime movement with solid pinions. The price of the former being 27s. 6d. while the latter sold for 55s. By the end of the year the toolroom and workshops were running smoothly and a chrome plating department had been added to the factory. Bezels were now made at Edmonton but gongs and dials were supplied by other English manufacturers. It would be interesting to compare a movement made at the beginning of the year with later models to ascertain whether the early examples using some German components were in any way different from those manufactured by the end of the year with all components made in this country.

Although prices were competitive the established German factories such as Kienzle and Junghans could still undercut prices and towards the end of 1933 it became obvious that amalgamation with another firm or the outright sale of the company was inevitable. Smith's Industries finally bought out the shareholders but left the existing directors C.T. Baxter, J.E. Roles and C. Schatz in control. Revitalised, the firm continued to expand and by the end of 1935 they were able to introduce grandfather and grandmother weight driven movements as well as strike wall regulators into their range. These were followed by a striking 14 day movement complete to sell in a Jacobean oak case for 10s. This particular model proved particularly successful as Dupon Bros. placed an order for 25,000.

With the outbreak of the Second World War in 1939 the factory was turned over to war work. A few clock movements were made as and when supplies of materials allowed. Most of these went to the NAAFI for Army and Navy use. After the war it was necessary to take stock of the state of the horological trade. It was decided to continue the production of the very successful 2ins. movement at the Smiths Industries assembly factory at Cricklewood. The Edmonton factory was sold in the late 1940s and production of the remaining "Enfield" movements was undertaken at the new Smiths Industries factory in Wales.

Further details of the history of this company and the movements made by them either while still under the management of the Enfield Clock Co. or Smiths Industries can be found in the May, 1980 issue of *Clocks* magazine in an article by Rita Shenton entitled 'History of the Enfield Clock Company — Battle against Foreign Competition'. It was an interesting company and was one of the pioneer ventures in this country to introduce mass production methods, etc. in an attempt to compete with foreign companies. See Appendix II for details of movements still in production in 1940.

Figure 158

Advertisement for Enfield Clocks dated August 1940. Note the trade mark — ENFIELD Guaranteed British Made Clocks. There was a great play on the customer's patriotism then and during the preceding few decades.

£50+ if with strike/chime

ENFIELD

movements now offered in a

NEW RANGE

of Cased Clocks...

In order to consolidate your connection with goods that are guaranteed to give reliability and service insist when ordering that the movement and dial are to be branded "ENFIELD"—the highest quality of British clock manufacture.

ALL striking clocks are fitted with 14-day, hour and half-hour strike movements, and all chiming clocks with the now famous 'underslung' Westminster chime movements.

Wholesale & Export only

ENFIELD CLOCK COMPANY (London) LTD.
PRETORIA ROAD, EDMONTON, LONDON, N.18

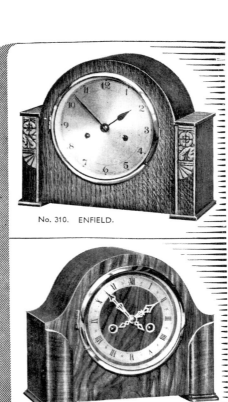

No. 310. ENFIELD.

No. 312. ENFIELD.

No. 334. ROYAL.

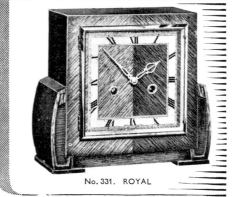

No. 331. ROYAL

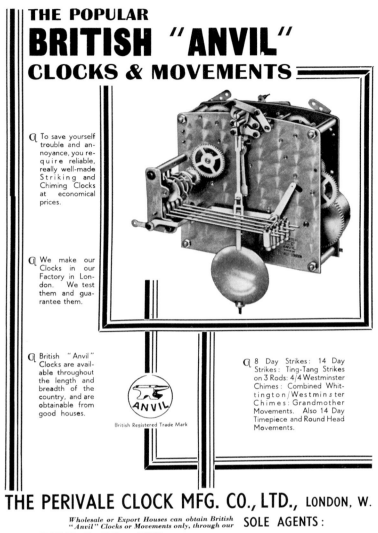
Figure 159

The "Anvil" trademark was that of The Perivale Clock Manufacturing Co. Ltd. of Wadsworth Road, Perivale, Middlesex. Their clocks were marketed by Andrew and Co. Ltd. 67-68 Hatton Garden, London E.C.1. Their range included movements for 8 day or 14 day striking/bracket clocks, ting tang striking on three rods, combined Whittington and Westminster chimes as well as modified versions of these for small longcase clocks (familiarly referred to as Grandmother Clocks) or just simple 14 day timepieces and dial clocks. The Britannia trade mark was in use for Perivale clocks in 1937 when they were being marketed by the Bentima Co. Ltd.

The Westminster quarter chime movement illustrated here was first marketed in 1934. The method of regulation peculiar to these clocks appears to be somewhat complicated but it enables the pendulum length to be adjusted by means of a small lever through a slot in the dial above the figure 12. Other features included a lead pendulum bob encased in brass; pillars secured by nuts; barrels easily detachable by letting down the spring, and removing two screws on the strap plate without further disassembly of movement; self correcting chimes and provision for Chime/Silent. The Anvil trademark appears on the bottom right hand corner of the back plate.

ENFIELD ROYAL
DE LUXE WESTMINSTER CHIMING CLOCKS
British Made

Montrose. 8½×12 inches.
Oak, Mahogany or Walnut with
Chrome and Ebony Relief.
Oak, £3 15 0F
Mahogany or Walnut, £4 5 0F

Dysart. 8¾×11¾ inches.
Oak, Mahogany or Walnut,
with Chrome Relief.
Oak, £4 15 0F
Mahogany or Walnut, £5 10 0F

Avebury. 9½×11 inches.
Oak or Walnut Case
with Simple Inlay.
Oak, £4 10 0F
Walnut, 5 5 0F

Devonport. 8¾×11¾ inches.
Walnut with Macassar and
Sycamore Inlay.
Walnut, £5 5 0F

Warwick. 8½×12 inches.
Oak with Macassar Relief.
Oak, £4 12 6F

Kinross. 8½×12½ inches.
Oak, Mahogany or Walnut
with Macassar Relief.
Oak, £4 17 6F
Mahogany or Walnut, £5 7 6F

Argyll. 8½×12¾ inches.
Oak or Walnut with
Macassar Relief.
Oak, £5 5 0F
Walnut, 6 0 0F

Buckingham. 8¾×12 inches.
Walnut with Thuya
and Ebony Inlay.
Walnut, £5 12 6F

Sutherland. 9×13 inches.
Walnut with Thuya
and Ebony Inlay.
Walnut, £4 17 6F

Enfield Royal clocks have high grade 4/4 Westminster Chime Movements and are sold under **5 years guarantee.**
Every care and every possible improvement to secure accurate performance is used in their construction, and each clock has special
packing and special safety devices to ensure safe delivery.
(F) *Minimum Retail Selling Prices.*

Figure 160

Part of the Enfield Royal de luxe range of Westminster Chime clocks that was introduced into the range 1936-7. This range was named after the Royal Dukes. The plates were damascened, the pendulum and gong rods chromed with the underslung chime allowing for a slim fashionable case. James Walker were one of the major stockists.

£50+

150

ENFIELD ROYAL
DE LUXE WESTMINSTER CHIMING CLOCKS
British Made

Portland. 8¾ × 13 inches.
Mahogany or Walnut with
Relief and Carved Mouldings.
Mahogany or Walnut, £6 6 0ꜰ

Stamford. 9¼ × 13¾ inches.
Mahogany or Walnut with
Carved Moulding.
Mahogany or Walnut, £6 17 6ꜰ

Grosvenor. 9¼ × 13½ inches.
Oak or Walnut with
Hand Carved Moulding.
Oak,   £7 10 0ꜰ
Walnut,   8 10 0ꜰ

Cumberland. 8¾ × 11½ inches.
Oak or Walnut with
Macassar Relief.
Oak,   £5 17 6ꜰ
Walnut,   6 12 6ꜰ

Cromer. 9 × 16 inches.
Oak or Walnut with
Macassar Relief.
Oak,   £5 12 6ꜰ
Walnut,   6 7 6ꜰ

Chandos. 8¼ × 11¼ inches.
Oak or Mahogany with
Carved Moulding.
Oak,   £5 17 6ꜰ
Mahogany,   6 12 6ꜰ

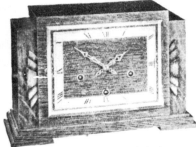

Melford. 8¾ × 11½ inches.
Walnut with Macassar Inlay.
£6 15 0ꜰ

Pembroke. 8½ × 13¼ inches.
Oak or Walnut with
Hand Carved Moulding.
Oak,   £6 5 0ꜰ
Walnut,   7 0 0ꜰ

Granville. 8¼ × 13¼ inches.
Oak or Walnut with
Ebonised Relief.
Oak,   £6 10 0ꜰ
Walnut,   7 5 0ꜰ

Enfield Royal clocks have high grade 4 4 Westminster Chime Movements and are sold under 5 years guarantee.
Every care and every possible improvement to secure accurate performance is used in their construction and each clock has special
packing and special safety devices to ensure safe delivery.
(ꜰ) *Minimum Retail Selling Prices*

£50+

WATCH AND CLOCK MAKER—NOVEMBER, 1936

EDWARD VIII CORONATION CLOCK

S. Davall & Sons Ltd.,
262. 264. 266. St. John St.,
London. E.C.1.

Coronation enthusiasm will become more intense every week from now. There will be an enormous demand for this clock. You cannot order too early. Be in time. Let your wholesaler know your requirements at once.

Figure 161

A special Coronation clock was manufactured by Davall & Sons in 1936 in preparation for the Coronation of Edward VIII as can be seen by this illustration taken from an advertisement of that date. A highly prized collector's piece if any were allowed to remain on the market. It is more than likely that the production run of dials had not been completed before the news of the Abdication broke so few examples were ever made of this particular design. The finalised product had a shield with the lions of England and Scotland and the harp of Wales together with the words "Crowned 12th May 1937" appearing in the centre of the dial with the initials "G R" and the figures "VI" either side. The side panels were engraved with lozenge ornamentation containing lions and fleurs-de-lys. The rear door was inscribed "Davall, 1937, London". Obviously these would make choice acquisitions for the collector of Coronation memorabilia as well as the clock fiend.

Collector's price £200+

Figure 162

Tower striking movement. Note cylindrical bob and strap method of affixing movement to case. The fact that these clocks were British made and "A Clerkenwell Product" was greatly stressed by their wholesalers.

Below, a few examples of the excellently designed London Made cases that house the "Tower" striking and chiming movements.

£40+

The formation of S. Davall & Son Ltd. of 262/266 St. John's Street, Clerkenwell, London was registered December 1931. By 1932 their "Tower" striking movement was being marketed by Dimier Bros. Ltd. 10-12 Charterhouse Street, London E.C.1. and subsequently shown at the 1933 Trade Fair "as the smallest striking clock on the market". It is as well to remember that the Trade Descriptions Act was not in force at this date and the accuracy of the statement has not been ascertained. A striking model with quarter chime and self correcting gear was introduced in 1932 (see Figure 163). This was a compact going barrel movement with rack striking mechanism. The plates measured 4½ x 3¾ ins. Some of the identifying features were:

Pillars riveted to back plate and secured by nuts to front plate.
Rack, detent and clicks blued.
The barrel and the great wheel machined in one piece.
Square winding square to barrel arbor.
Pallet arbor and pallet could be dismantled as a entity through opening in back plate.
Click springs were formed in one piece and attached to plate by means of a screw.

1937 also saw the 'new' night-silent Westminster chime movement (see Figure 164). They were also by this date making a 6ins. 8 day Shipclock striking on a bell, longcase, dial clocks and specimens in ornate finely veneered cases. In most instances the hands on the higher quality cased movements were hand pierced as opposed to pressed out.

Production appears to have ceased with the beginning of the Second World War.

Figure 163

The new "Tower" Clock in a Modernist case.

£40

Figure 164

Night silent Westminster chime by Davall & Sons Ltd. Note underslung chimes and circular pattern of snailing on back plate and visible wheelwork.

JAZ 8 DAY CLOCKS

in moulded Jazolite cases

(NOT ALARMS)

MOVEMENT

Solid plates, detachable escapement, jewelled. Conical staff pivots. Safety wheel to prevent over-winding. Telltale on dial to indicate winding. Easy to wind.

Send TO-DAY for illustrated LIST OF ALL MODELS to

JAZ CLOCK CO., LTD., 6, HOLBORN VIADUCT, LONDON, E.C.1

Phone: CITY 7414

Figure 165

The name "Jaz" is usually associated with alarm clocks (see page 334) but in 1934 the Jaz Clock Company Limited launched a 8 day clock with some interesting features. The case was of red and white Jazolite (bakelite). It was moulded in three sections — the base a spacing piece and the main upper portion, with the designs being considered very avant garde! A two coloured flag (red or white) indicated (through the triangular aperture immediately below the figure 12) whether the clock would shortly be in need of rewinding. A moulded depression with a light spring fixed within was incorporated in the design of the back of the case for holding the key. The movement was surprisingly small but with thick plates and comparatively sturdy. The pillars were of steel, springs were open, pinions were of the lantern variety and the escapement a platform pin pallet type.

£100+ due to case style and material

Figure 166a

Just a few of the case styles being advertised by
Garrard Clocks Ltd. in 1932.

£75+

The 14-day Garrard Striker.

Figures 166b, 166c and 166d

The Garrard spring driven chime clocks first appeared on the market in mid 1931 and had been discontinued by 1940. It was a three train full Westminster quarter chime fitted with self correcting gear, detachable barrels (greatly facilitated assembly and subsequent repair as barrels could be changed without need to dissemble main frames; recoil escapement and a patented pallet depthing device. A flanged screw was mounted on the pallet clock. As the screw was turned the clock was raised and lowered and finally kept at the required position by tightening two fixing screws. The pendulum rod was of Invar. The block holding the chiming rods was unusual in so far as instead of each rod being inserted in a screw and this in turn being screwed into the block the rods were inserted directly into slots in the block and kept clamped in place by two bolts. The barrels were turned from the solid as were the bezels holding the glass. The backplate was snailed and lacquered as were all their subsequent models. Generally speaking all the movements made by this firm were of a high quality — thick plates, stout pillars with all parts well machined and finished commensurate with their founders' aim of producing a quality clock.

£75+

In 1932 a 14 day striking clock was added to the range (see Figure 167b). It bore a strong resemblance to their first product being built on the same sturdy lines and where possible with the same components. The timepiece that appeared later was of a somewhat lighter construction. Other models introduced included a chime clock with the choice of Westminster, Winchester and Whittington quarters with the chime rods mounted below the movement to allow for a narrow clock case. The previous escapement was replaced by a platform lever (full jewelled with elinvar balance spring) escapement.

Garrards also went into the market of synchronous movements, (see Chapter on Electric Clocks) but in order to soothe those who were still a little dubious of these 'new' timepieces they manufactured a small eight day lever timepiece housed in a bakelite drum of identical dimensions to that housing the synchronous movement. The argument being that if you moved to an area where the electricity supply was either unsuitable or unreliable the movement could be readily exchanged. (See Figures 167c and 167d).

Case styles were prolific with one advert claiming a range of over 500 styles. Garrards did produce a longcase — the cases continuing the tradition for quality in so far as the backs were veneered and well finished unlike a number of manufacturers who felt that the heart would not greave about what the eye could not see! See Appendix II for details of models still in production in 1940.

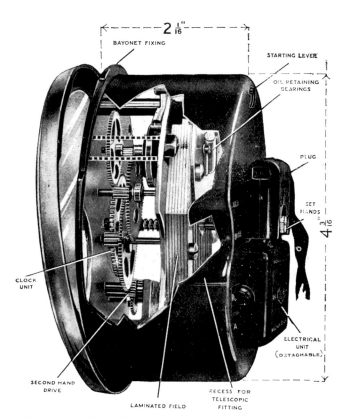

The Garrard Electric Timepiece.

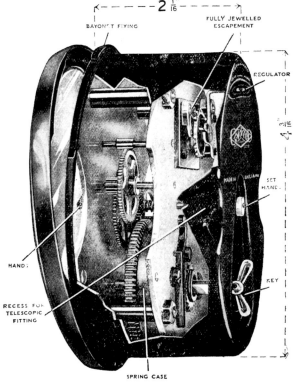

The Garrard 8-day Lever Clock.

The Clarion Clock Co., 213-4 Upper Street, Islington, London made both movements and cases for their clocks, and according to contemporary advertisements "Aims at Super Quality". Special features included:-

Thick plates (7/64th of an inch thick)
Screwed sturdy pillars
Detachable mainspring barrel
Chiming barrel and hammers on separate frame
attached to back plate by screws
Self correcting chime
Solid chromium plated lenticular pendulum bob
with central rating nut

It was marketed at the time with instruction book and a ten years warranty against mechanical failure. The cases were made by the Clarion Clock Co.'s associated company — Messrs. B. Petronzio & Sons — a firm noted for their clock cases especially those using fine veneers. The movement illustrated in Figure 167a is the Westminster quarter chime model that was sold in 1935.

£50+ Three train

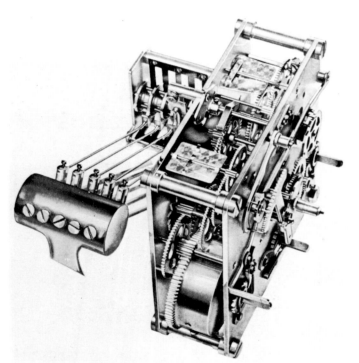

Figure 167b

An example of one of the 40 Clarion Models to be had in modern and ultra-modern styles.

CLARION

The finest Clock in the World

100% British and made at the Clarion Works.

Has many exclusive improvements including extra thick and heavy plates, specially designed pillars and screws; solid steel pinions, hardened, tempered and polished; pivot holes jig-drilled, broached and burnished; perfectly designed escapement adjustable at every necessary point; detachable barrels for easy spring replacement. Catalogue and full particulars will be sent on application.

THE CLARION CLOCK Co., 213-4, Upper Street, London, N.1

Figure 167a

The Clarion Clock Company were advertising this striking clock in 1935 as 100% British made. Note that even the flies were snailed.

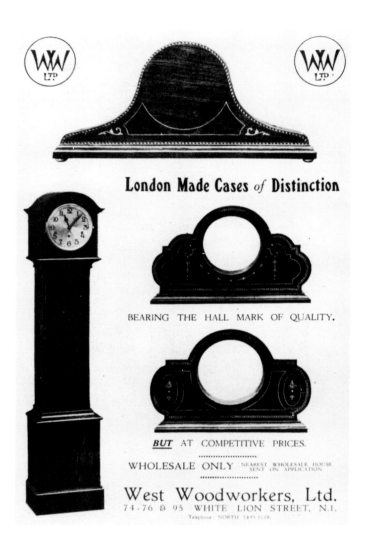

Figure 168

West Woodworkers Ltd. White Lion Street, London N.1. were but one of the London case makers active in the 1920s. The two smaller cases would have been made from the solid wood with the centres removed to house Swiss, French or English small round movements.

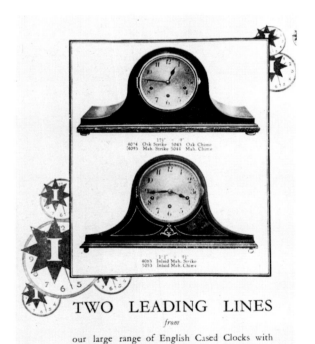

Figure 169

Typical Napoleon Hat styled case of the 1930s. Although the movements were from the Junghans factories in Germany the cases were of English manufacture. These clocks often have extremely melodious chime/strike and there has been a marked increase in the desirability both by home and overseas buyers. Modern reproductions have also been noted in the High Street jewellers windows with quartz movements. Still necessary to be selective and only acquire examples with good quality cases and undamaged dials, etc.

£30+

Figures 170a and 170b

The case of this example is oak veneered on soft wood, with added oak beading. The silver anodised dial has a small Chime/Silent lever to the right of the numeral three. As can be seen in the back view the foreign-made (? German) three train movement chimes the quarters and strikes the hours on rods mounted on the base of the case. As the demand grew in the 1930s for smaller cases the hammers were often slung from the base of the movement, instead of, as illustrated here, on the backplate.

Three train £35+

No. 373

23 in. × 15 in. × 10½ in.
Oak
Brass Dial
2 Gongs
294/-

Silvered Dial
9 and 5 Gongs
Fusee
600/-

No. 374
21½ in. × 14 in. × 10 in.
Oak
Brass Dial
8 Bells and 5 Gongs. 560/-
9 and 5 Gongs. 560/-

Courtesy of F.W. Elliott Ltd.

Figure 171

Two examples of clocks manufactured by Winterhalder and Hofmeier of Neudstadt, Germany and imported by Grimshaw, Baxter and J.J. Elliott Ltd. sometime after 1909. At the prices shown here they must have been a serious challenge to the English clockmakers. English clocks of a similar nature were being offered in the same catalogue at approximately twice the price.

Figures 172a and 172b

17½ins. high. Typical example of a bracket clock manufactured by the Winterhalder and Hofmeier. Anton Winterhalter started a workshop in Schwarzenbach, Germany. He had four sons Herman, Ludwig, Linus and Bernard. The factory closed in 1930 upon the death of the last surviving son Linus. The case is oak, the dial conventionally styled brass dial, silvered chapter ring with small seconds dial and Strike/Silent in the top corners. It is interesting to note that this clock is identical to that advertised in the catalogue of the D.N.N. & Co. Clocks, Bronzes and Optics published immediately prior to the First World War.

Note the substantial movement with coiled gong. W.H. & Co. at the base of the plate indicates maker.

£200+

Courtesy of Reeds Rains

Figure 173

Early Victorian mahogany cased bracket clock by Charles Skinner, London. The estimated price was £150 — £300 when offered for sale in late 1982. It realised £160.

£250+

Figure 174

Selection of brackets designed to accompany clocks from the range being produced by the Hamburg American Clock Company in 1910-12.

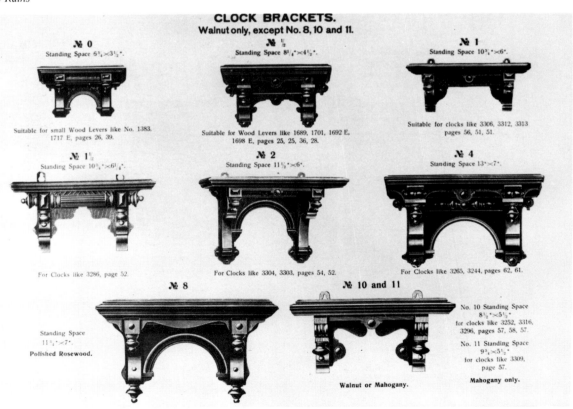

163

Trademark

Figure 175

Two examples taken from the catalogue of J.B. Rombach offering a range of clocks made by the Baden Clock Co. (factories at Furtwangen and Gutenbach, Germany). The Badische Unrenfabrik founded in 1889 was the largest firm in Baden.

Both these clocks are described as "Marble Imitations". This description could just be accepted for that shown on the right, but is hardly an apt description for the other clock!

£50+

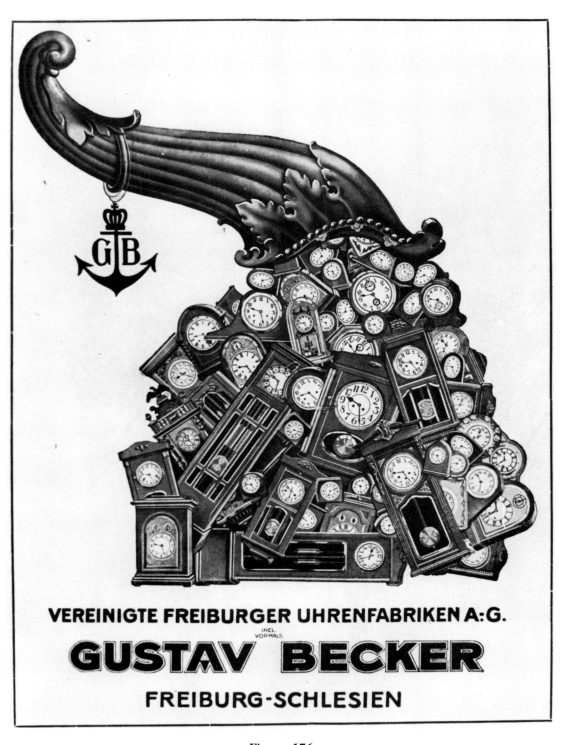

Figure 176

1939 Gustav Becker advertisement appearing in the Deutsche Uhrmacher Zeitung, *included here as it so succinctly demonstrates the variety and quantity of clocks flooding out of the German factories at this date. Note trademark of anchor and GB.*

WELL FINISHED CASES.

These 14 Day Half Hour Bracket Striking Movements have solid plates, hardened polished pinions and polished pivots.
EASY WINDING.

№ 3059   № 3061   № 3062

Height 18¾ inches.   Height 18¾ inches.   Height 18¾ inches.
Walnut.   **Walnut.**   **Walnut.**

14 Day Half Hour Gong Strike. Ivory Dials, pierced gilt centre.

WELL FINISHED CASES.
Reliable Movements.

№ 7181   № 7190   № 7191

Height 22½ inches.   Height 22½ inches.   Height 22½ inches.
Walnut or dark Oak.   **Walnut and Brass Mounts.**   **Walnut and Brass Mounts.**

1 Day Time, Alarm or Strike or 14 Day Time. Ivorine Dials, gilt centre. Imitation mercurial Pendulums.

Figure 177

Two pages from the catalogue of Niehus Bros, 40 Bridge Street, Bristol, as supplied to them by the manufacturers Hamburg American Clock Company in 1912, showing a few of the styles of the bracket clocks offered by them at this time. The illustration at the top of the page shows those with 14 day half hour striking movements with solid plates, hardened polished pinions and polished pivots in walnut cases. The other examples came from their range of Cottage pendulum clocks, are more in character with similar clocks made in the United States and would have obviously been cheaper.

£100+ largely depending on condition

Pendel-Uhr.

Pendulum clock
Walnut brown

Echappement a balancier
Noyer brun.

Brittannia.

Height 18 inches
Scale ¹/₄

Hauteur 455 mm
¹/₄ de la grandeur naturelle

	1 Day	30 heures	1 Tag
Nr. 360	alarm,	a reveil,	Wecker.
Nr. 360/1	time	sans reveil,	Gehwerk.
Nr. 360/2	alarm & strike,	sonnerie avec reveil,	Wecker mit Schlagwerk.

Courtesy of August Schatz & Sohne

Figure 178

Two pages from the 1895 catalogue of August Schatz of Triberg, Germany showing shelf clocks manufactured by them at this date.

Pendel-Uhr.

Pendulum clock
Walnut brown

Echappement à balancier
Noyer brun

Wellington.

Height 16¹/₂ inches
Scale ¹/₄

Hauteur 415 mm
¹/₄ de la grandeur naturelle

	1 Day	30 heures	1 Tag
Nr. 361	alarm,	a reveil,	Wecker.
Nr. 361/1	time,	sans reveil,	Gehwerk.
Nr. 361/2	alarm & strike,	sonnerie avec réveil,	Wecker mit Schlagwerk.

WELL FINISHED CASES.

These 14 Day Half Hour Striking Movements have hardened polished pinions and polished pivots. EASY WINDING.

The Orchestra Gong Clocks.

№ 3291   № 3311

Height 16½ inches.   Height 16¾ inches.
Engraved Silvered Dial.   Engraved Silvered Dial.
Carved Smoked Oak.   **Polished Mahogany Inlaid.**
                                      Bevelled Glass.

The Orchestra Gong Clocks.

½ Hour Strike with 4 hammers on 4 Silver Gongs
¾   „   „   „   2 sets of 4 hammers each.

Figure 179

Two of the range of large Boardroom clocks being manufactured by the Hamburg American Clock Company in 1911-12. Note the reference to Orchestra Gongs ("the sweetest and most distinct gongs on the market").

£300+

Figure 180

View of the movement of one of the Hamburg American Clock Company Orchestra Gong Clocks. The half hours and hours were sounded with 4 hammers. These four hammers were fixed on the same arbor and according to the manufacturers "so they fall at the same time on the four toned (steel) rods and the result is a beautiful sonorous tone". In the three quarter striking clocks there were eight hammers. Four hammers struck the quarters, halves and three-quarters in four different notes and the hour was again sounded by the set of four hammers as in the half hour striking model. This movement was also used in their regulator clocks.

WELL FINISHED CASES.

These 8 Day and 14 Day Half Hour Bracket Striking Movements have solid plates, hardened, polished pinions and polished pivots.

EASY WINDING.

CASES OF SUPERIOR FINISH.

№ 3252   № 3309   № 3296

Height 11½ inches.
Real Mahogany inlaid.
Antique Ivory Dial.
Round Plate Movement.
8 Day Pendulum Time
or 8 Day ½ Hour Gong Strike.

Height 12½ inches.
Real Mahogany inlaid.
Antique Ivory 4½ inch Dial.
Solid Bezel. Bevelled Glass.
14 Day ½ Hour Gong Strike.

Height 11¼ inches.
Real Mahogany inlaid.
Antique Ivory Dial.
Round Plate Movement.
8 Day Pendulum Time or 8 Day ½ Hour
Gong Strike.

Figure 181

Examples of some of the mantel clocks appearing in the Hamburg American Clock Company catalogue for 1912. The annotation appearing with each model is self explanatory. Note the crossed arrows trade mark prominently appearing on each of the examples shown in top illustration. The case styles were typical of the period and one finds everything from electric movements to high grade mechanical movements housed in similar cases.

£75+

EIGHT DAY FANCY WOOD PENDULUM CLOCKS.

Polished Full Plate Movements. Spring in Barrel. Good Time Keepers.

№ 1746 E   № 1747   № 1748 E   № 1749 E

Height 10½ inches.
2⅜ inch Plain silvered dial.
Carved Dull Walnut.

Height 10 inches.
3¼ inch Plain silvered dial.
Smoked Oak inlaid.

Height 11 inches.
2⅜ inch Garland dial.
Polished Mahogany with Cast Brass Mounts.

Height 11 inches.
2⅜ inch Garland dial.
Polished Mahogany inlaid.

8 Day Pendulum Time.
No. 1747 is also made in 8 Day Gong Strike.

Figure 182

Although appearing in an English stockist of the Hamburg American Clock Company's products these 8 day fancy wood pendulum clocks do not appear to be typical of the clock case styles sold in this country at that time (1912). It is doubtful if they would have been to English taste and so it is more than likely that few if any were imported.

£75+

Cuckoo Clocks

The cuckoo clock originated in the Black Forest area of Germany. Legend has it that clocks were first introduced into the area by glass hawkers from Bohemia. Local farmers turned their woodworking skills to the art of clockmaking to augment their income during the hard winter months. The early movements were of wood apart from lantern pinions and some use of wire. The theme of the cuckoo is said to have been invented by Franz Anton Ketterer in or around 1730. Later brass wheels were introduced but these were mounted on wooden arbors and set in wooden plates until the end of the nineteenth century. To most people the typical clock is wall hanging and has a weight driven movement, weights cast in the form of pine cones, carved bone hands and a wooden case heavily embellished with further carvings. This is not completely accurate as can be seen from the following illustrations. Some of the more elaborate examples have a quail as well as the cuckoo — the former sounding the quarters and the cuckoo the hours. The trumpeter clocks have a trumpeter sounding and appearing instead of the cuckoo. These clocks date from around 1857 and came from the Furtwangen area. They are considered choice collectors' pieces.

Courtesy of Camerer Cuss & Co.

Figures 183a and 183b

18ins. high. This is an unusual German cuckoo clock — most examples are wall hanging with weights. Until the second half of the nineteenth century, few bracket cuckoo clocks were manufactured. The innovation of the spring driven cuckoo clock is attributed to Johann Beha (1815-98) dating from around 1850. The movements are either thirty-or fifty-hour with a going barrel, with some examples combined with a double fusee. As can be seen from the back view the plates are wooden, the pendulum and wheelwork brass and the circular round wire gong is fixed to the door of the case and not the back of the movement. Similar clock with fusee sold in 1977 for £450. Without this feature the price would be in the region of £300.

£600+

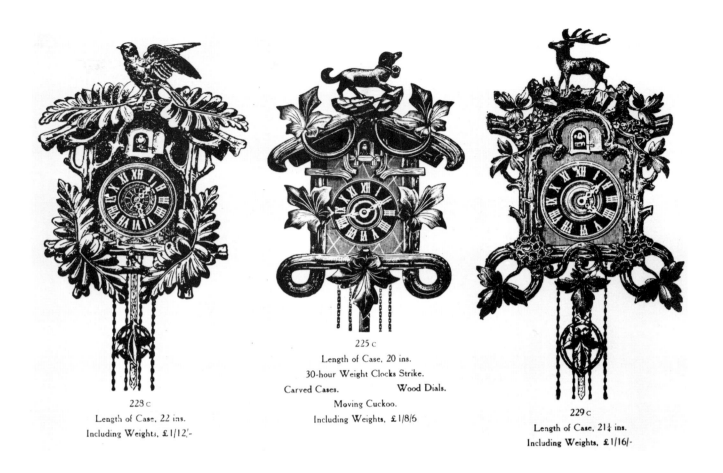

228 c
Length of Case, 22 ins.
Including Weights, £1/12/-

225 c
Length of Case, 20 ins.
30-hour Weight Clocks Strike.
Carved Cases.   Wood Dials.
Moving Cuckoo.
Including Weights, £1/8/6

229 c
Length of Case, 21¼ ins.
Including Weights, £1/16/-

192 c   17½ by 11½ ins.
£2/2/6

193 c   16¾ by 10½ ins.
Best Carved Cases.   Solid Movements.
Moving Cuckoo.
30-hour Bracket Clocks Strike.   £2/1/0

Larger Range of Standing Cuckoo. Lever and Pendulum. from 12/6 each.

Courtesy of Bout Time, Oregon, USA

Figure 184

Two pages of illustrations taken from the catalogue Clocks and Bronzes
published sometime between 1914-1915 by an English stockist.
£200+ weight driven
£300+ spring driven

American Shelf Clocks

Although clocks of varying types and complexities had been manufactured in America for many decades, the tribute for initiating the manufacture of movements on a large industrial scale must go to Eli Terry (1772-1852) of Connecticut. With the aid of water power, machinery and standardisation of parts, he accomplished the apparently impossible task of producing some four thousand wood movements for grandfather clocks within four years. The next stride forward in this growing industry was the introduction by Chauncey Jerome (1793-1868) of a *cheap* brass movement. So successful was his method that he was forced to find new and larger markets for his wares. This combined with the fact that these movements could be transported long distances far more successfully than the older wood movements led him to consider England as a possibility. In 1842 he sent his son and Epaphroditus Peck to England with a shipment of wooden cased shelf or wall clocks similar to those seen in Figures 186 and 188. They were invoiced at a dollar and a half each with twenty per cent duty payable on arrival, and were intended to sell in England at twenty dollars each. The British Customs decided that the shipment had been deliberately undervalued and according to the then current law purchased them for the declared value plus ten per cent, thereby hoping to teach Jerome a lesson. He sent a further shipment

Figure 185a

10ins. high. This is an extremely stalwart-cased American movement and it can only be assumed that the movement was imported and cased here. Apart from the door the case is of solid oak, veneered with oak and with small gilt metal feet. The bezel is thin pressed brass, the dial painted zinc and the hands thin pressed metal with regulating arbor above Figure XII. The door has the original labels — the first stating that the clock was of American manufacture (by Waterbury Clock Co., Waterbury, Conn., U.S.A.) and that the model was called the 'Enfield'. The other label gives the Directions for Setting the Clock Running and Keeping it in Order, i.e.

"Remove the paper, or packing wire holding the Pendulum Rod and hang the Ball on the Rod. Be sure that the paper or packing wire is taken out of the Clock. The best should be equal and regular, and will be so if the Clock is set in a LEVEL position.

This Clock is fitted with our "No 30 A-2 MOVEMENT" — EIGHT DAY SPRING, STRIKING THE HOURS AND HALF-HOURS.

TO SET THE CLOCK, use the long or MINUTE hand only; NEVER TRY TO TURN the short or HOUR hand. The MINUTE hand can be turned in either direction without injury to the Clock.

The clock can be made to strike to correspond with the position of the hands by carrying the minute hand forward to the figure. If then backwards to the Figure VIII and back and forth between these figures until the right hour is struck.

THE CLOCK IS FITTED WITH OUR PATENT REGULATOR, whereby it can be regulated without touching the Pendulum by means of the small arbor just over the Figure XII. If the Clock goes SLOW turn it to the RIGHT. If FAST turn to the LEFT.

If for any reason the Clock should not regulate satisfactorily from the front it can be regulated the same as any ordinary clock by raising or lowering the Pendulum ball itself. Raise the ball to make the Clock go faster, lower it to make it go slower.

The clock should be wound REGULARLY once a week."

Intact labels would enhance the price £50+

Figure 185b

Movement of the Waterbury clock. The points to note are the Brocot-type suspension adjustment, open springs, thin wheels, round wire gong (hours are struck on the gong and half hours on the bell), and fancy gilt lead bob. Also shown is the original double ended key to both wind the movement and adjust the suspension.

185a

185b

Figure 186a

15ins. high. This is an example of a typical Gothic twin-steeple American shelf clock in a veneered softwood case with a fully glazed door. This example has decorative gold tracery on a black background but other examples are common with a tablet in the lower half of the door containing a scene, spray of flowers, bunch of fruit etc. The dial is zinc painted white with black numerals.

£100+

Figure 186b

Movement of American clock shown in Figure 186a. The points to note are the shape and method of fixing the dial, alarm disc surrounding the centre arbor for the hands, the position of the alarm bell (screwed to the back of the case), the general mass produced appearance of the movement and the label pasted to the back describing the clock as a 'Small Sharp Gothic One Day Timepiece' and stating that it was manufactured by Jerome & Co., New Haven, Connecticut, USA. This was the trade name used by the New Haven Clock Company in the second half of the last century.

Figures 187a and 187b

15ins. high. Further example of a small Gothic styled American clock in a veneered softwood case. This example, however, has a decorative tablet with a rural scene. The name on the dial is merely that of the retailer — the manufacturer being the Waterbury Clock Company of Connecticut, USA. Details of the movement can be seen in Figure 187b. A painted tablet would add a further £10 to the basic price of a clock of this type.

£85+

Figure 188a

18ins. high. This is an example of another American thirty-hour alarm clock in a veneered softwood case manufactured by Jerome & Co. of New Haven, Connecticut. Again the decoration takes the form of gold tracery on a black background but in this instance an aperture has been left through which can be seen the mock mercurial pendulum bob. This has been simulated by a piece of silver-coloured metal tube in place of what would, in a genuine mercurial pendulum, be a glass jar partly filled with mercury.

£100+

Figure 188b

View of the interior of the clock shown in Figure 188a. The points to note are the alarm bell that has been brightly gilded, the mock mercurial pendulum and the fact that, in place of the manufacturer's label, the interior is lined with black paper to set off the two features already mentioned. The small label on the back of the case describes the clock as 'One Day Foutenoy, Time Piece Alarm, Jerome & Co., New Haven, Connecticut, U.S.A.'

£100+

Figure 189

8ins. high. Veneered softwood clock manufactured by
Seth Thomas, Plymouth Hollow, Connecticut, USA with
a small mirror in the lower third of the glazed door.
Most American clocks have paper labels pasted in the
back giving instructions for use and incidentally
providing the name and address of the manufacturer.
The latter can often assist in dating a clock. It is known
that Plymouth Hollow was renamed Thomaston in 1866
and so this particular example must have been made
prior to that date. This is a thirty-hour timepiece, but
other examples have been seen with the additional
feature of an alarm mechanism.

Early example £75+

which was treated in the same manner, but the Customs allowed the third cargo to pass through. Other American factories followed suit and for some time their clocks flooded our market. Until around 1850 these were nearly all weight-driven one-or eight-day movements, but some smaller and more compact types were also manufactured with brass springs. After this date steel springs proved more suitable and became cheaper to produce.

Although scorned by many collectors, there is a great deal of interest to be found in American clocks. Unfortunately we do not see many of the rare examples in this country, but only those imported for the cheaper end of the market. Once it is realised, however, that the attraction of these clocks lies in studying the production methods employed by the various manufacturers, as they devised ways of making them at economical prices, so the interest grows. They also have the added virtue of being of sufficiently sturdy construction to allow the amateur repairer a certain amount of scope he could not hope for in more delicate mechanisims. The immense variety of designs of these clocks and the large number of manufacturers appearing and disappearing in what was an extremely competitive business makes it impossible to cover the subject here. However, a few typical examples of American clocks manufactured after the mid-1800s appear in the illustrations here and in the following chapters. Examples are usually readily identifiable and a list of suitable reading material for anyone wishing to pursue the subject has been included in the bibliography.

CHAPTER VIII

Small Clocks

The following brief selection shows clocks that have little in common apart from the fact that they are not Bracket Clocks or Shelf Clocks. Since 'Fancy' or 'Boudoir' are the words used to describe them in contemporary catalogues, possibly their manufacturers had the same difficulty over classifying them. It is felt that they all have some attractive characteristic to offer the collector. Apart from the Strut Clocks of Thomas Cole, only passing reference is made to them in modern literature, but they are all worthy of closer examination and are not always quite so straightforward technically as external appearances would imply.

Figures 190a and 190b

10ins. high. This is a compact and attractive clock in a solid mahogany case which has been decoratively veneered, with added cast brass mounts. The white porcelain dial is marked to imitate a thirteen piece dial of French design. When enamelling was first introduced it was a technically difficult and expensive process and so each segment was painted and fired individually. Long after it was a necessity the custom continued as it was considered decorative. The name on the dial is that of the retailer 'Howell & James Limited, To the Queen, London'.

According to J.B. Hawkins in his book Thomas Cole and Victorian Clockmaking *this firm was referred to in most Directories as 'Warehousemen'. He goes on to say that they were styled Howell & Co., from 1836 to 1840, after which they became Howell, James & Co. He makes no mention of Howell & James Limited. It had been hoped to be able to trace the years that the Company were suppliers to The Queen, but it has not been possible to ascertain that they actually ever held a Royal Warrant! It is known that they marketed the clocks manufactured by Thomas Cole until his death in 1864. The example here has 'Made in Paris' stamped on the movement and other examples of clocks imported from Germany have been seen bearing their name. It would appear that they were retailers of a wide variety of goods.*

Note the transformation in Figure 190b. The mounts have been removed cleaned and lacquered.

General elegance of case, together with good quality French movement would bring the price to

£350+

Figure 190a

Figure 190b

Figures 191a and 191b

6¾ins. high. An Edwardian timepiece in a solid mahogany case with a little decorative inlay. The pillars and small bun feet are cast brass. The dial is of a high quality white enamel, with black Roman numerals and blued steel hands. The features to note in the view of the movement are:

a) *the cylinder escapement*

b) *the externally mounted barrel of an easily detachable type to facilitate simple repair*

c) *the Japy trade mark and 'Made in France' stamped on the backplate (Figure 191b). (Further details on this manufacturer can be found in Chapter II with regard to the manufacture of marble-cased clocks.)*

Until recently these clocks had little commercial value but with the prevailing fashions in interior decorating they are slowly rising in price. There are many similar examples on the market and when contemplating a purchase only those of a high quality should be seriously considered, i.e. solid case, any decorative feature on the case which would have taken a little more labour or material at the time of manufacture and a French, Swiss or English movement.

£100+

Figures 192a and 192b

These two clocks have identical 8 day movements. Both appear in the 1911/12 catalogue of the Hamburg American Clock Company. The four glass example was also offered in a white wood case. It is interesting to note how in order to maintain a production run columns, finials etc. are fully interchangeable.

£90+

Figure 192c

View of movement. Note the barrel is easily removable without the need to dismantle the movement. This design also has the advantage of allowing for a barrel of a larger diameter (longer duration) than would have been possible if the barrel had been placed between the plates. The 'mercury' pendulum would have been non-functional.

193b

Trade mark of
Horstmann Gear Co. Ltd.
of Bath.

Courtesy of the Horstmann Gear Co. Ltd.

Figure 193a

This is an example of one of the clocks manufactured for the domestic market during the 1920s by the Horstmann Gear Co. Ltd. of Bath. This particular example, the 'Montmorency', is a mahogany inlaid case — had a fourteen day striking lever movement (as shown in Figure 193d). Other models would possibly have had the alternative pendulum movement shown in Figure 193c.

This history of this firm is not without interest and it is felt that any of their clocks would be extremely desirable examples of this era of English clockmaking. Gustav Horstmann, after serving an apprenticeship with Dejean, a pupil of Bréguet, came to England in 1853 and one year later established a clockmaking business in Bath, which continued as a retail establishment until 1925. Joined by his three sons, and aided by his inventive genius (he had some hundred patents to his credit) the firm prospered. After his death in 1893 the firm strayed a little from the field of horology and finally, in 1904, the Horstmann Gear Company Limited was launched to manufacture and market a car gear box. This did not prove to be a commercial success, but the youngest son designed and the firm produced the Horstmann car which continued to be manufactured until about 1930. This firm is known, however, the world over, for their screw gauges and time controllers used mainly in the gas industry. They are still one of the most important manufacturers and many a street light or domestic boiler has a timer by this Company.

It is, however, the clocks manufactured by this Company in the 1920s that are of interest to the collector. Having decided to apply the same methods of producion as they did to their timers, they started in January, 1921, to manufacture their pendulum movement. By the end of the year a striking version had been added, and by early 1923 a further model with a lever escapement was included. It would appear that, having made some 3,500 movements, production ceased around 1928 with the tooling and remaining parts sold to a London firm sometime during the Second World War.

From a contemporary catalogue it is possible to ascertain that they manufactured some thirty case designs for shelf clocks, one drop dial wall clock ('The Bungalow' Miniature Clock Dial with a pendulum movement) and one round dial wall clock. The case designs were typical solid Edwardian mahogany cases, oak pseudo-Jacobean and others more reminiscent of earlier bracket clock designs, including some with Oriental lacquer cases. These were the most expensive and were sold to the trade at £5. 18s. 4d. with strike or £4. 13s. 4d. without. Additional features, again at trade prices were:

Solid Bezel	*5s. 0d.*
Enamel Dial (standard one being silvered)	*1s. 6d.*
Ball Feet	*1s. 6d.*

Exceptional quality of case and movement as well as documentation of manufacturer would make the valuation of this clock £100+.

193c

193d

Courtesy of Horstmann Gear Co. Ltd.

Figure 193c

This is an illustration of the pendulum movement manufactured by the Horstmann Gear Co. Ltd. under the tradename of 'Newbridge' between 1921 and 1928. The official specification states:

"The movement is made to fit cases 4½ins. aperture. Owing to precise production methods all parts are interchangeable, spares being sent by return.

DIALS. — Flat or raised zone, silvered, engraved with either Arabic or Roman figures. Enamelled dials fitted at slight extra cost.

PLATES. — Solid brass with pillars screwed at back.

WHEELS. — Machine cut of substantial thickness.

PINIONS. — Solid steel, cut, hardened and polished.

BARREL. — Cut wheel, polished brass tube, fitted to wheel with splines.

ESCAPEMENT RECOIL. — French type, eccentric adjustment, solid steel pallets.

SUSPENSION SPRING. — Permanently held in crutch.

PENDULUM. — Brass weighted bob, steel rod regulated by nut. Can be hooked on by amateur.

STRIKING MECHANISM. — Of the rack type, so that the hours always follow the hands, and can be repeated without disturbing the sequence. The hands can be turned back without disarranging the striking mechanism. This is invaluable for summer time. The hammer lifting is from a star wheel, instead of a pin, and the strike is on a deep-toned cathedral gong."

Figure 193d

This is an illustration of a lever movement manufactured by the Horstmann Gear Co. Ltd. under the tradename of 'Newbridge' between 1923 and 1928. The official specification states that "this movement is similar to the pendulum model, but the pendulum is replaced by a Swiss straight line double roller lever escapement, having a non-magnetic compensating hair spring. The escapement is fitted with a shield to exclude dust. This model can be fitted in all cases". This particular example is a striking model, but the movement could be just a timepiece. The points to note are the distinctive lattice finish on the backplate and the trade mark, of which a larger example can be seen in Figure 193b.

Figure 194

13ins. The case and figure of the little Dutch girl are of a non-specified metal that has been left to acquire an imitation bronze patina. The pin pallet movement has an off-centre wind as described when discussing Figure 203b. There are no maker's marks but it is probably of German manufacture. Judging by the sentimentality of the design and the nationality of the figure it would be logical to assume a date in the late 1920s or early 1930s. Although of no great horological merit, it is a pleasing piece and worth acquiring as typical of the period.

£120+

Figure 195

6¾ins. high. The great advantage of any clock cases with substantial silver mounts or decorative panels is that they carry a hallmark which provides a date for the clock. It is doubtful if these small clocks would qualify for a replacement movement having been substituted. This example is in a silver and tortoiseshell case (Birmingham 1912) and sold in 1977 for £120. Although an attractive timepiece the price would have been enhanced by the value of silver.

£200+

Courtesy of Keith Banham

Figure 196

These clocks are a selection of small clocks attributed to the London maker Thomas Cole (1800-64). It is known that he supplied a number of the leading jewellers with finished clocks, but as was the custom it was the name of the retailer that appeared on the dial of these pieces as demonstrated by the example illustrated. This has the name 'Tessier & Sons, London' appearing on the dial but the name 'Thos. Cole' on the backplate. The clocks constructed by Cole after 1855 usually carry the punchmark 'Thos Cole'. He is best known for his small thirty-hour oval or rectangular strut clocks which often include additional features like thermometers, calendars, etc. As most examples of his work are numbered (commencing c.1846 with number 500 and ending upon his death in 1864 at 1900) it is possible to gain some indication of the volume of his output. His life and work have been exten-sively documented by J.B. Hawkins in his book entitled Thomas Cole and Victorian Clockmaking.
A wide price range spanning several hundreds of pounds depending upon documentation of provenance and complications of movements.

£2,500+

Courtesy of King and Chasemore

186

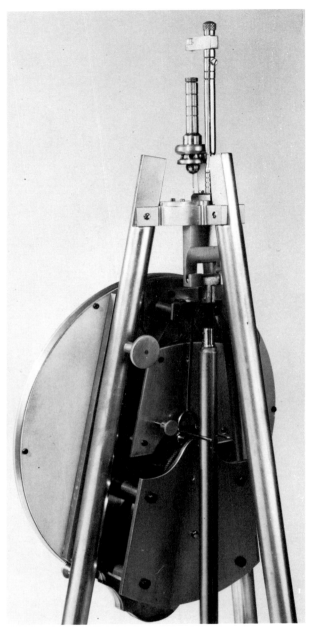

Figures 197a and 197b

Thomas Cole gilt-brass tripod clock, with engraved silvered dial blued steel moon hands. The movement has a deadbeat escapement with the pendulum independently suspended from the top of the tripod surmounted by a plumb bob (see Figure 197b). There is a thermometer, aneroid barometer and beat scale raisable for locking the pendulum. Estimated in 1981 at £1,200 — £1,400 the realised price was £1,800. **£3,500+**

No. 1533/9676.—White and Gold, with Colours. Fancy Dial.
Height of Clock, 6¾ins.   Height of Vases, 6ins.
1-day Lever Time   ...   **10**/- the set of 3 pieces.

No. 242/1031 - Green and Gold.   Fancy Dial.
Height of Clock, 8ins.   Height of Candelabra, 7½ins.
1-day Lever Time   ...   **10**/- the set of 3 pieces.

No. 1533/5236.—Royal Blue and Gold.   Fancy Dial.
Height of Clock, 6½ins.   Height of Vases, 6½ins.
1-day Lever Time   ...   **14**/- the set of 3 pieces.

No. 1533/5615.—Light Green, Brown Edging, Coloured Flowers.   Fancy Dial.
Height of Clock, 6¾ins.   Height of Vases, 5¾ins.
1-day Lever Time   ...   **9**/- the set of 3 pieces.

It might be said that the clocks in the illustration were intended as the poor man's answer to that shown in Figure 309 in a fine Coalport case. Those shown here are from a page of a catalogue of Grimshaw, Baxter and J.J. Elliott some time after 1909 and the cases would have been the product of one of the Continental factories that specialised at this time in cheap ware for exporting. Though having said that, an advertisement dated 1906 was noted wherein the St. Louis Fine Art Pottery Co., Argyle Works, Fenton, Staffs. were informing manufacturers that there would be a travelling display of fancy clock and clock sets at Saracens Head Hotel, Snow Hill, E.C. between 19th February and the 2nd March. Obviously not all cases were imported. These clocks vary in quality and amount of decoration and can be extremely attractive. The movements were thirty-hour pin pallets although some of the larger cases did have an eight-day movement. Many of the movements have not worn well and it has been noted that the original movement has frequently been replaced by a small mass produced modern Swiss movement obtainable from any clock and watch material dealer. So long as the purchaser is aware of the substitution it is possibly more beneficial to have a non-authentic working clock than a fully authentic non-working example! However, untampered with examples can still be found, with and without their side ornaments. It may be noted from the 1914 catalogue of the Ansonia Clock Company that they also offered similar clocks. Their models, however, include hour and half-hour striking on gongs and/or visible Brocot escapements. It is possible that a few may have reached this country, but as they carry the Ansonia trade mark, as shown on page 316, they would be readily identifiable.

£50+

188

199a

199b

Figure 199a

8¾ins. high and 6¾ins. high. Superficially these two French timepieces appear to be similar; both have veneered drumhead cases, white card dials and Bréguet-style hands with spun brass press-on covers. Both movements are stamped with the initials 'V.A.P.' and 'Breveté S.G.D.G.' Many writers refer to 'V.A.P.' as being an abbreviation for Valogne á Paris — a maker known to have been working there after the mid-1800s. However, Tardy in his Dictionnaire des Horlogers Français *says that 'V.A.P.' was the trade mark of Maison Pierret-Borel et Blin of Paris around 1900. In view of the fact that the movement with an alarm in Figure 199d and similarly marked is identical to that in* La Pendule Francaise, *also by Tardy, and attributed to Victor-Athanase Pierret (1806-93), the latter assumption is more likely to be correct, the trade mark having been taken from Pierret's initials. He was an important maker of alarms and a further example of his work can be found in the skeleton clock illustrated in Figure 381. He came to Paris in 1800, and eventually opened a factory to manufacture clocks of his own design; he retired from active business in 1865 and sold a half-share to Borel but continued his researches until he died in 1893. The movements of these two clocks are, however, completely dissimilar, Figures 199b and 199d. The third example, Figure 199e has the additional feature of an alarm mechanism. Although basically using the traditional drumhead case, the materials used vary and other examples have been seen in white alabaster, onyx or brass, see Figure 199c. The dials in the examples shown here are of card, but in some instances the dials were of white enamel. Although it does not carry the mark 'V.A.P.' the carriage clock in Figure 82 is most certainly a 'V.A.P.' type of movement.*

Even though they are not of superb workmanship these movements are basically sound and have the virtue of offering some variations to the collector. It is doubtful if there is any difference in value between a basic pendulum type and a balance wheel example, but obviously a more decorative case or the additional feature of an alarm would increase the price accordingly.

£90+   *depending on quality of case*
£15 *additional feature of an alarm*

Figure 199b

The movement of the larger of the two clocks shown in Figure 199a has a 'V.A.P.' movement but with a lever escapement. The escape wheel and lever are between the plates, but the balance is on a small separate platform on the back plate. The double ended key fits the winding arbor and the square for setting the hands.

Figure 199c

The movement of the smaller clock in Figure 199a has a deadbeat anchor escapement which was widely favoured in French drum clocks from the middle of the nineteenth century.

Although frequently referred to as 'tic tac' escapements this is not in this instance true. A 'tic tac' escapement has pallets embracing two teeth only and these movements most certainly do not conform to this definition. The solid spherical bob is screwed on to the pendulum rod which is in turn attached directly to the pallet arbor, i.e. no crutch. As the pendulum is self setting this should on no account be firmly fixed but left just friction tight. The original paper label with instructions for use adds to the intrinsic and academic interest. The annotations are those of the repairers.

Figure 199d

The movement in this black ebonised case of similar design to those shown in Figure 199a is of standard 'V.A.P.' balance wheel pattern with the additional feature of an alarm mechanism. An identical movement is illustrated in La Pendule Française, by Tardy. As with many alarms of this period the pointer was set to indicate the number of hours you wished to elapse before being awakened and not set to point to the hour at which you wished it to ring. The small bell is set between the plates behind the balance wheel. The spun brass cover is pierced in order to allow the bell to sound freely but this has been covered by a piece of thin silk to prevent the intrusion of dust.

Figure 199e

Further case example with brass drum and good enamel dial.

Figure 200

13ins. high. The frame of this clock is a good quality, heavy brass casting, with an ivorine dial, recessed gilt-metal centre and delicate hands. The movement is German having as it does the Junghans 8 star trade mark stamped on the backplate. The layout of the movement is identical to that shown in Figure 201b. As the unusual arrangement of spring and winding necessitated further labour and parts it must be assumed that it was carried out in order to infer that it was an eight-day movement and not a thirty-hour.

£50 — £110

201b

Figures 201a and 201b

3½ins. high. An identical clock to the example in this illustration appears in the 1913 catalogue of Gamages of Holborn, London, priced 3s. 11d. The case is of nickel with a coating of copper, the dial of celluloid with Arabic numerals. Looking superficially at the back it could be assumed that the movement was an eight-day variety with concentric centre wind. It is, however, a thirty-hour movement with the spring positioned off-centre with gearing to enable it to be wound centrally, as can be seen in Figure 201b. Although not marked with any trade mark or manufacturer's name, the layout of the movement is identical to that in the cast brass frame shown in Figure 200.

£50+

Figure 202

3½ins. high. "This clock is unusual in so far as it has a centre seconds hand. The cast iron base, supporting a gilt metal drum movement with a decorative finish, makes an attractive contrast. The dial is celluloid. The thirty-hour movement has off-centre wind (as is to be expected), but instead of a centre hand-set, this is off-centre concentrically with the winding arbor, yet another small but interesting variation in design."

When described thus in the first edition of this book the opportunity for minutely examining the movement had not occurred. Later this was possible and immediately a lesson was learnt. Here was an excellent example of the trap that one can fall into by making assumptions without detailed examination. The true nature of the movement had not been fully appreciated. Upon examination it became apparent that it did in fact have a most interesting escapement of a type attributed to Claude Saunier (1816-96), (see Gros — Echappements D'Horloges et de Montres pp.84-86). This clock was therefore greatly undervalued in its first assessment.

£200+

Figure 203a

7ins. high. This clock gives the impression upon first being handled of having a heavy cast brass case; however, on closer inspection this is not true. The case is of spelter and has been filled with plaster prior to being gilded in order to add to its overall weight and to retain its shape. The little drum movement is held in place by the plaster only, as can be seen in Figure 203b. The celluloid dial has an attractive gilt-metal centre.

Appearance would attract a price of £50+

Figure 203b

The movement of the clock in Figure 203a was never intended to be repaired, and it was only with care and tenacity that it was possible to remove the back cover without causing damage. However, to do so was rewarding as it was then possible to note that the movement which gives the appearance of having an eight-day movement with a spring across the entire width does, in fact, only have a thirty-hour spring that has been positioned off-centre with a concentric hand-set and wind.

3½ins. high. This small timepiece, in the form of a chariot pulled by a dog driven by a winged cherub, was manufactured by the Ansonia Clock Company of America. Originally founded in 1850 in Connecticut, they moved in 1879 to New York, where they had their second disastrous fire. A new factory was built in Brooklyn in 1881 and within two years they had expanded sufficiently to open sales offices in New York, Chicago and London. Around this date they started to produce large numbers of novelty and figurine clocks, as illustrated in Figure 267. By 1914 they were exporting to over twenty countries, but after the First World War they found that the demand for their products was declining. In 1929 they went into voluntary liquidation and production ceased. Most of their machinery together with some of the dies and patterns were sold to the Russian Government.

In this example of one of their clocks, the drum holding the movement is nickel, but the little dog and the cherub are gilt metal. The dial is celluloid and has the Ansonia trade mark as shown on page 316, together with the name of the Bournemouth retailer. The words 'The Ansonia Clock Co., New York, United States of America' appear on the lower perimeter of the dial. This address would indicate a date after 1879. The backplate carries the date of the patent covering the design of the movement (23rd April, 1878).

Novelty value £65+ purely on appearance

Figure 204b

The movement of the timepiece shown in Figure 204a and manufactured by the Ansonia Clock Company of America. Note the spring across the entire width of the movement but observe that, unlike the examples manufactured by the British United Clock Company (see Figure 206c), the arbor remains still while the spring barrel rotates. It is important not to wind these clocks in the wrong direction as this sheers the two small tabs completely off in some instances and, at best, irretrievably damages them (see Figure 204c). Provision was made for their replacement as can be seen by their presence in the clock and watch material dealers' catalogues of the day. These thirty-hour Ansonia movements come in a wide range of cases of varying qualities.

Figure 204c

This illustration shows in detail the inside of the back cover housing the spring. As this is rotated, the slots engage with the tabs on the movement (small detached tab shown beside the spring barrel).

Figure 205

5½ins. high. Small timepiece from the American factory of the Waterbury Clock Company in a gilt spelter case of a somewhat rococo style. The detail is far from crisp and rather coarsely cast. The paper dial has become stained over the years but some collectors would feel it was a pity to renew this and lose some originality of the clock although it might be helpful to be able to read the time more easily!

The 30 hour movement has off centre key wind at the 12 o'clock position with the escape wheel at the 9 o'clock position. This was obviously intended to be a 'poor man's' carriage clock.

£55+

Figure 206a

These three small clocks were manufactured by the British United Clock Company of Birmingham between 1885 and 1909. The largest example shown here has a good cast brass frame and is possibly an early example as the dial shows the patent number (Patent Number 13538) and not the name of the manufacturers. This patent was taken out in 1887 by Mr. Edward Davies, the Manager of the Company. The other two examples have similar movements. The case of the imitation carriage clock is of nickel and is of poor quality when compared to many of the other clocks produced by this Company. The method of retaining the glass in the metal frame was patented in 1891, which provides some indication of when production of this particular case design commenced. All three dials are of white card and carry the Company's trade mark. An enlarged view of this is shown in Figure 206d.

The history of the Company is not without interest as they played an important role in the history of the changes occuring in the manufacturing methods of English clockmakers during the closing years of the nineteenth century. The Company was formed in 1885 — Manager Edward Davies and Secretary John Fisher — with their factory first at York Terrace, Hockley Hill, Birmingham, and after 1891 at Leamington Road, Gravelly Hill, Erdington, Birmingham. They were one of the few manufacturers who realised that factory methods were here to stay and that the demand for cheaper clocks had to be met and not ignored in the pious hope that it would go away! The factory at Gravelly Hill was three-storeys high, covered 1,250 square yards and employed 250 hands, of whom 150 were girls. Everything, from their tools to the balance springs for the clocks, was made by them. Machinery was installed and the standardisation of parts was introduced — a selection of which can be seen in the pages of the Watch and Clock Materials Catalogue *in the Appendix. Their products received awards in Adelaide (1887), Melbourne (1888), Sydney (1888) and, at the French International Exhibition of 1889, they received a Prize Medal and the following comment from their French competitors: ''Good conception of calibre, sound workmanship in their tools, and excellent tase in the decoration — qualities which place them at once above their American rivals.''*

Unfortunately they did not survive. There are several possible reasons for this. Although they were sufficiently perceptive to recognise that changes in production methods must come and had adapted accordingly, they failed to realise that quality must also be lowered if they were to compete successfully with the ever increasing flow of cheap clocks from Germany. They introduced other and larger movements and cases. ''All kinds of plain and fancy lever clocks'' appears in their advertisements and other contemporary catalogues have been seen with English dial wall clocks carrying their trade mark.

However, by 1909 they had succumbed and there is no further record of the Company. It is considered that any of their clocks would be of interest to the collector, but perhaps the most desirable are the small clocks housing the movement patented in 1887 (see Figures 206b and 206c) as these are the clocks most typical of the Company's aims and ambitions.

Price would increase with vendor's realisation of technical interest of these movements.

Carriage £50 Round £75 Large £75+

Figure 206b

Side view of the movement manufactured by the British United Clock Company in direct competition with those manufactured by the Ansonia Clock Company of America, but of a much superior quality and to a better design. This can be seen by comparing this movement with that in Figure 204b. The British United design is superior in so far as the arbor rotates while the spring barrel remains fixed, whereas in the Ansonia design the arbor is fixed and the barrel rotates. However, as it rotates it twists askew and, as it does so, loses much of its power with friction against the sides of the case. A full view of the spring in the British United clock movement can be seen in Figure 206c.

Figure 206c

View of the back of the clock manufactured by the British United Clock Company showing the spring taking up the complete width of the barrel. Note the clumsy jobber's repair using a screw to fix the outer end of the spring in place of a barrel hook.

Figure 206d

Trade mark of the British United Clock Company.

198

Figures 207a and 207b

This further example is of noticeably better quality than the other 'carriage' clock shown on the left of Figure 206a. The four glass case is heavily gilt, the dial ivorine (plastic imitating porcelain) with blue tracery, circular background to numerals and markings.

This particular clock is in mint condition, with original gilt and as such would command a higher price than a poor unkempt example.

£100+

Figures 208a and 208b

La Belle was manufactured by the E.N. Welch Manufacturing Company of Forrestville, Connecticut, USA, c.1880. Elisha Niles Welch originally worked with his father in Bristol, Connecticut, producing weights and bells for clocks. He formed the E.N. Welch Manufacturing Company in 1884 and eventually they became one of the largest American manufacturers. Clock production ceased between 1893 and 1895 through financial problems. Unfortunately within two years of recommencement of work the movement workshops caught fire (March 1899) with the case-making department sustaining a similar disaster two months later. Subsequent rebuilding was in brick! In 1903 the management was taken over by the major stockholders — William and Albert L. Sessions and the corporate name changed to Sessions Clock Company. This particular model was a nickel cased 36 hour movement with pointed tooth lever escape wheel and solid pallets. The skeletonised dial with white card chapter ring. Note the uncased spring, stamped strap plates and folding winding key. This case style was noted throughout the 1880 catalogue housing a variety of movements.

£80+

Figure 208a

Figure 208b

Figure 209

4ins. high. Small clock with 30 hour pin pallet movement. Transferred Dutch scene on dial with automata in the form of a windmill — the vanes are attached to the pallet arbor. A simple method of motivating still widely used in similar novelty clocks and modern children's alarm clocks. This particular example dates from the 1920s.

£40

Figure 210

4½ins. high. This small clock appears in the Junghans catalogue for 1905 under the name of Hermosa. The case is of olive wood, with brass wreath encircling dial, caps, etc. on side pillars and ball feet. The plaque on the lower section depicts a child and heron playing at the waterside. The 30 hour movement has a pin pallet escapement with end stone to balance. The back cover is stamped with the Junghans five pointed star in a circle trademark.

£40+

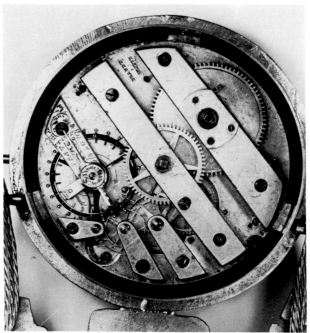

Figures 211 and 212

4ins. high. Small attractive gilt clock with case style obviously influenced by the 1920s interest in the discovery and opening of the Tomb of Tutankhamun.

The movement is actually a watch movement manufactured by Brandt of Geneva. Brandt was the founder of the firm that later became the Omega watch manufacturers. The ébauche (rough unfinished watch movement) used here was similar to that used in their bar movement watches with cylinder escapement. This watch has a single roller lever with side pallets — an early type of lever escapement that did not always prove successful. The balance wheel is unconventional. It is of the three arm variety usually associated with cylinder watches. In cross section it is like a flat bottomed U and loaded with extremely fine ornamental screws. The purpose of these refinements is obscure. There are no practical reasons — the balance is of steel and in no way a compensation balance. The detail is too fine to see without recourse to magnification so hardly an added attraction for the customer. Production cost would have been increased by the added complications. It has obviously added interest for future collectors!

£40+ Novelty value

Figures 213a and 213b

Small nickel cased ''Bee'' timepiece with similar movement to that shown in Figure 204, but of a later date (note addition of further patent date — 1884). An example of this movement with alarm mechanism can be seen in Figure 376. Illustrations of components obtainable from material houses in order to execute repairs can be seen in Appendix I.

The addition of the small tin container as originally supplied by the manufacturers still with legible labels is a decided asset. ''Prize Medal Awarded at Paris Exposition 1878'' and ''No key required to wind the clock, Turn the back.'' are but two of the discernible slogans. The base of the tin carries a label depicting the Ansonia Clock Factory buildings.

£45+
£55+ with original container

Figures 214a and 214b

This is actually a large watch (4ins. diameter) suitably cased to serve as a small travelling or desk clock. It could conceivably be referred to as the forerunner of the calotte clocks made later by such firms as Smiths, etc. (see Figure 219).

Figure 214a shows the watch in situ encased in a velvet lined dark green leather case with silver banding — hallmark London 1902. Access was gained through a sliding panel in the back of the case. This particular clock formed part of a desk set with matching blotter, pad and so on. The nickel cased watch is of Swiss manufacture. The bar movement is jewelled with club tooth lever escapement, compensated balance and overcoil. Hand setting was by means of a rocking bar. Dial is white enamel as can be seen with good bold clearly read markings protected by a thick bevelled glass.

These large watches appeared in a number of cases in order to make them acceptable as small clocks. Other examples frequently seen are in leather cases with the front totally covered with decorative silver work or silver cases of varying qualities and detail of decoration. Obviously the latter price is greatly governed by silver content as well as craftsmanship of the casemaker.

£75+

Figure 215

4ins. high. Small nickel cased 30 hour travelling timepiece of American manufacture in red leather outer carrying case. The pin pallet lever movement has offset key wind. The interest of this item is enhanced by it being complete with carrying case.

£50

Figures 216a and 216b

In no way an important high grade horological specimen but interesting as being one of the Big Ben — Baby Ben 'family' of clocks manufactured by the Westclox manufacturers. This little timepiece was called Tiny Tim. The case is die cast and the case is nickel. The Fast/Slow adjustment is enclosed behind a shutter. This is opened by means of the top knob being moved to the right hand of the slot. Winding is by the small knob at the bottom of the case, while hand set is achieved by the same knob but with it being pressed to engage with the setting mechanism. Designs for these innovations were patented in 1927.

£20+

The Zenith Watch Co., Le Locle, Switzerland was founded in 1865 by Georges Favre-Jacot. Their watches quickly established a reputation for quality — a gold medal was awarded in 1896 at the National Swiss Exhibition in Geneva, the Grand Prix at the Paris Exhibition in 1900 and in 1926 a Zenith watch gained 97.3 out of 100 marks at the National Physical Laboratory tests. This was the highest score attained by a watch at that time.

Gradually the range of production was extended to include wrist and pocket watches, table clocks, wall clocks and telephonmeters (for timing telephone calls on a switchboard). Quality prevailed throughout the entire range. The small boudoir clocks (as shown in Figures 217a and 217b) ably demonstrate this. The solid brass cases were usually engine turned, plain, or as in this instance red and green enamel in a floral design. The feet are characteristic — well finished in decorative style. The example in Figure 217a is a timepiece only, that in Figure 217b has an alarm mechanism. The small travelling cases are often worn — the silk linings have disintegrated with age. Jarvis of Clerkenwell was but one of the many firms making these cases.

Figures 217a, 217b and 217c

The movements were mainly diecast and must be one if not the first example of this method of manufacture used in clockmaking. Note the ratchet teeth moulded into the backplate in Figure 217c. Unfortunately these tended to wear quite quickly and obviously this involved replacement of the complete unit. The Zenith Watch Company provided an excellent materials service and no great inconvenience was encountered in their repair. These are most assuredly one of the better quality small travelling clocks.

Timepiece £125+
Alarm £75+
+ £15 if with case

Originally in business as jewellers with premises in the Strand, London, Smith's later developed into one of the largest manufacturers of instruments and accessories for the motor trade. Following the success of their car clocks they developed an 8 day spring driven timepiece for the domestic market. To date the ramifications of this company — which movements were actually manufactured by them; which movements were imported and retailed under their name; the dates of the changes in their title, i.e. S. Smith & Sons, Smith's English Clocks Ltd. or which firms were absorbed by them but clocks still marketed under the original trade name, etc. have not been satisfactorily untangled. Subsidiary factories at Cricklewood, one making bakelite cases and the other escapements (ABEC Ltd., All British Escapement Co. Ltd., Chronos Works, North Circular Road, N.W.2.) were all part of the Smith empire. Some of their movements still in production in 1940 appear in Appendix I and on the following pages there are just a few of the small clocks made by them — they most certainly dominated the market in the 1930s — bracket clocks, shelf clocks, longcases, boudoir clocks, alarms, etc.

The Smith's 8 day calotte clock made its debut at the British Industries Fair in 1934 and as the first of its kind to be produced in this country it was warmly received by both trade and public. Prior to this date calottes had been exclusively of foreign manufacture. Figure 218 provides the specifications of the various models exhibited. Although possibly an oversimplification ''callotte' although generally used when describes the small folding travelling clock more accurately only describes the movement i.e. a small lever timepiece resembling a large watch. The special features of those designed and manufactured by Smiths included the replacement of the usual watch type winding stem and button by a wing winder on the back, a fixed bezel in place of the more usual method of utilising it as a means of fixing the callotte into the case as well as variations in lay out of the movement and the use of quality materials, such as elinvar for the balance spring, five jewelled escapement with steel club tooth escape wheel, and a easy letting down of the mainspring by means of a special lever on the click.

1937 saw the introduction by Smith's of the Chronos alarm movement. It was the natural progression from the timepiece introduced three years earlier. It was marketed in similar styles — small fancy boudoir type cases, fitted into leather travelling wallets or fitted into a leather wallet but detachable for use subsequently as a strut clock. The movement was also 23 calibre, platform and bridges finished in a small circular grain pattern and with 5 jewels. See accompanying illustrations for styles, etc. and also the British Clock Manufacturer supplement in Appendix II (p.7 and back cover) for further views of movement and case styles as produced in 1940.

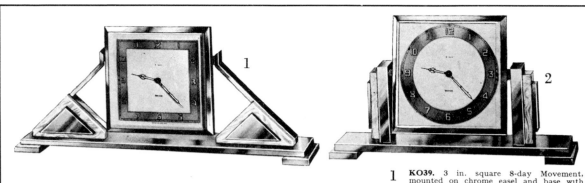

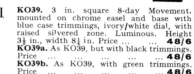

EXPLOITING the success of the Smith 8-day "Chronos" Movement, these new designs are essentially smart, up-to-date and efficient. Particular attention is directed to the Smith Two-in-One Clock, combining a travelling wallet model with a dainty boudoir timepiece — a most useful, desirable and acceptable gift.

1 **KO39.** 3 in. square 8-day Movement, mounted on chrome easel and base with blue case trimmings, ivory/white dial, with raised silvered zone. Luminous. Height 3½ in., width 8¼ in. Price **48/6**
KO39a. As KO39, but with black trimmings. Price **48/6**
KO39b. As KO39, with green trimmings. Price **48/6**

2 **KO11.** 3 in. square 8-day Movement, mounted on chrome easel and base, with blue case trimmings; raised silvered zone; luminous. Height 3⅝ in.,width 6¼ in. Price **42/6**
KO11a. As KO11, but with green trimmings. Price **42/6**

3 **TWO-IN-ONE CLOCK.** Comprising a wallet-type case from which a complete little 3 in. Strut Clock can be instantaneously detached or replaced. This exceedingly useful double-purpose Clock is ideal to have at hand, and also makes an excellent present. 8-day movement. In real Morocco leather case, obtainable in the following colours: navy blue, brown, green, light blue, black or red.
KO6z. With square raised silvered zone. Price **45/6**
KO6zr. With round raised silvered zone. Price **45/6**

Figure 218

1934 saw the introduction by Smith's of "the first popular priced all-British callotte Clock. 8-day Wallet movement, fitted with A.B.E.C. 5-jewelled lever escapement. Highly polished hardened steel roller pin".

£50
increasing interest in typical 1930 style

Specifications OF SMITH'S CALOTTE MODELS

MOVEMENTS

KL3 —2¼″ Square Calotte Movement with shallow fixing ring. (For Wallet Cases, etc.) Chrome Bezel and Case, Ivory/White Dial.

KL3A —Ditto, but with Gilt Bezel and Case.

KL1 —2½″ Round Calotte Movement with shallow fixing ring. (For Wallet Cases, etc.) Chrome Bezel and Case. Ivory/White Dial.

KL1A—Ditto, but with Black Dial.

KL1B—Ditto, but with Silver Dial. (Engine turned dial).

KL1C—Ditto, but with Gilt Bezel and Case.

KL1D—Ditto, but with Gilt Bezel and Case and Silver dial (engine turned dial).

KL2—2½″ Round Calotte Movement with deep (½″) fixing ring. Chrome Bezel and Case, Ivory/White Dial. (As fitted to KL7 Clock. Suitable for similar cases and narrow wood struts).

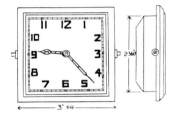

KL4 —3″ Square Movement, without swivel, but with ½″ fixing ring. (For wooden Strut Cases). Chrome Bezel and Case, Ivory/White Dial.

KL4A—Ditto, but with Gilt Bezel and Case.

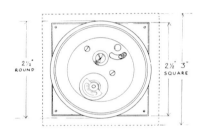

BACK VIEW
Common to all Models.

KL5 —3″ Square Movement with swivelling studs. (As fitted to types KL11, 12 and 13). Chrome Bezel and Case, Ivory/White figures.

KL5A—Ditto, but with Ivory/White Chaplets.

APPROXIMATE SIDE VIEWS OF COMPLETE CLOCKS

KL6

3″ Sq. 1″ Deep

KL7

Height — 3.187″
Depth — 1.375″

KL9

Height — 4″
Width — 4″
Depth — 1.500″

KL10

Height — 4″
Width — 4¾″
Depth — 1.500″

KL11-12-13

Height — 3.625″
Width — 6.375″
Depth — 1.500″

KL15 & KL16

4 ins square
when closed.

2½″ Round Movements in 3-fold Leather Wallet Case.

KL15A—In "Pearmak" No. 3 Crushed Skiver Case (Finishes—Dark Blue, Brown, Dark Green, Maroon). Chrome Bezel and Case, Ivory White Dial.
Minimum Retail Price **31/6**

KL15B—In "Pearmak" No. 3 Crushed Skiver Case (Finishes—Dark Blue, Light Blue, Dark Brown, Light Brown, Dark Green, Light Green). Chrome Bezel and Case and Ivory/White Dial.
Minimum Retail Price **33/-**

KL15D—In "Pearmak" No. 4 Crushed Calf Case (Finishes—Blue, Dark Brown, Light Brown, Green, Red). Chrome Bezel and Case, Ivory/White Dial. Minimum Retail Price **37/6**

The above can be supplied with gilt movements. Minimum retail price 8/- extra nett. (Please quote "gilt" when ordering).

2½″ Square Movements in 3-fold Leather Wallet Case.

KL16 —As KL15, but fitted with KL3, 2½″ Square Movement. Chrome Bezel and Case, Ivory/White Dial, Minimum Retail Price **33/-**

KL16A—As KL15A, but fitted with KL3, 2½″ Square Movement.
Minimum Retail Price **34/6**

KL16B—As KL15B, but fitted with KL3, 2½″ Square Movement.
Minimum Retail Price **36/-**

KL16D—As KL15D, but fitted with KL3, 2½″ Square Movement.
Minimum Retail Price **40/6**

(When ordering please quote KL16 and letter "A," "B," or "D" according to case required.)

Also supplied with gilt finished movements at minimum retail price 8/- extra nett. (Please quote "gilt" when ordering).

KR6. *This attractive set consists of a well-made Chapelle real leather case fitted with Smith 8-day timepiece. Case colours: Blue, Red and Green. Height 3¼"; Width 4¼"*

CAMERA TYPE TRAVELLING CLOCK

KCE1. *2⅜" square 8-day movement also 48 hour alarm. The case is in morocco grained skiver in the following colours: Navy Blue, Brown, Green, Light Blue, Red and Black.*

KL15. *Morocco Grain Skiver Case Travelling Clock with 2½" round Calotte Movement. Luminous.*

KL16. *As KL15, but with 2½" square Calotte Movement.*

Figure 219a

Smith Callotte clocks as advertised in Horological Journal *in 1939.*

£50+

Figure 219b

(*Left*) Smith calotte movement, showing unusual layout.
(*Right*) Complete Smith calotte showing external case thread and fixing ring.

CHAPTER IX

Wall Clocks

Vienna Regulators

The original Vienna regulator manufactured by the master clockmakers of Austria from the end of the eighteenth century and well into the nineteenth, were excellent timekeepers and, therefore, frequently used in public buildings. The cases of the early examples made during the Biedermeier period (1815-45) were simple and followed the classical lines popular at this time. The white enamel dials were plain and one piece, although some had a decorative brass edge. The movements, apart from the wheels, would have been hand made and examples can be found that run for a year, a month or a week. Any strike work was usually on gongs which were fitted to the backplate of the movement. This could be full strike with repeat or just striking on the hours and half hours. Full strike was with the quarters *first* (four blows at the hour, one blow at a quarter past, two blows at half past and three blows at quarter to the hour), with the hours struck *after* each appropriate quarter strike. Some four thousand makers' names have been listed but to date little has been written on the subject. It is doubtful whether many examples of the very early Vienna regulators are to be found in this country, although some excellent later pieces made by these makers do appear occasionally. The original Vienna regulator continued to be made in diminishing numbers until the end of the last century. Those manufactured after about 1850 had far more ornate cases with additional carvings, pillars, etc. The early elegant hands were now replaced by ornate pierced examples and the dials were manufactured in two pieces with an intervening brass ring between the two sections.

What do appear frequently, and are quite erroneously referred to as 'Vienna regulators', are the later German made copies. They neither come from Vienna nor are they precision timepieces. From the 1850s clock factories began to develop in several areas of Germany and inevitably copies were made of the superior Austrian clocks. The quality of these clocks varies — those made by the Lenzkirch factory in the Black Forest area for example are held in high esteem, while many others bear all the signs of mass production. One peculiarity of these German made 'regulators' is that the seconds dial although marked for sixty seconds, actually has the hand calibrated to complete its circuit in forty five seconds. Apart from a very few exceptions these German clocks have two piece dials.

Two manufacturers known to have made these clocks are Aug. Schatz and Sohne of Triberg and Gustav Becker of Freiburg. Becker was a prolific manufacturer and according to the history of the firm as recounted by Karl Kochman, he originally manufactured these timepieces for offices, etc., but about 1860 began to produce more ornate striking models for domestic use. Although he received many awards for his achievements, he found that the fierce competition of the cheaper spring-driven 'regulators' produced by the makers in the Black Forest area forced him also to manufacture movements of this type. His clocks prior to 1880 had been weight-driven. His trade mark was an anchor with the letters GB. This either appeared on the dial or stamped on the backplate.

The original Vienna regulators were precision timepieces and as such are finely poised and adjusted.

It is, therefore, necessary to ensure that the movements are kept well cleaned and polished to eliminate any friction caused by dust or dirt. The German factory-made examples are far more robust as can be seen from the pages shown in the Appendix devoted to parts for them.

It is as well to be aware of the fact that there have been kits available in America for some time for assembling your own Vienna regulator, and it has been recently noted that reproduction movements and dials are being imported from Germany into England. One manufacturer of these modern copies being J. Kieninger of the Black Forest area of Germany.

Figure 220

6ft. 1in. This large example of one of the comparatively early Vienna regulators has an elegant mahogany case with one piece dial and centre sweep seconds hand. The graduated enamel regulating plate can be seen at the base of the case on the back board, but it is not possible to discern the two small screws at the sides of the case for levelling the clock against the wall.

Courtesy of Derek Roberts Antiques

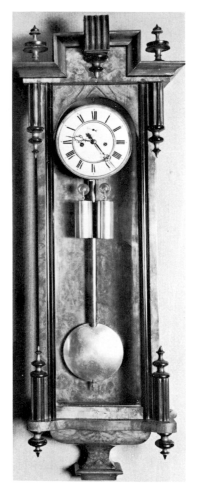

Courtesy of Derek Roberts Antiques

Figure 221

This is a Vienna regulator of a later date as indicated by the shaped case and two piece dial. Note the attractive use of the walnut veneer both on the back board and the bottom section of the case. These clocks have frequently lost their top pieces.

Figure 222

This example is a timepiece only. The shaped door and two piece dial indicate a later date, with overtones of art nouveau styling.

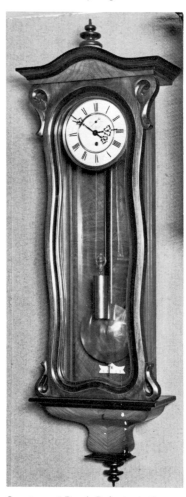

Courtesy of Derek Roberts Antiques

Colour Plate 16

French striking (bell) movement in gilt spelter case which reflects the sentimentality of Victorian/Edwardian taste.

£450

Figures 223a and 223b

4ft. 1in. The German weight-driven clock (left) is in a relatively elegant mahogany case with decorative turned finials, carved mask and pillars reminiscent of the style used on the later Vienna regulator cases. Note the two piece enamel dial with subsidiary seconds dial. The pendulum has the conventional wooden rod and spun brass bob. Striking is on the hours and half hours. Note that the gong is not mounted on the backplate of the movement, but on the back board of the case.

Examples with elaborately turned finials and superstructures like this are less popular than those with cleaner lines, such as Figure 221.

£550+

Figure 224

The similarities between the cases of the clock shown in Figure 223 and this illustration are interesting. The ebonised pendulum rod and parts of the turned components contrast well with the particularly attractive walnut graining in this clock. The estimated price at the end of 1981 was £180/£220 and it actually realised £240. Not a high price for an attractive two train example.

£500+

Figure 225

Some readers might find this case too ornate but it must be admitted that the veneer on the backboard is most attractive while the carved design is of a pleasant foliate effect. Not surprisingly when sold at the end of 1981 it realised £320 although it is only a timepiece.

£800+

Colour Plate 17

22½ins. high. The case of this clock is extremely ornate and with its ebonised wooden case and lacquered brass cast mounts and brass dial is reminiscent of the design of much earlier bracket clocks. The clock is of Swiss manufacture. Striking is at the hour and half hour on a single gong and provision has been made for repeat at will. See Figure 139b for details of the movement.

Colour Plate 18

10ins. high. This is a compact and attractive clock in a solid mahogany case which has been decoratively veneered, with added brass cast mounts. The white porcelain dial is marked to imitate a thirteen piece dial of French design. When enamelling was first introduced it was a technically difficult and expensive process and so each segment was painted and fired individually. Long after it was a necessity the custom continued as it was considered decorative. The name on the dial is that of the retailer 'Howell & James Limited, To the Queen, London'. See Figure 190 for details of the retailer.

Figures 226a and 226b

3ft. 10ins. A German spring-driven eight-day regulator in an ornate walnut veneered case. In place of the more usual fully glazed door only the dial has a convex glass with another small glazed aperture through which to view the pendulum bob. The conventional pendulum has been replaced by a mock compensated one. The two piece dial is of cream porcelain with a subsidiary seconds dial. Compare this good quality spring-driven movement with that shown in Figure 229b and note that the movement in Figure 226b is fastened on to a removable wooden base board, has solid plates, an encased spring, ting-tang striking on two gongs and bears a German patent number stamped on the backplate.

£450+

No. 1747.—Brass mounted Dial.
Length 55½ins.   Width 19ins.
8-day Strike   ...   68/-

No. 1641.   Length 52ins.   Width 16½ins.
8-day Strike   ...   87/6

No. 1798.—Length 55½ins.   Width 19⅜ins.
Brass-mounted Dial, Ivory Circle.
8-day Strike on Steel Rod Gong.

Courtesy of F.W Elliott Ltd.

Figure 227

*The page shown in this illustration is from a post-1921 catalogue of Grimshaw, Baxter and J.J. Elliott
Ltd. and shows a few of the 'Vienna Weight Regulators' sold by them at this time. 'G.B. movements'
would indicate that they had been manufactured by Gustav Becker of Germany.*

£400+ each

Colour Plate 20

Above, a small decorative Zenith boudoir clock manufactured by the Zenith Watch Co., in Switzerland. See Figure 217 for details of the carrying cases and movement.

Colour Plate 19

Above, mahogany pagoda type clock with an 8 day movement, which appeared in the 1911/12 catalogue of the Hamburg American Clock Company. See also Figure 192b.

Colour Plate 21

Cylindrical brass cased pigeon clock (6ins. x 5ins.) with keywound jewelled lever movement. The patents (No. 24905) were taken out in 1905 by Alec Turner of 30 Craven Street, Coventry, a watch and clockmaker. He had taken out a patent (No. 1885) two years earlier with his brother William Henry, which had been the first pigeon clock patent. The case in the early models was also of brass but rectangular.

£300+

Colour Plate 22

12ins. high. An exceptionally good quality French mantel clock in a solid gilt brass case with solid silver plaque, pillars, gallery and dial (weight of silver 15oz). Dial signed Maples Paris. Maples commenced trading in this country in 1840 with their Paris establishment in Rue Bondreau opening in 1896. The movement is with striking on a gong. Complete with dome and base (removed for ease of photography) this clock would be priced somewhere around £1,000+

Colour Plate 23

An example of a battery driven clock, where the coil oscillates between the two poles of the magnet. This model has a veined marble base with the letters LR on a silvered dial. (See also Figure 417).

Colour Plate 24

This small carriage clock was manufactured by the British United Clock Company, with a four glass case in heavy gilt. A good example in mint condition. See also Figure 207.

№ 2587

All of these either Eagle, Horse or Knob Tops.

Length 34¾ inches.
Ivorine Dial, gilt centre.

№ 1011

Length 39½ inches.
Ivorine Dial, gilt centre.
The Monk tolls the Hour and Half Hour.

№ 788

Length 43 inches.
Ivory Dial.

These Regulators have a very powerful **Strike on a large bell**, so that, used as "Hall Clocks" they can be heard all over the house.

Figure 228

Other examples in the same catalogue (Hamburg American Clock Company 1911-12) had a gnome in place of the monk or a flat bell in the same position but with the hammer obscured by the top crestings of the case. The manufacturer's comment was that as these clocks had a very powerful strike on a large bell if they were placed in a hall they could be heard all over the house. Each case design could be varied by an eagle, horse or knob top piece. It would be interesting to know how popular these Bell Striking Models were in this country or whether they were ever imported as few examples have been noted.

£400+   *(more for examples with automata)*

2ft. 8ins. This illustration shows an example of a German spring-driven regulator from the Black Forest area. The case, although attractively carved, is made of stained and varnished soft wood. The dial is of zinc and brass sheet overlaid with painted card. The pendulum is of a mock compensated design.

£300+

Figure 229b

Although the case is attractive (see Figure 229a) the 'pressed-out' movement of this clock is immediately seen to be from the lower end of the range of mass produced German regulators. Note that the gong (and indeed the movement itself) is mounted to the back board of the case. The coarse wheelwork, thin plates and open spring, together with the imitation grid-iron pendulum are all typical features. Compare this movement to that shown in Figure 226b. Although both mass produced the difference in quality is immediately noticeable.

Colour Plate 25

A Nightwatchman's clock. Very little original research has been carried out on these clocks and most authorities attribute their invention to John Whitehurst, FRS, of Derby. However, an article in the Antiquarian Horological Society's journal written by Adrian Birchall and entitled 'The Noctary or Watchman's Clock, Its Introduction and Development' provides substantial evidence that the Whitehurst in question is in reality John Whitehurst III (great nephew of John Whitehurst, FRS). Spring driven examples such as the model illustrated here date from the mid-nineteenth century with many of the movements coming from the workshops of Thwaites & Reed. The name of Bellefontaine (who worked in the Islington area 1832-75) is also associated with their manufacture. The Science Museum, London have one clock bearing this name.

This particular example has passing strike on the bell on the hour and half hour. The watchman was required to prove his wakefulness by being sufficiently alert to depress the knob on the top of the clock as the clock struck and thus drive in the pin that was at that moment immediately below the plunger. The padlocked front door would prevent any tampering. The quality of the movement is excellent and it would have been difficult for any nightwatchman to substantiate the statement that the clock was inaccurate. There are two dials — the main one remains static while the pins are read off against the outer smaller dial.

This type £1,000+

Colour Plate 26

At first glance this would appear to be a typical English bracket clock with inlaid Lancet case, however, the words on the dial 'Patent Moeller' indicate otherwise. This is actually a self winding electric clock manufactured to patents taken out between 1901 and 1908 by Max Moeller of Altona (Germany). The later patent refers to a striking model. These are very rare clocks and it has not been possible to trace the existence of sufficient examples to study the variations. There was one in the James Arthur collection in the United States of America similar in appearance to a skeleton clock and with the movement under a glass dome (see The Lure of the Clock *by D.W. Herring). The movement is completely self contained in so far as once disconnected from the batteries housed in the base of the case it can be drawn out with the seatboard (see Figure 429). Difficult to put a value on such an item but it must be in excess of £1,000.*

£3,000+

225

Figure 230

This is an unusual brass dial weight-driven regulator with a calendar aperture, seconds dial, date and phase of the moon. Striking is on a gong. A similar styled clock was made by Lenzkirch of Germany in 1926. There are no identifying marks on the movement apart from a repairer's date of 1921. It is nevertheless an interesting and attractive piece well worth acquiring.

£350+

Figure 231

This illustration has been included in order to emphasise the importance of studying fully any timepiece available and not to make sweeping assumptions. This movement comes from one of the cheap mass produced German 'regulators' made by Junghans`(name appears stamped on movement), but it is noted that it has an extremely unusual rack striking mechanism. In place of the more usual toothed rack, this example has pins arranged at right angles to the arm. Although not adding greatly to the monetary value of a piece with such a feature it would make it a desirable collector's piece.

LENZKIRCH, BEST QUALITY REGULATORS.

No. 856/243.
Length 41¾in.   Dull Walnut.
Strike on Cathedral Gong.
Timepiece £4 4 0
Plain Enamel Dial ...   £5 2 0

No. 859/244.
Length 56½in.   Dull Walnut.
True Centre-seconds Pendulum.
Price, £14 0 0.
Plain Enamel Dial.

No. 860/245.
Length 48in.   Dull Walnut.
Strike on Cathedral Gong.
Timepiece £4 15 0
Plain Enamel Dial ...   £5 13 0

A. M. & S. L.

Figure 232

Selection of Lenzkirch 'regulators' being marketed about 1905. Note the restrained lines of the example on the left. Attention is drawn to the fact that the large clock in the centre has 'true' seconds pendulum hence therefore the concentric seconds hand. Not a cheap clock (retailing at £14. 0s. 0d.) at about three times the price of its two companions. It is interesting to see that those with a plain enamel dial are priced higher than those with a two piece dial.

Today they would be about £450+

Figure 233

Two examples of the medium size regulators being marketed c.1911-12 by the Hamburg American Clock Company. The Silver Gongs referred to in this illustration were made 'from the best quality steel rods and are a harmonious combination of steel rods and of a bell'.

The 14 day movements had hardened polished pinions, polished pivots and springs enclosed in barrels.

£400+

MEDIUM SIZE WALNUT REGULATORS.
Glass sides. 5⅞ inch Dials, gilt centres. Movements fitted on sliding racks.
EASY WINDING.

№ 976   № 977

Length 39 inches.   Length 37 inches.
Ivory Dial.   Ivory Dial.

These Regulators can be fitted with the H. A. C. Silver Gong.

14 Day Half Hour Strike, also in 1 Day Strike and 14 Day Time.

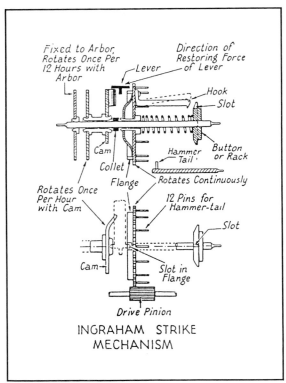

Colour Plate 27

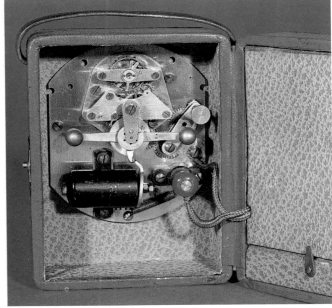

Colour Plate 28

4ins. x 3ins. x 2½ins. Small Swiss travelling battery electric clock. The red leather case has a compartment in the back to take a flat Ever Ready (4½ volt) battery. The gilt dial has luminous arabic numerals with the name Perpetua and Electric on the lower section. A removable sheet of celluloid protects the movement. The Swiss patents were taken out in 1928-9 by A. Schild of Grenchen, Switzerland, (see also Figure 434).

10ins. high. American synchronous electric striking clock (self starting) manufactured by E. Ingraham Company, Bristol, Connecticut prior to 1934. This example in a Gothic style walnut case with cream dial and centre seconds has a movement running on 240/250 volts 2/3 watts. These are interesting clocks in so far as the striking mechanism has been specially designed around the synchronous motor rather than the conventional type used in mechanical clocks.

Ingraham Strike mechanism: "In the Ingraham strike mechanism, all the striking wheel-work is carried on a single arbor. Two wheels mesh with the motion-work, that rotating once per 12 hours being fixed to the arbor. The once-per-hour wheel carries a plane-snail and off-set tip cam. This lifts, and then sideways-displaces, the universally-pivotted lever, lowering it on to the flange of the continuously-rotating strike-wheel. The cam is shaped to prevent the lever's return to its normal position, except when it drops through a slot in the flange and pumps the strike-wheel into hooked-on locking on the counting-button. Rotation, and hammer-operation, take place until the lock-hook escapes through the button-slot. Every hour the button has rotated a further 1/12th rev. A masterpiece of ingenuity! The drive-motor in this clock has a 6-pole hysteresis rotor, so rotor-speed = 6,000/6 = 1,000 r.p.m." (Diagrams and information by courtesy of H. Stott.)

£350+

THE ORCHESTRA GONG WALL CLOCKS.

14 Day Half Hour Strike, or 8 Day Three Quarter Strike.

For description of this Gong see opposite.

№ 616   № 617

Length 31 inches.
Dull Walnut or Fumed Oak Carved.
Plain 7" Silvered Dial.
Arabic or Roman Figures.

Length 30 inches.
Dull Walnut or Fumed Oak Carved.
Bevelled Glasses.
Plain 8" Silvered or Ivory Dial.

The Pendulums are behind Glass Doors.
Half Hour Strike with 4 hammers on 4 Silver Gongs.
Three Quarter Strike with 2 sets of 4 hammers.

Figure 234

Two weight driven 8 day regulators marketed by the Hamburg American Clock Company in 1912. There were also a number of models with spring driven movements with similarly glazed doors. Figure shows basically the same clock but placed on a plinth to resemble a longcase clock.

£200

NEW VIENNA STYLE.

THESE TWO DESIGNS
APPEAL TO THOSE WHO DESIRE
GRACEFUL OUTLINE AND
SIMPLICITY IN DECORATION.

KOSSOUTH. /249.
Length 35¼in.
Price, £3 0 0.

14-day Gong Strike, Brass and Silvered Dial,
Dull Walnut Case with Bevelled Plate Glass.

LUEGER. 250.
Length 31in.
Price, £3 7 0.

SPRING REGULATORS.

A. M. & S. L.

Figure 235

The case styles here are definitely showing signs of being influenced by the art deco trends. A factor that could enhance the price although many of these German spring driven clocks have particularly sonorous chimes and are sought after for this feature alone.

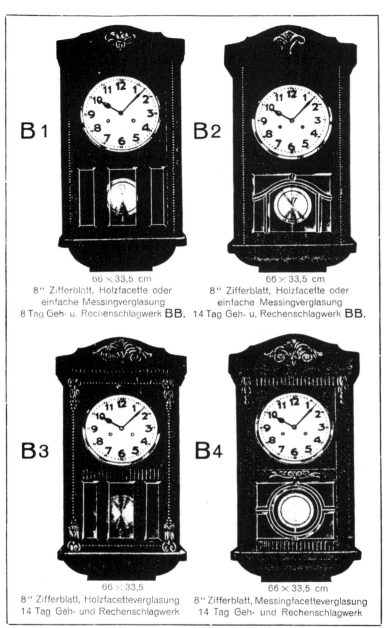

B1

66 × 33,5 cm
8" Zifferblatt, Holzfacette oder
einfache Messingverglasung
8 Tag Geh- u. Rechenschlagwerk **BB.**

B2

66 × 33,5 cm
8" Zifferblatt, Holzfacette oder
einfache Messingverglasung
14 Tag Geh- u. Rechenschlagwerk **BB.**

B3

66 × 33,5
8" Zifferblatt, Holzfacetteverglasung
14 Tag Geh- und Rechenschlagwerk

B4

66 × 33,5 cm
8" Zifferblatt, Messingfacetteverglasung
14 Tag Geh- und Rechenschlagwerk

Courtesy of Aug. Schatz & Sohne

Figure 236

These four wall clocks appeared in the 1936 catalogue of Aug. Schatz & Sohne of Germany and were typical examples of the clocks being imported at this time. With few exceptions the cases were of oak, with solid oak mounts, bevelled glass panels of varying shapes and sizes, silver anodised dials and striking on gongs. The movements were German — Gustav Becker being one of the most prolific manufacturers. Further details relating to this maker can be found on page 211. As with any of these late clocks, to be considered as a collector's item they must be in good condition and in their original state.

Becoming more sought after £100+

Figure 237a

25ins. This is a particularly attractive weight-driven clock made by Seth Thomas of Connecticut in a well preserved veneered softwood case with fully glazed door and colourful tablet in the lower portion.

Prices in England for these clocks bear no relation to the figures they fetch in America!

£300+

230

Figures 237b and 237c

Movement of clock shown in Figure 237a. The points to note are the original cone-shape weights suspended either side of the movement, the generally poor but serviceable quality of the mass produced movement, and the circular wire gong mounted on the back of the case. In common with most American manufacturers of this period the maker's label is pasted in the lower half of the case. From this it is possible to discern the name Seth Thomas, Thomaston, Conn. As the name of the town was changed from Plymouth Hollow to Thomaston in 1866, the date of this particular clock must be between 1866 and 1888. It is wise however to bear in mind that there is now a vast trade in reproduction labels and indeed stencils for the tablets. It would be dangerous to totally rely on a pristine label for purposes of identifying and dating!

Figure 238a

17ins. This is an exceptionally small American clock in an ogee veneered softwood case with a fully glazed door. The scene appearing in the tablet on the lower half of the door is entitled 'View in Rome, Italy'.

£250+ because small, compared to £200+ for standard size

Figure 238b

View of the movement of the American wall clock shown in Figure 238a. The interesting points to note about this thirty-hour spring driven movement are that the springs are brass (later examples used steel), the thin brass plates have been ribbed to give added strength and the name of the maker on the label is that of Brewster and Ingrahams of Bristol, Connecticut, USA. (1840-50).

Dial Clocks

Although there were various wall clocks prior to 1770, the English dial clock as we know it today dates from about this time. The earliest cases were solid mahogany, with a concave surround and flat bottomed box behind the dial to house the movement. A small door at the side was made to enable easy access to the fixed pendulum and verge escapement, with another at the base giving access to the pendulum bob for any necessary adjustment. Except for a few scattered examples the anchor escapement replaced the verge, which led to some changes in the shape and design of the cases. In order to accommodate a larger pendulum the base of the box was curved and in some instances lengthened. These are generally referred to as 'drop dial' or 'trunk clocks'. The lower section can be decorated with stringing, inlay, veneering and in some instances the pendulum is visible, but in other examples it is enclosed by a small mirror or panel.

After 1850 the wooden surround tended to be convex, although by this date other shapes, for example hexagonal or octagonal, also appeared. About the same date the cheaper spun bezel (the earliest examples had been cast brass) also appeared. Until 1875 the glass was retained in place by plaster, but after this date a sight ring (an inner brass ring) was used. However, it was still necessary to use plaster to retain the glass in place. Early this century the spun bezel with a sight ring that could be sprung in became more common. The dials of the clocks manufactured after the mid-1800s were of iron — painted white — either flat or convex, although a few examples can be found with the older-type silvered brass dials. Boldness and clarity would be the most noticeable characteristics of the painted dials. At this date the names on the dial will be those of the retailers — the makers of the movements being content to omit their names or at best stamp them on the dial plates. Thwaites and Reed were one supplier to the trade during the nineteenth century, details of others will appear with the appropriate illustrations.

With very few exceptions these clocks manufactured after 1850 will have anchor escapements, although some can be found still using a verge or deadbeat escapement. In typical English tradition the movements will have a fusee and chain or steel wire. It is possible to ascertain whether the movement has the original line by counting the fixing holes in the barrel. One indicates that a chain was used, whereas the presence of three infers a steel wire or gut. It would not necessarily be true to state that one was any earlier than the other as it was noted, for example, from a 1921 catalogue: "New feature — the 8-day movement now made with chain". Although there are some examples with strike, seconds, alarm or calendar work the commonly found dial clocks are timepieces.

So far the clocks described have been *English* dial clocks, but inevitably they were copied in large numbers by the American and German manufacturers. German manufacturers such as Winterhalder and Hofmeier included them in their range and made some models with fusees, but almost without exception those manufactured by other makers omitted the additional mechanical refinement of the fusee. Many German and American movements were imported and cased here and therefore outward appearances can be deceptive. Movements, backplates and, if possible, dial plates should be noted to arrive at a positive conclusion!

One gains the impression from the catalogues of American manufacturers that the majority of their movements were housed in cases of the drop dial pattern with varying shaped surrounds. They did not favour the simple English-style dial clock, but added either on the same dial or on a second dial calendars, barometers, etc. Strike work was far more common than on the English counterparts.

Further detailed information concerning these clocks can be found in the book on *Dial Clocks* written by R.E. Rose. Although the late standard movements are of no great horological merit they are good solid clocks that deserve higher praise than being referred to as 'just' schoolroom or kitchen clocks.

Figure 240

An example of an English dial clock in an attractive case, with a carved mahogany surround and trunk. The painted iron dial is signed Grosvenor, Ellesmere. The date of this piece is c.1870.

£250+

Courtesy of Strike One Ltd.

Figures 241a and 241b

This is a typical mahogany dial clock as used in offices, shops, etc., at the beginning of this century. This example has the added interest of having been part of the shop fittings of Playle Bros. of 137 Northcote Road, London, S.W., who were in partnership as watchmakers until the end of 1905. In the movement (241b) note the fusee and that although the rest of the wheels are crossed out, the hour wheel is not — this was common practice as it was not usually visible. Although it cannot be said that this clock is horologically exciting, it is an excellent example of a good sturdy English fusee timepiece and can only appreciate in value. Price depends on size of clock. A small (8ins.) example £200 would be twice the price of large (12-18ins.) with ebonised examples even less (£60+).

£200+

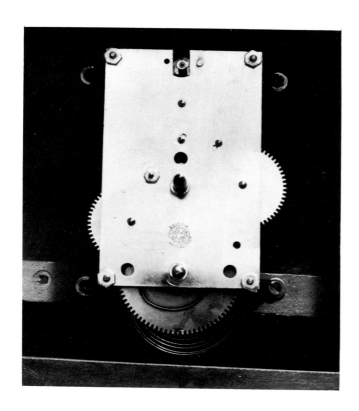

Figure 242

This is an illustration of a rather curious movement viewed from the dial plate (i.e. dial has been removed). Note the uncrossed wheels, open spring, only means of access through dial and lastly the trade mark. A movement of this quality could only be American or German and this is the trade mark of the New Haven Clock Company of Connecticut, but it has also had the words 'British Manufacture' stamped around it. Possibly the case was made in this country.

Figure 243

This appears to be another example of a plain timepiece manufactured at the beginning of this century similar to that shown in Figure 241. However, on closer examination it is realised that it is a highly sought after collector's item as it is of one month duration. Note the convex surround, spun brass bezel and sight ring.

£650+

Courtesy of Strike One Ltd.

No. 501.—Oak, Walnut. or Mahogany Case.
8-day Time. 10 and 12in. Dial. **19**/-. 8-day Strike, 12in., **26**/-
8-day Time, 14in. Dial. **32/6.** 8-day Strike 14ins., **39/6**

No. 503.—Oak, Walnut. or Mahogany Case,
12in. Dial.
8-day Time, **27**/- 8-day Strike, **34**/-

G. B. & J. J. E. Ltd.

No. 502.—Oak, Walnut. or Mahogany Case,
12in. Dial.
8-day Time, **21**/-. 8-day Strike, **28**/-

Figure 244

This illustration shows three styles of dial clocks manufactured by the British United Clock Company (for further details of this maker see page 197). They appeared in a trade catalogue published c.1910.
Price here depends on size and whether the case dial surround is ebonised or wood. Ebonised would fetch a lower price.

top illustration £200+
other two examples £350+

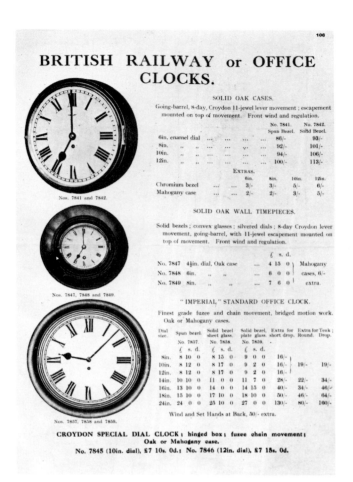

Figure 245

This illustration shows a selection of dial clocks manufactured by F.W. Elliott Ltd. sometime between 1932 and 1946.

Figure 246

William Haycock, Ashbourne, Derbyshire is generally associated with the manufacture of skeleton clocks in the late 1800s but they were also extremely active in the manufacture of many other types of movements including that illustrated for dial clocks or the longcase just discernible in the right hand bottom corner of this advertisement dated October 1930. As their outlet was through the wholesale and export trade it is extremely rare to find a clock bearing their name. They are still in business and have in recent years produced a small range of reproduction longcase and bracket clocks.

Wm. Potts and Sons Ltd. Leeds was founded in 1833 by William Potts, who began his career in the horological trade as a maker of small domestic clocks. By 1840 the business had expanded to include the manufacture of large movements for churches, factories, etc. He became acquainted with Lord Grimthorpe when competing for the contract for the new turret clock for the parish church at Ilkley, Yorkshire. The acquaintanceship ripened into friendship and the two men worked on a large number of similar projects over the years. One of the most important installations carried out by Potts was that for the Cathedral in Lincoln. The range of clocks advertised by them in their 1888 catalogue included a wide range of domestic clocks, regulators, longcases, chime and strike bracket clocks, (see Figure 157), dial clocks for factories and schools, etc. The firm became a limited company in 1906 and continued to manufacture a wide range of clocks until well into this century. The depression of the '30s and the fierce competition from overseas manufacturers saw the closure of their domestic clock workshops. Their work with turret clocks continued. The firm was absorbed by John Smith Ltd. of Derby.

Going Barrel Timepiece

The illustration above shows our new GOING BARREL TIMEPIECE—Case, Movement and Dial.

Figure 247

Illustration of dial clock manufactured by Wm. Potts and Sons Ltd. As they were at one time officially appointed clockmakers to the railways there should be a number of these clocks generally available especially in the Leeds and Newcastle area.

£100+

THE "POTTS" TIMEPIECE

Finest Quality throughout, and
the BEST ENGLISH TIMEPIECE MOVEMENT IN THE WORLD.

SPECIFICATION

Hard Brass Plates, of heavy gauge, and Brass Pillars. Wheels of Hard Brass, with teeth machine-cut from the solid to mathematical accuracy and correctly depthed. Pinions of Finest Steel, hardened, tempered and polished. Well turned and correctly grooved Brass Fusee, with strong Steel Click, Brass Ratchet and Click Spring. Specially strong Steel Chain. Steel stopwork. Bridge pattern motionwork. Pallets of Solid Steel, hardened and highly polished, recoil pattern. Best Steel Spring, in well-made Barrel. Brass Pendulum, with heavy lens shaped bob. Screws of Nickel Silver (unrustable). Brass Winding Key.

FINISH. All parts highly polished; and Brass parts lacquered.

PRECISION WORKMANSHIP THROUGHOUT. All the depths are run individually by hand, ensuring the greatest possible accuracy, long life and perfect timekeeping.

DIAL. Our Standard Pattern dial is of best Hard Enamel on heavy gauge Mild Steel (Baked White) with bold figures and minutes. Hands of Blued Steel. Special Dials, in Brass, Silvered and engraved, etc.. with any type of figuring—Block, Roman or Arabic—to order.

CASE. Every type of case supplied—Bracket, Wall Hanging, etc., etc.

We make a feature of providing suitable Movements and Dials to specially designed fitments. We also make and supply complete installations to be in keeping with Architects' Designs.

DOUBLE DIAL CLOCKS

Either self-contained Double Dials ("Cheesehead" Cases) or Clocks to shew the time in two places—the dials of which may be of different sizes—supplied to meet requirements.

In the case of the "Cheesehead" the second dial is fixed to the back of the case, and the whole Clock is then capable of being fixed in almost any position.

With the other type the second dial is mounted in a well-made wood rim, fitted with bezel and glass, for securing to the wall of the building, complete with set of suitable dialworks, connecting rod, universal joints, etc., to connect up with the movement.

The gears are machine-cut from Hard Brass, and the universal joints are of a particularly efficient pattern.

On large dials the minute hand is correctly counterpoised. For the second dial, in the larger sizes, the bezel is usually of cast iron, but may be of brass if required.

When the second dial is to be larger than 14" we supply a more powerful movement in every way, to meet the extra requirements.

GUARANTEE. EACH CLOCK IS GUARANTEED FOR A PERIOD OF FIVE YEARS against faulty workmanship, materials or design.

Figure 248

*Further example of one of the wall clocks manufactured
by William Potts & Sons Ltd.*

£150+

Figure 249

*Dial movement made by H. Williamson Ltd. of
Coventry. Note Astral Trade mark.*

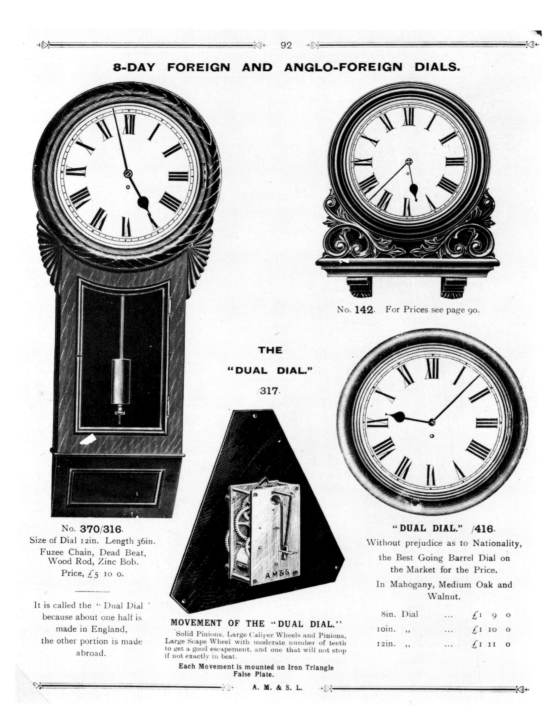

8-DAY FOREIGN AND ANGLO-FOREIGN DIALS.

No. **142**. For Prices see page 90.

THE

"DUAL DIAL."

317.

No. **370/316**.
Size of Dial 12in. Length 36in.
Fuzee Chain, Dead Beat,
Wood Rod, Zinc Bob.
Price, £5 10 0.

It is called the "Dual Dial"
because about one half is
made in England,
the other portion is made
abroad.

MOVEMENT OF THE "DUAL DIAL."
Solid Pinions. Large Caliper Wheels and Pinions,
Large Scape Wheel with moderate number of teeth
to get a good escapement, and one that will not stop
if not exactly in beat.

**Each Movement is mounted on Iron Triangle
False Plate.**

A. M. & S. L.

"DUAL DIAL." **/416**.
Without prejudice as to Nationality,
the Best Going Barrel Dial on
the Market for the Price.
In Mahogany, Medium Oak and
Walnut.

8in. Dial	...	£1 9 0
10in. ,,	...	£1 10 0
12in. ,,	...	£1 11 0

Figure 250

The value of this particular illustration lies in the reference to the "Dual Dial". Firstly it provides evidence of the practice of foreign movements being cased here and secondly shows details of the movement mounted on an iron Triangle False Plate. It is interesting to note that the pinions are solid. Other features commented on are large caliper wheels and pinions, large scape wheel with moderate number of teeth to get a good escapement and one that will not stop if not exactly in beat. The initials on the base of the plate are those of the wholesale firm marketing these clocks, A. Mayer and Sons Ltd.

£100 — £250

ANGLO-AMERICAN 8-DAY DIAL.

12in. DIALS, INLAID CASE FITTED WITH AMERICAN MOVEMENTS.

SETH THOMAS
STRIKING
MOVEMENT,
7/- EXTRA.

No. **18**.
Scroll O.G. Bottom Walnut.
Time, £1 8 0.   Strike, £1 12 6.

No. **17**.
Double Hollow String and Dot.
Time, £1 5 0.   Strike, £1 9 6.

SETH THOMAS
TIMEPIECE
MOVEMENT,
3/- EXTRA.

No. **11**.
Long Line and Dot in Walnut.
Time, £1 0 0.   Strike, £1 4 6.

No. **13**.
Long Stringed and Dot.
Time, £1 1 0.   Strike, £1 5 6.

On Orders of Six of the above Clocks, a **Nett** Allowance of Sixpence per Clock will be made.

A. M. & S. L.

Figure 251

The term "Anglo-American" indicates that the case was made in this country while the movement came from the United States. It was common practice for American firms to purchase movements from each other for casing hence it was only logical especially when considering transportation limitations that there was a large movement trade between the two countries as well as fully cased clocks. It does not state in the catalogue (a page of which is illustrated here) whether the movements originated from the E.N. Welch, Ansonia or New Haven manufactory. Obviously not from Seth Thomas as these are offered as alternatives at extra cost. 7/- extra for a striking movement and 3/- for a timepiece only. The cases shown in this illustration are typical of those made in this country, i.e. with "ears" and bottom of case curved back to wall. There is some evidence to indicate that mainly Italian workmen were employed in making these cases. One well documented firm was Holloway & Co. (c.1842 to c.1883) of London. Their trademark was the letters H & Co. under a crown. This mark is often found stamped on the backs of cases. Further details regarding the firm can be found in American Clocks for the Collector by E.J. Tyler. There is a strong possibility that cases for the Jerome & Co. Ltd. and New Haven Clock Co. were also made in Liverpool between 1894 and 1932 by J.C. Plimpton & Co.

£150 — £250 depending on case decoration

Courtesy of Strike One Ltd.

Figure 252

This is an extremely attractive timepiece with 12ins. painted iron dial, octagonal surround and elaborate trunk. The case is part oak and part paper-mâché inlaid with mother-of-pearl. It is an unusual design for an English dial clock, but one which was used extensively in America.

£175+

Figure 253

2ft. 4ins. Anglo-American 12 drop dial clock with eight- day striking movement. This is a typical case style for these clocks.

£150+

Courtesy of Ruislip Antiques, Alfies Antique Market

Cuckoo Clocks

The clockmaking industry in the Black Forest area of Germany had originated as a secondary winter occupation for the farming community, but in 1850 a great spur was given to the industry by the establishment at Furtwanger of a clockmaking school. The following account was written by Miss Seguin:

"Formerly, and until within the last fifteen years, every portion of the works of these Black Forest Clocks was made by hand, and each workman began and finished his own Clock in his own cottage, being assisted in his labours by the different members of his family. Now, this hand and individual labour is, to a great extent, done away with, being supplemented by large establishments, where a hundred or more men are engaged, in which machinery is employed, and the labour is subdivided into at least a dozen processes. The men work twelve hours, are paid from a shilling to half-a-crown a day, and women are employed as polishers of the cases. The old hand-labour system is maintained only in a few remote villages, and for the inferior kinds of Clock.

"Since the introduction of machinery the Black Forest Clocks have been, it is said, not only cheaper but more accurate, although it is certain that some of the old wooden Clocks, made a hundred years ago, are still in use, having withstood the various changes of temperature, and the wear and tear of a century, with scarcely any diminution of their powers.

"One peculiarity of the Black Forest Clocks, is that they are almost all made to be fastened against a wall, not as chimney timepieces, and thus they are used throughout Germany, where, in truth, it would be difficult to find a place for timepieces, as chimneys do not exist there.

"The favourite form is the cuckoo-clock, and a variety of other mechanisms is also introduced in the more elaborate specimens. It would be difficult, indeed, to say that any result was impossible to the inventive genius of these Black Forest Clock-work makers.

"In the ninety-two parishes which form what is called the Clock-country, are over 1,400 master Clockmakers, who employ some 6,000 workmen. Altogether, about 14,000 people, including women and children, are occupied in this one industry. The number of Clocks manufactured yearly in this district is calculated at two millions, valued roughly at one million sterling."

The Americans, with their cheaper factory-made shelf clocks captured for a while at least the English market that had previously been supplied by the German clocks. However, the Germans rallied and began to use factory methods themselves, and, as well as manufacturing copies of the American, English and French clocks, adapted their own traditional designs to the new methods. The Black Forest cuckoo clock was, according to legend, invented by Franz Anton Ketterer of Schoenwald about 1730 to 1740, and is still popular to this day. Apart from the pinions and perhaps a few pieces of wire, the entire clocks were made of wood, although gradually brass wheels were introduced and by the end of the nineteenth century only the wooden plates were retained. About 1850 the cuckoo clock as we think of it today was introduced — the case having a sharp-angled gable roof, and being in the form of a box with a flap at the back for access to the movement housed within. The variations that are shown in the illustrations in Figures 254 and 255 are not without interest.

The points to note when considering a purchase are first to inspect the case carefully for damaged or missing parts. Frequently made of pine and having become brittle with age only too often the carvings have been broken. Secondly, ensure that your cuckoo can still 'cuckoo' and the bellows have not been perforated. Although the Appendix lists a comprehensive range of parts for these clocks it is doubtful if they can be obtained today and patching bellows can be a tedious task. Lastly, never be tempted to improve upon the basic design of the movement as, although to all intents and purposes it appears a very bodged affair, it has evolved over many years of trial and error to become perfectly satisfactory. For example a flat strip on the pendulum with a hole to place over the suspension hook could possibly stop the clock as the two hooks in the original design roll within each other and reduce friction. Similar refinements also bring problems in their wake.

Figure 254a

This is a fine example of a cuckoo and quail wall clock, dating from around the end of the last century. The quail giving the quarters and the cuckoo the hours. It is far more common to find examples with a cuckoo alone.

The double automaton make this a particularly attractive piece.

Figure 254b

Movement of cuckoo and quail clock shown in Figure 254a. Note the steel arbors and lantern pinions with wooden plates, and the pipes and bellows above the movement.

£500+

Figure 255a

An attractive and slightly unusual cuckoo clock of around 1850 to 1875. The points to note are the gabled roof of the case, painted mother and child scene (after 1860 transfers were commonly used), enamel dial and steel moon-shaped hands and brass cased lead weights. Later examples have cast pine cone weights (see Figure 254a).

£900+

Figure 255b

Movement of clock shown in Figure 255a. The points to note in this view are the gong secured to the door of the case, wooden plates and general appearance of string and bent wires! The two supporting pillars on either side are covered with coloured marbled paper. The bellows and cuckoo are housed at the top of the case. Later examples (after 1900) are more likely to have brass movements and plates.

Various other types of Wall Clock

Figure 256

A typical exmple of a Black Forest 'postman's alarm'. These were manufactured up until the First World War and maintained a steady popularity. The dial is glass, with the numerals and markings painted on the obverse. In this example the alarm is set by positioning the third hand at the required hour. In this instance the alarm would ring at about quarter past seven. The arrow indicates the counter-clockwise setting of the alarm. Other examples have been seen with a centre disc for setting the alarm mechanism.

Depending on decorative quality of dial surround

£250+

Figure 257

Good selection of Postman's Alarm clocks being offered for sale about 1911. Pine cone weights were an innovation of the English market. Value depends upon the decorative quality of the dial surround, size, style of weights, and whether pendulum bob and weights are of brass or brass covered. Inlay on the example shown here is brass.

Figure 258a and 258b

14ins. high. Bronzed metal case with white enamel dial with blued steel hands. The cylinder movement (note easily detachable barrel) is of French origin. The trade mark adjacent to the barrel is that used by Japy Frères et Cie in the late nineteenth century. See the chapter on Marble Clocks for further details of this manufacturer. This example has been included to demonstrate the great variety of clocks made by this firm as well as being an amusing case style with obvious nautical influence.

£150+

Figure 259a

Having stated that these clocks are valued by their decorative appeal, an exception is immediately made! The case of this example is attractive — Napoleon III style case c.1855 with copper repoussé dial, enamel plaques with black Roman numerals and pewter mounts, etc., but it is the position of the winding hole that is of great interest. It is above the centre arbor. The reason for this will be seen in Figure 259b. A clock in an identical case was noted in the 1908 catalogue of Chateau Frères & Cie, 125 Boulevard de Grenell, Paris. A barometer in a matching case was available for an additional 68frs.

£300+

Figure 259b

View of the movement to the clock shown in Figure 259a. As access to the movement is difficult an aperture has been left in the wooden backboard for any necessary adjustment to the effective length of the pendulum. Upon looking at the movement it is immediately noted that it has an inverted pin pallet escapement. The other point of interest lies in the trade mark stamped on the back-plate. It is identical to that on the Swinging Cherub clock illustrated in Figure 275b. To a collector these technical curiosities will greatly enhance the value.

Figure 260

25ins. The cartel clock originated in France about 1750, but whereas the English developed their counterpart — the dial clock — on simple lines, the French continued to utilise ornate highly decorative surrounds, usually of ormolu. The case styles of these early examples were repeated and copied through the years. The example in this illustration is reminiscent of the designs used during the Louis XVI period. The movement is stamped with the name of the maker L. Marti et Cie together with the fact that they received a Médaille d'Argent in 1889, thereby indicating that the clock must have been manufactured after this date. It is an attractive piece — well proportioned and with fine detail. As the value of these clocks lies in their decorative appeal this is important.

£400+

Courtesy of Kingston Antiques

Figure 261

24ins. These clocks come from the Franche Comté region of France and are referred to as 'Tableaux Comtoise'. They are spring driven and although of simple construction are extremely robust and durable. The shape of the case can be square, oval, scalloped or plain and decorated by brass or mother-of-pearl inlay. The name on the dial is Claude Mayet à Morbier. The name 'Mayet' is associated with the origins of the clockmaking industry in this area. Legend has it that three brothers of this name founded the industry in the seventeenth century after successfully replacing a worn out public clock. This Claude Mayet would possibly be the name of a nineteenth century retailer. Due to a certain amount of renewed interest in these clocks, both as decorative and practical pieces for a small modern home, reproductions are being manufactured by firms on the Continent. Several books have been written on the subject including: The Morbier 1680-1900 *by Steve Z. Nemrava;* La Comtoise, La Morbier, La Morez, Histoire et Technique *by Francis Maitzner and Jean Moreau; and* Die Comtoiser Uhr *by Gustav Schmitt.*

£200+ if dial painted £150+

Courtesy of Sotheby's

CHAPTER X

Mystery and Novelty Clocks

In the early days of horology many clocks were the products of fertile minds seeking to improve upon the then limited knowledge of the art and mystery of horology with experimental devices. The clocks of Nicholas Grollier de Serviere (1596-1689), soldier, mathematician and inventor being one example, with the concepts and ideas suggested by the Marquis of Worcester (1601-67) in his patent of 1661 being another. This patent (Patent No. 131) was the first patent relating to clocks and watches to be taken out. Further details of the Marquis and his horological inventions can be found in the Winter issue of *Antiquarian Horology,* 1976. As the centuries passed and for all practical purposes the technicalities and basic design for a clock were mastered, the necessity arose to please the wealthy customer and satisfy his desire for an unusual timepiece. So the clockmakers strove to provide such items by adding complicated chimes, automata, elaborate cases and dial decoration, etc. Still later the factory methods of production, which had brought the clock within the price range of the working man, also created the problem of uniformity of design. Individuality had to be achieved in some aspect in order to overshadow competitors. Some manufacturers relied upon being the first with some new innovation or being able to add some gimmick to their range. This created the need to protect new ideas by filing applications for a patent. The flood of applications for patents to be granted for items being shown at the Great Exhibition led, in 1852, to the reorganisation and simplification of the patent system. Even amateur inventors tended to protect their devices in this manner, many of which were ludicrous and never reached the production lines, while others only enjoyed a short period of popularity, with a residue of commercially acceptable ideas remaining. In some instances the bases for these were reintroduced from previous areas and adapted to current manufacturing techniques and materials. One example of this is the clock in Figure 281a illustrating the rack or gravity clock. Others were considered by their inventor to fill "a long felt want" – one example being the Memorandum Clock of John Davidson shown in Figure 278.

Frequently the patent dates or numbers appear on the clock and provide an invaluable piece of documentation, for these can be traced at the Patent Office in order to ascertain the name of the inventor, occupation, location and, if not already quoted, the date the invention was patented. Although this can only establish the earliest date from which an article was manufactured it is nevertheless invaluable.

As the following clocks are so diverse in concept, quality and design it is impossible to provide any specific guidelines as to features to seek or note except in the accompanying text with each illustration.

Figure 262

The mystery of this clock lies in the fact that there is no apparent connection between the pendulum held firmly in the hand of the bronze statue and the movement in the marble pedestal. With a high power lens it can be noted on careful inspection that in this example the bronze ring upon which the figure stands rotates by 1/40th of an inch either right or left with each impulse of the escapement. A full description

262

263

Courtesy of Keith Banham

of this interesting mechanism can be found in the Horological Journal *for August and September, 1948. The movement is French. Similar clocks can be found with extremely ornate ormolu and coloured marble bases, and in some instances a skeletonised dial. The figures can be bronze or spelter. It will be the quality of these figures that determines the price. A good bronze figure fetches several hundreds of pounds in its own right, whereas one of spelter would be a fraction of this. It must also be remembered that any restoration needed to the movement of any of these mystery clocks could be a problem as only a handful of restorers are prepared to overhaul and clean, let alone repair, this type of mechanism. It is therefore not particularly wise to purchase other than a working example unless provisional arrangements have been made together with an estimate of the cost that will be incurred with a restorer.*

£2,000+

Figure 263

Approx. 12ins. by 18ins. An extremely rare novelty clock. Its invention is attributed to Nicholas Grollier de Serviere (1596-1689). He spent much of his life serving as a French soldier during which time he lost the sight of one eye. During this period he applied his mechanical and mathematical genius to the designing of fortifications, etc., but upon retiring from the army he turned his attention to a wider field which included various ingenious timekeepers. A few reproductions were made during the later part of the nineteenth century and the beginning of this. Their value would lie in their decorative quality and the general principle that lies behind the mechanism rather than being efficient timekeepers. This particular example has the 'fan' mounted behind glass in recessed picture frame. The recess is lined and has two decorative mouldings in the top corners which have been gilded to match the outer frame. The fan itself is highly decorative. The time as shown in the illustration is 6 o'clock as indicated by the small pointer in the shape of a serpent. The fan snaps shut when fully unfolded at 6 o'clock and then slowly commences to unfold from left to right — the time always being that indicated on the left hand side. The movement in this example was mounted externally at the back of the frame.

£1,750+

Figure 264

The principle upon which this particular mystery clock works is attributed to Robert-Houdin (1805-71) the French conjurer. He is not to be confused with Houdini the escapologist of this century. Houdin, born Jean Eugene Robert, was first trained as a lawyer but prevailed upon his father to apprentice him to his cousin, a clockmaker. In 1830 he married the daughter of Jacques Françoise Houdin, a watchmaker who worked with Bréguet, and continued his career of showman and conjurer as Robert-Houdin. With his own knowledge of clockwork, and no doubt with the assistance of his father-in-law, he devised many automata and mystery clocks for which he received several medals at the various exhibitions. There were two models for this clock — in one the column between the plinth holding the clock movement and the dial is of clear glass (see Figure 265) where in the example illustrated it is opaque. In both there is no visible method by which the drive reaches the hands. In the example depicted here there are actually two glass dials within the bezel — one is fixed and has the hour and minute markings while the other carrying the hand is free to rotate. The periphery of this second dial has a finely toothed rim and is rotated by a carefully contrived series of rods and worm and bevel gears connected to the movement housed in the base. This example carries the name of 'Promoli A Paris' and has an anchor escapement with a silk suspension. A similar model appears in La Pendule Française by Tardy. The value of these clocks is governed by virtue of their being a novelty, and so it is of no great importance as to whom the makers of the movement and case are.

£2,500+

Courtesy of Keith Banham

Figure 265

This is the model with the column and dial in clear glass. A clockmaker's label inside the case stated that the clock was 'cleaned and repaired for Ld. Lington at Kiloran Castel, May 16th '08'. It is assumed that the clock's present owner has investigated this helpful clue as to the clock's provenance! The movement was described as a twin going barrel with countwheel strike on a gong. When sold in 1982 it realised £3,672 including 8% buyers premium.

£3,500

Courtesy of Christie's

Figure 266

19½ins. high. These clocks are usually of French origin and this is no exception. Although a late nineteenth century copy of those made over a century earlier it is a highly desirable piece. The figures of the three Graces support a globe with two rotating bands — one showing the hours and the other the minutes. The arrow held by the figure of Cupid serves as a pointer. Other similar examples would have an urn, or vase and be with or without figures. The quality of the figures (bronze signed or unsigned, or spelter) or the manufacturer of any porcelain decorations would be influencing factors when estimating the value of such a piece. This particular piece fetched £1,050 in 1976. It would command a higher price now.

£2,000+

Courtesy of Sotheby's

Figure 267

1ft. 9½ins. high. The relevant catalogue entry for this attractive mystery clock reads "A French Bronze Patinated Spelter Mysterieuse Timepiece in the form of a scantily dressed young girl holding a timepiece aloft. . . . signed Louis Moreau and with founder's mark, on a turned socle, c.1900". The movement is housed in the globe held by the figure and both globe and pendulum swing in toto, a small pendulum within the globe being the true regulator. (It is interesting to compare this with the example shown in Figure 268 which is a later example and would have been made sometime in the 1930s; this spelter figure does not have the sharp finish enjoyed by the example shown here and in place of the globe there is a normal dial.) The movement is that of a watch and runs for a week. The timepiece and pendulum swing fifty times per minute. It realised £300 in 1976. As early as 1886-7 the Ansonia Clock Company of America were advertising copies of these clocks in 'Japanese or French Bronze' finish but with eight-day movements. Later, in 1914, their catalogues listed swinging ball clocks, with 'real bronze finish', coloured balls and again, eight-day movements.

Courtesy of Sotheby's

£250+ if spelter, £400+ if bronze

Figure 268

This example of a Diana swinging clock was manufactured during the 1930s but similar clocks appeared in the Hirst Bros. catalogue for 1910 retailing at 30/-. The German manufacturers Junghans and T. Haller took out a patent in this country for "a clock of the kind which the movement is built into the pendulum so as to oscillate with it, the pivotal supports of the pendulum are two pins the points of which rest in recesses in a bracket". Patent No. 4812, Convention date 1909.

The Ansonia factories in America also produced similar clocks and in fact it is still possible to obtain figures to this pattern in the States. An alternative model is that of an elephant holding the timepiece aloft in his trunk.

£150+ with the quality of the figure largely determining the price.

Figure 269

8ins. high. The case of this clock is in the form of a gaily painted pottery dwarf. It was introduced in 1929 when Walt Disney films were making their début and there is a label pasted to the base stating that the design was 'By permission of Walt Disney. Mickey Mouse Ltd.' The clock was, however, made in England and has a basic uncomplicated movement — the eyes rotate as the clock beats the seconds. Unlike the example in the following illustration the dial below is the true time teller.

Comic character clocks and watches have seen a marked price increase £200+

Figures 270a and 270b

6ins. high. Unlike the pottery dwarf in Figure 269 the eyes of these little dogs actually tell the time. The white line running from the centre of each pupil points to the appropriate marking at the edge of the eye socket. The hours are marked around one orbit and the minutes the other. The original patent taken out in this country in 1926 by J. Oswald, Gartenstr 30, Frieburg, Germany envisaged a man's face flanked by two girl's faces. The girls' eyes indicating the hours and minutes while the man's showed seconds or alarm mechanism. Both 8 day and 30 hour models were made as were little Pekinese Dogs. The cases simulated carved wood.

The advertisements that appeared in 1928 and 1929 English periodicals stated that this was "The greatest novelty on the market, of finest workmanship and only to be obtained from the Patentee and sole manufacturer. Infringements of Patent proceeded against." They were shown at the Leipzig Fair and were an instant success. Further models included owls and turbanned women.

£150+

Figure 270b

View of the movement — note cheap quality generally and method of fixing. The larger dog had an eight day movement while that in the smaller was only thirty hour.

£50 — £80 depending on duration of movement

Figure 271

Alternative model advertised in Die Uhrmacher Woche *in 1935.*

£50–£80

Figure 272

13ins. high. This is an illustration of an American 'Blinking Eye Clock'. The eyes flick back and forth with each beat of the pendulum. Several factories produced these clocks, but this particular example in a black cast iron case was made by Chauncey Jerome, Connecticut, USA, around 1870. They are eagerly sought after by collectors both here and in America.

£900+

Figure 273

This is an early twentieth century replica of a seventeenth century clock in the form of a clock pedlar with a clock on his back. In this example the movement within the painted spelter figure has the mark JVE stamped on it. The duration of the movement is thirty hour. The model manufactured by Hamburg American Clock Company and sold by Hirst Bros. in 1910 carried a pendulum in one hand and a drum alarm in the other as well as the longcase movement strapped to his chest. The head turned slowly with the movement of the pendulum. They sold for 31/4d. each. Many replicas have appeared on the market including one kit that was on the market to make your own 'clock pedlar' complete with aluminium body and epoxy cement. The latest versions seen needless to say had quartz movements!

Early original examples with movements by known Austrian makers can realise as much as £1,000 but those to which reference is made here would be £125+ depending on when they were reproduced.

Figure 274

This is an excellent example of the reproduction of Sir William Congreve's (1772-1828) Rolling Ball Clock, based on the John Wilding interpretation which was serialised in the Horological Journal, *1975 and later published in book form. Although best known for his achievements in the development of rockets for military use, Congreve was also interested in devising new principles for the measurement of time. He lists in* A Second Century of Inventions, *written about 1796, some five improvements to clocks — none of which were apertaining to this particular example. He presented in 1808, an example of his Rolling Ball Clock to the Prince of Wales and this is now on view at The Rotunda, Woolwich. Reproductions of this clock have been made in limited numbers during the last fifty years by such makers as Geoffrey Bell, E. Dent & Co., and Thwaites and Reed. They were never produced on a commercial scale during the lifetime of Sir William Congreve. Price dependent on quality of workmanship put into the finish.*

£800+

Courtesy of Kingston Antiques

275a

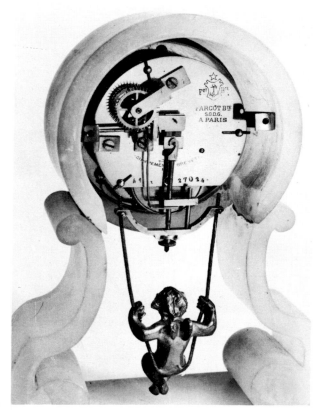

275b

Figures 275a and 275b

An interesting and attractive clock with the novel feature of a cherub seated on a swing acting as the pendulum, which swings from back to front instead of the more conventional side to side action. To enable this to occur a special form of anchor escapement, patented in 1862 by Farcot of Paris, is used (see Figure 275b). The trade mark is that of Farcot. This particular example is housed in a white alabaster case, with some ormolu trim and a gilt-metal cherub. The dial is attractively skeletonised. These clocks have great decorative appeal depending upon condition of case. (See also Colour Plate 32.)

£300+

Figure 276

The principle upon which this tortoise (or possibly more correctly turtle) clock works is attributed to Grollier de Sevriere of whom mention has been made in the text accompanying Figure 263. A simple timepiece is housed in the base, to which is fixed an arm carrying a magnet. As this revolves throughout the twelve hours it attracts with it, by means of the small piece of iron embedded in his stomach, the small turtle floating in the bowl of water above. The hours are marked on the chapter ring around the rim of the plate. This particular example has a handsome marble base or plinth, a silver dish and a horn tortoise. The price is dependent, to some extent, on date of manufacture, but far more emphasis is placed on the aesthetic appeal and value of the materials used.

£100 — £400

Figure 277

7½ins. high. This is an illustration of a modern reproduction of a rotary pendulum clock patented by John C. Briggs of Concord, New Hampshire, USA, in 1855 and 1856. The two patent numbers (13,451 and 15,356) are stamped on the top of the plates. Although the principle behind them was ingenious they were far from a commercial success. Several models were manufactured, the first of these being in 1858 and 1860 when Abel Chandler, a clockmaker and instrument maker, also of Concord, became interested in them. They were manufactured by another American Clock Factory (E.N. Welch) during the 1870s when catalogues offered them at $2.50 for models with brass finish and $3.50 for models with nickel finish. Further details appear in The American Horologist and Jeweler, *August, 1946, Vol. 13, No. 8, entitled 'The Brigg's Rotary' by J.E. Coleman.*

£90+ for reproduction

Figures 278a, 278b, 278c, 278d and 278e

14ins. high. The clock in this illustration is a perfect example of Victorian ingenuity and would have appeared to have a great many practical applications. It had, however, only a modest commercial appeal and was manufactured for about five or six years. The clock was patented by John Davidson in 1891. His invention was primarily the mechanism for the automatic memorandum device for use with any ordinary clock movement, so one finds examples in a variety of cases. This example is housed in a well-made oak case with brass finials and frets. The square dial is similar in concept to that of a longcase or bracket clock, being of brass with a matted centre, brass spandrels and silver chapter ring. It was intended that the revolving drum at the top should have forty eight slots and that an ivory tablet bearing the relevant message should be placed in the appropriate slot. At the appointed hour a bell would ring and the tablet drop into the small brass box seen below the dial. John Davidson foresaw his Memorandum Clock being a boon to busy financiers, or nurses in the sickroom, and five years later in 1896, he took out a second patent for 'Improved Automatic Memorandum Clock'. In the second version a bell rang and the tablet had to be removed from the slot in order to read the reminder. John Davidson had come from Scotland with his wife and children about 1895 but by 1900 the automatic Memorandum Clock Company was wound up and the family returned in 1900 to Wick, Caithness. Further information regarding this clock can be found in Antiquarian Horology, *June, 1975, in 'The Automatic Memorandum Clock' by R.K. Shenton.*

£275+

Figure 278b

View of the movement of the Memorandum clock. The extra wheel and rod connected to the revolving drum seen in Figure 278d is just discernible above the pendulum crutch. The battery to the electric bell seen in Figure 278c was housed in the lower section of the case.

Figure 278c

View of the electric bell in the base of the clock that is activated whenever the 'reminder' is due.

Figure 278d

View of the revolving drum showing hour and quarter markings and slots for placing reminder markers.

DIRECTIONS FOR MANAGEMENT OF THE
(PATENT)

Automatic Memorandum Clock.

To Unpack.—Lift off the drum and remove the paper underneath, then place back the drum. When the centre tube is put on the spindle, move round the drum until it drops into a fixed position.

Remove the paper from under the tray, and inside the clock. Wind weekly.

To use the Memorandum Apparatus.— Write nature of engagement in *pencil* on ivorine tablet, then place it in the slot representing the time you wish to be reminded of your engagement ; when the desired time arrives, the tablet will discharge into the tray in front of the clock, and the bell will ring until the tablet is removed from the tray.

All your engagements may be placed in the clock at once, or as each engagement is made.

To use as an Alarum.—Place a tablet in the slot representing the time you wish to be called in the morning.

To use in the Sick Room or Hospital, as a reminder of each period when medicine, &c., should be given. If the bell should be considered objectionable, place the small piece of wood (supplied with the clock), under the tray, when the tablet will discharge into the tray, without ringing the bell.

6937

Figure 278e

Instruction label pasted into the door of the Memorandum clock.

Figure 279

13½ ins high. Alternative style of the Keyless Clock illustrated in Figure 281. Winding in this instance is by means of the lever visible at the left hand side of the case. The dial is matt black with gold arabic numerals and white painted hands.

£120+

Figure 280

7ins. high. The concept of a 'flying pendulum' was not new, but the version shown in this illustration was invented by Alder Christian Clausen in 1883. The clock has been aptly referred to as 'the craziest clock in the world'. The small ball on the end of the thread wraps and unwraps itself first around one post and the the other — a procedure that either enthralls or infuriates the onlooker! Originally these clocks were manufactured for one year only (1884-5) and were produced by the New Haven Clock Company of New Haven, Connecticut, USA, although sold under the name of Jerome and Company also of Connecticut. The clock in the illustration is a replica of the clock patented in 1883 and made in the 1960s in West Germany for the Horolovar Clock Company of the USA. The case is of wood, with matt black finish and brass trims. The original had an oak case with brass trims and retailed in 1885 for $5.18. It would appear that there was also a similar version of these clocks manufactured in France. The cases of the French models were similar, but of ebonised wood, with the central pillar carrying an umbrella from which hung the thread and ball. A small cherub sat crossed legged under the umbrella.

£90+ for reproduction

281a   281b

Figure 281

9½ins. high. The principle of utilising the weight of the clock movement, as it slowly descends a notched rack, to drive the clock is not a new one. This type of clock was made as early as the seventeenth century. The clock in these illustrations, however, was manufactured during the 1920s and was first introduced at the Crystal Palace British Industries Fair in 1920. Patents had been taken out the previous year by Thomas Watson and Christopher Frederick Webb. It was manufactured and marketed under the name of the Watson Clock Company and then under the name of the Kee-Less Clock Company, both of Kentish Town, London. At this time the clock was described as "The Silent Kee-Less Clock. A revolution in Clock Construction. No Keys. No Springs. Moving Parts reduced to a Minimum giving Greater Reliability at Less Cost. Price 50/-". The example in this illustration is of brass, with a clear glass dial with white figures painted on the obverse. The front plate being painted matt black in order to make the figures and hands more readily visible. Later advertisements indicate that gilt figures, or plain or luminous dials, were also to be had. Other examples have been seen on marble or onyx bases, and one with the brass base embossed with a leaf pattern and a wreath of brass foliage surmounting the bezel. These clocks needed winding daily by pushing the drum containing the movement back to the top of the columns. Figures 279c and d provide detailed illustrations of the movement and case. (Horological Journal December, 1959 provides some repair details.)

In 1921 patents were taken out in America and some examples have been seen which were manufactured by the Ansonia Clock Company of New York (1879-1930).

£150+

281c

281d

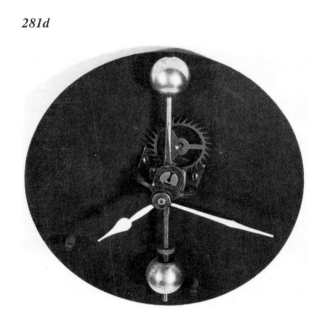

Figure 281e

Escapements of Keyless Clocks as shown in Abridgments of Specifications Class 139, Watches, Clocks and Other Timekeepers 1901-1930 for Patent No. 140,668 as taken out by T. Watson and C.F. Webb.

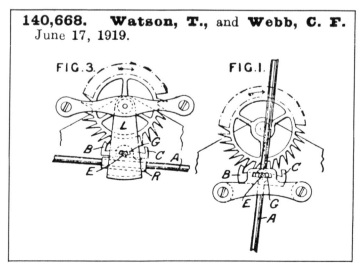

Figure 282

5ins. high. These clocks are generally referred to as 'Ticket' or 'Plato' clocks. The first name was intended to be a reference to the fact that, in place of the orthodox dial, small plates or tickets bore the figures for the hours and minutes, but the general public preferred another interpretation in that the cylindrical case was reminiscent of the lantern which Plato carried while looking for an honest man. The top ticket indicates the hours and the lower minutes. The patent for this clock was taken out in 1902 by Eugene Fitch of New York, USA. The American Electrical Novelty and Manufacturing Company, also of New York, sold some 40,000 of these clocks, although the horological trade refused to recognise them as timepieces and hence marketing and repairing difficulties arose. The advertisements of 1904 and 1905 refer to them as the " 'Every-Ready' Plato Clock, an ideal timeteller without hands or dial. The change of the figure every minute on the lower plate is very catchy and impressively suggestive of the flight of time. The plates can be had in any colour — white, red, blue or dark green. Price 6 dollars.''

A conventional pin pallet movement is in the base of the clock. In common with most of the novelty clocks these cannot be expected to keep atomic time and it is common practice for the tickets to fall in multiples instead of one at a time, thereby really making time fly! It usually takes but a minor adjustment to the small hand or pointer holding the bottom edge of the tickets to rectify this. There were four models, three on circular bases and the fourth on a rectangular base, see Figure 283. A later model was made by Junghans of Germany and the words 'Made in Germany' appear on the base. In the 1920s a further model was manufactured in France.

£150+ if good condition

"EVER-READY" NOVELTIES.

THE "CHRONOS" CLOCKS.

No. 300 —Nickel-plated or Gilt Case.
Height 4¾ins.
1-day Lever Time, **34/-**
Similar Clock with Alarm, **37/6**

No. 314.—Mahogany Case with Gilt
Columns.
Height 6½ins. Width 6½ins.
8-day Lever Time, **82/-**

No. 230.—Gilt Case, Height 5⅜ins.
8-day Lever Time, **62/-**
No. 302.—1-day Lever Time, Black
Oxidised Steel, **49/-**: Gilt Finish, **42/-**

Courtesy of F.W. Elliott Ltd.

Figure 283

This advertisement appeared in one of the trade catalogues of Grimshaw Baxter and J.J. Elliott Ltd. at some date after 1909.

The **CHRONOS CLOCK**
EVER READY PRODUCTION

300 OXYDISED SILVER
301 OXYDISED COPPER
302 POLISHED NICKEL PLATE
303 LACQUERED BRASS
304 MATT SILVER

NOW ONLY **30/-**

The Chronos Clock is a finely finished utility article which settles everybody's gift problem. It is interesting and unique, and the neat design and handsome finishes harmonise perfectly with modern decoration.

The Chronos is wound, adjusted, and regulated as an ordinary clock. The usual dial is displaced by ivorine leaves with clearly marked figures which are mechanically operated to record the hours and the minutes. The 30-hour movement will give accurate and reliable service.

Figure 284

1930 advertisement listing the case finishes available at this date. Here it is referred to as the Chronos Clock.

285a

Figures 285a and 285b

8ins. high. This is an extremely curious item and provides an excellent lesson on the importance of being able to document a clock in order to enhance its potential value. This particular clock is believed to have been manufactured in China in the early 1900s.

The veneered clock case is typical of that period. The main dial indicates the time, with the smaller apertures showing the sun or moon (whichever is appropriate) and the character indicative of the time according to the Chinese system. The old way had been to divide the day into twelve parts, each of which had their own Chinese character. The left hand panel indicates the day of the week, date and month.

The right hand panel carried a thermometer, while the bottom panel has a hygrometer and an open aperture through which the movement of the pendulum is visible. A rough translation of the instructions pasted on the door read as follows:

1. To start the clock — wind it by the key once a week.
2. If you need to adjust the time or the date, use the key and turn it until the correct reading appears.
3. In the short months (those with 30 days) turn the date to the 31st in the evening of the 30th day.
4. To adjust the month. Use the key and turn it until it shows the correct month.
5. To regulate the clock, just adjust the hand.
6. If the hygrometer is not accurate, use the lever at the back to adjust it. The clock should stand on a small metal stand with adjustable feet so that it can be moved in order to ensure a good circulation of air. In good weather the hygrometer should read between 70 to 50.

The most interesting horological feature of this clock is the unusual construction of the escapement which automatically adjusts to keep the clock in beat even if the case is tilted through ten to fifteen degrees either side of the vertical. Apart from this the movement is perfectly straight forward and purely functional.

£90+

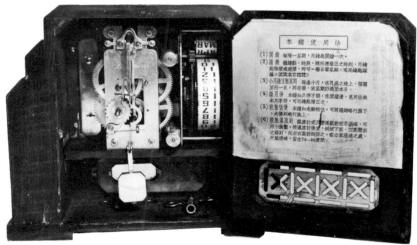

285b

Figure 286

These clocks are commonly referred to as Tape Clocks but contemporary (c.1930) editorial comment indicates that the maker's name was The Wywurrie. The overall quality of case is mediocre, being thin spun brass painted brown with silvered chapter band. The time is indicated by a small pointer fixed to base of case. The movement is a pin pallet. Made in USA appears on the bottom of the case together with the Patent No. D-95 184.

£60+ depending on state of case

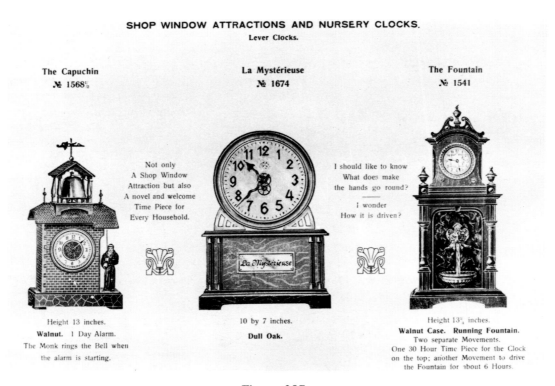

SHOP WINDOW ATTRACTIONS AND NURSERY CLOCKS.
Lever Clocks.

The Capuchin
№ 1568½

La Mystérieuse
№ 1674

The Fountain
№ 1541

Not only
A Shop Window
Attraction but also
A novel and welcome
Time Piece for
Every Household.

I should like to know
What does make
the hands go round?

I wonder
How it is driven?

La Mystérieuse

Height 13 inches.
Walnut. 1 Day Alarm.
The Monk rings the Bell when
the alarm is starting.

10 by 7 inches.
Dull Oak.

Height 13½ inches.
Walnut Case. Running Fountain.
Two separate Movements.
One 30 Hour Time Piece for the Clock
on the top; another Movement to drive
the Fountain for about 6 Hours.

Figure 287

Three novelty clocks manufactured by Hamburg American Clock Company in and around 1912. La Mystérieuse has no visible means of powering the motion work clearly seen in the centre of the dial but is actually utilising the same technique as that described for Figures 264 and 265. The Fountain has a standard 30 hour movement for the clock and another movement to drive the glass rod that simulates the fountain for 6 hours. The shorter period of duration for the second movement is quite a common feature of these clocks and appears not to be influenced by the overall quality of the clock itself.

£100+   **£200+**   **£150+**

Figure 288

10ins. high. A mantel clock in the form of a lighthouse with the light forming the balance with glass rods. The movement had a double wheel duplex escapement. When sold in 1982 it realised £320.

£750+

> *The variation between the price of these two examples must be attributable to the fact that one had two trains, was documentable and was of better workmanship and materials throughout.*

Courtesy of Sotheby's, Chester

Figure 289

20ins. high. A gilt brass lighthouse clock inscribed "Presented by the Commander, Officers and Crew of the RMS 'Adriatic' to Mr. T.H. Russell as a mark of their High Esteem and in Appreciation of his Uniform Courtesy and Kindness to All". Some collectors feel that inscriptions detract from the value of an item but if the people and circumstances of the presentation can be traced it can often greatly enhance the value. The two train French movement had an adapted pendulum attached to faceted glass rods and helical balance spring contained within the brass domed light, with striking of the half hours on a gong. The date given for this piece was 1885.

£1,500+

Courtesy of Phillips Son & Neale

Courtesy of Sotheby's

Figure 290

21ins. high. It is worth remembering that not only the specialist clock sales contain items of significance to the horologist. The clock shown in this illustration was sold in a sale of Fine Marine, Paintings, Drawings and Watercolours, Navigational Instruments etc. This brass lighthouse clock was embellished with applied lions' masks, and window frames and the figure of a sailor on the 'balcony' around the revolving glass drum painted with the numerals 1 to 12. The time was indicated by means of a brass pointer. The movement was English and the clock was thought to date from the mid-nineteenth century. Estimated at £200 to £300 it realised £187. This price was influenced by the fact that the movement was in need of restoration, and no doubt in perfect order and condition would have attracted a higher bid.

£500+

Figure 291

Early mantel clock with 8 day movement, silk suspension, and half hour strike on the bell. A separate movement activates the sailing boat on the top of the clock case. The nautical theme was a popular one and the gentle rolling movement of a ship on the high seas was readily simulated. A further example of a clockmaker's interpretation can be seen in Figure 445.

The overall quality of the mounts on the case, decorative bezel, etc., would make this an expensive example.

£500+

Courtesy of Auktionshaus Peter Ineichen

Figures 292a and 292b

An unusual timepiece with a separate barograph in a four glass case. There is a remarkable lack of maker's marks but this is obviously an interesting and desirable piece.

£400+

Figures 293a and 293b

Small floral clock c.1930. The pin pallet 30 hour movement is housed in the metal flower pot with the motion work behind the small dial that acts as the centre of the flower. A cable runs from the movement up through the stem of the plant to the motion work. Apart from noting that it is of German origin it has not been possible to ascertain the manufacturer. Other models have been seen but with other species of flower. The original flowers were of stiffened cloth and silk and it is as well to ascertain that any examples contemplated for purchase have not been refurbished with modern counterparts.

£60+ purely as novelty value

Figure 294

American synchronous electric twin faced bedside clock. The 1920s and '30s saw a marked increase in the popularity for twin beds hence a market for twin faced clocks! This example is in a black wood case, chrome bezel and white dial with viewing aperture to allow rotating disc to be seen to indicate whether the clock was working or not. The lack of sound would have been another asset when contemplating using these clocks in bedrooms.

£50+

Figure 295

An example of an 8 day Keyless Clock as first patented by A. Phinney in 1907 and manufactured by Phinney-Walker Co. The eight day movement is wound by rotating the bezel that is in the style of a ship's wheel. Hand set is by unscrewing the bezel and turning the hands by means of the knurled screw on the centre arbor.

This is a nice solidly made little clock with good thick brass case. Car clocks made on the same principle by this company have also been seen.

£65+

Jean Leon Reutter a French engineer had commenced working on his concepts for utilising the change in temperature and atmospheric pressure to wind a clock early in the 1920s but it had not been possible to design and manufacture the first model until 1926. Patents were taken out by him in this country (Patent No. 331,764) November 1928 and (Patent No. 356,216) in 1930.

The first models had a glass tube in the shape of a U which held mercury, a liquified gas and a saturated vapour. One arm of the tube was encased in a Dewar vacuum flask and hence temperature changes had no effect on the contents within that arm, only on the other uninsulated arm. The resulting difference of pressures between the two arms caused a displacement of the mercury which in turn caused a rocking movement of the tube. This movement was communicated to the clock movement by means of a ratchet wheel and thus the clock was wound. A very small variation in temperature was sufficient to wind the clock. These early models did not prove entirely successful commercially and modifications were made. Ethyl chloride was substituted as a reagent. This was contained in metallic bellows, the expansion and compression of which caused by the changes in temperature and pressure activated the rewinding of the movement. Further details of these clocks can be found in *Some Outstanding Clocks Over 700 Years* by Alan H. Lloyd and a two part article in the *Horological Journal* for May and June, 1934. The latter provides some practical technical data on the early models.

Figure 296

One of the first production Atmos clocks dating from 1929.

Figure 297

Page from Goldsmiths and Silversmiths Co. Ltd. illustrating three styles of Atmos clocks available. Many of the case styles remained constant hence it is dangerous to assume the type of movement by external appearances. It could be earlier than you think, or vice versa.

£900+ would not be unreasonable

Figure 298

Horological advertisement for May 1939. The Atmos clocks were manufactured by the Swiss firm of Jaeger le Coultre and marketed in this country from about 1934 by De Trevars Ltd. of Regent Street, London, W.1.

Figure 299

Diagram of Atmos movement taken from 1950 catalogue.

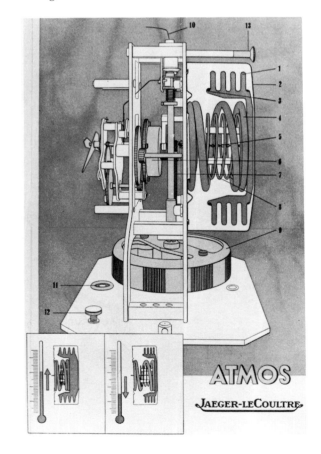

Figure 300

8ins. Calendar clock and money box invented and patented in 1929 by Hans J. Jepson and L.C. Nielson of Copenhagen. The steel case has a dark blue crackle finish, with brass bezel and silvered dial. The date can only be advanced after a one inch pin has been inserted into the slot in the back of the case. The lever operating the date can then be depressed. The clock has a simple pin pallet movement. Apparently some of the Savings Companies in Denmark introduced such money boxes with various devices to ensure that their members were constant in their contributions whether it be personally motivated or for a specific purpose such as an insurance premium. The value to the horologist is possibly lower than a money box collector would consider equitable.

£90+

Figures 301a and 301b

6¼ x 6 x 2½ins. Small portable Night Clock with white opaline dial, blue/gilt numerals and black spade hands. The dial is inscribed American Waltham Watch Company. The black leather carrying case is velvet lined. As shown in the illustration the folding bracket when in the open position supports the clock in an upright position and provides a small shallow cup to hold the night light in the correct position for the dial to be illuminated from behind. The movement is a standard Waltham movement Size 14, Mode 95 with 7 jewels. From the serial number on the movement it is possible to establish that this particular example came from the batch manufactured between 1895 and 1897.

£200+

Figure 302

Night clock and alarm in the style of an oil lamp with a frosted glass bowl within which there is provision for a night light to be housed and kept lit. The movement is that of the pin pallet variety. This design was patented by W. Soutter in 1886 (Patent No. 2248). The pointer is fixed with the globe being carried round with the clock casing. Originally designed to be wound by rotating the casing the final design manufactured had conventional key wind, alarm set, etc., which was accessible on the bottom of the movement. The quality and materials used of many of these Victorian and Edwardian novelty clocks are poor and the movements far from robust. No doubt this is one factor explaining why so few examples have survived.

Good quality example with VAP type movement £120+
Illustrated quality with original glass bowl £100+
Without original glass bowl £50+

Figures 303a, 303b and 303c

This small Night clock was manufactured by A. Shuker of Birmingham according to the design patented by W.H. Anderson in March, 1892. The clock movement is housed in the small mahogany case mounted behind/below the dial surround with the motion work behind the opalescent dial, (see Figure 303b). A bracket on the rear of the case allows the whole clock to be hung on a gas bracket in order for the gas burner to illuminate the dial. The 8 day movement has centre wind (see Figure 303c).

Although gas was available for domestic lighting in some areas as early as 1813 it was an expensive commodity and out of the reach of the general public. As technical skills and better understanding of the commodity progressed production was improved and the cost decreased dramatically so that more and more people were able to light their homes by gas. Naturally the manufacturers of gas brackets, mantles, etc. were quick to seize the new market and exploit it. One of the novelties introduced was a variety of timepieces with the dials illuminated by means of their being placed on a gas bracket. The example described and illustrated here is but one such clock. A further example can be seen in Figure 301.

£95+

Figures 304a and 304b

As batteries became more portable and manufacturers looked for methods by which to popularise them and to demonstrate their potentiality for a wide variety of uses. The Ever Ready Company was particularly adept at promoting their batteries by offering a wide range of novelties in which the use of a battery was necessary! These ranged from the large electric clock powered by small dry cell batteries to designs patented by Herbert Scott in 1902 (see Figures 403 and 404) to small portable timepieces and alarms illuminated by means of a small bulb and battery. The clock seen here is one such example. The movement is that of a small keywound 8 day watch with a pin pallet escapement. Adjustment depthing is provided by a slot in the top plate. This can be closed by a clamping screw in the thickness of the plate. Illumination is achieved by means of a bulb placed behind the transparent celluloid dial.

£75+

Figure 304b

View of the red leather carrying case fully opened. The battery is housed in the centre section. Note neat loop for retaining key.

Figures 305a and 305b

A further example of a Night clock in a blue leather case marketed by Messrs. Mojon, Manger & Co. — Watch Importers — in the early 1900s. The patents for this design were taken out in 1894 (Patent No. 10292). This model had the additional feature of an alarm with the Silent/Alarm operated by a lever at the 11 o'clock position on the bezel. The central area of the dial (a 1 inch plastic disc) is raised and slightly larger than the aperture trepanned in the main dial. A small tubular low voltage bulb behind the raised disc sheds light on to the remaining encircling dial.

Figure 305b

View of case opened. The battery is housed in the rear section of the case behind the velvet covered pad. The movement is Swiss made and carries the trade name Logic.

£60 but cases must be in good original condition

£85+

Figure 306

This Night Projection clock appeared in the 1913 Gamages Catalogue. It was manufactured to the design patented by A. Loebl (Patent No. 9462) in 1909. A 4½ volt battery was housed in the mahogany base upon which the clock movement pivots. The time is told either by means of the small dial as seen in the accompanying illustration or when the 'barrel' housing the movement is swung vertically so that the magnifying glass fixed into the other end of the barrel points towards the ceiling. By means of the pear switch a magnified illuminated image of the dial and hands is then projected on to the ceiling.

8 day movement with platform cylinder escapement £250+
30 hour movement £170+

Figure 307

30 hour pin pallet movement mounted on a wooden box housing the battery that powers the illumination for the dial. The dial is opaline glass with the bulb placed behind. This is magnified by a large bull's-eye lens in place of the conventional glass. Examples have been seen to this pattern with either green or red leather covering to the wooden box. Note the original wooden pear switch and flex.

£100+

Figure 308

The De Luxe model of this Ever Ready night projection clock had the wood base covered in "Green crushed morocco for 63/- or Violette de Parme crushed morocco 73/-". The projection apparatus is mounted on the top of the clock movement ready for use. The movement in these clocks is either a 30 hour cheap pin pallet movement or in a few instances a better quality 8 day movement. Either are quite adequate for the purpose and seem to have worn well over the years.

30 hour variety £130+
8 day variety £200+

CHAPTER XI

Clock Cases

Decorative Clocks

The majority of the clocks in the following illustrations are valued primarily for their decorative appeal and are of little interest to the horologist who refers to them as 'furnishings'. However, even a hardened clock collector may occasionally wish to purchase a piece for its decorative appeal and not solely for technical interest.

One of the factors influencing the price of this category of clock is the type of metal used for the cases or mounts.

a) Bronze — originally a mixture of copper and tin. Figures in bronze are much sought after by collectors today and it is an accepted fact that it is particularly suitable for producing sharp castings which can be finished later to a high degree of perfection by chasing, chiselling or engraving. Although by the Great Exhibition of 1851 opinion was deploring the quality of the mass produced figures, statues, mounts, etc., being made by the foundries using the sand casting methods, they are still highly desirable pieces.

b) Brass — originally a mixture of copper and zinc. Although this does not produce quite such a sharp casting it has most of the virtues of bronze and was used extensively for mounts. In fact the two medals have now become so adulterated and the proportions of their basic components so altered to facilitate casting that there is often little difference in their composition.

c) Spelter and other similar alloys — basically mixtures of tin and zinc. Although used extensively during the latter half of the nineteenth century for the sake of cheapness and the ease with which it was cast (its melting point is much lower than that of brass or bronze), it was too soft to be finished in any way. Therefore, the quality of the detail depends entirely upon the soundness of the mould. It also has the added disadvantage of being less durable than bronze and brass and cases or figures made in this material need to be closely examined. If too much zinc has been used the alloy will be soft and articles made from it will readily dent; whereas if an excess of tin has been used the alloy will be brittle and extremities may be broken or damaged. Repairs are difficult.

To distinguish between the first two metals mentioned and spelter is relatively simple as a piece made of spelter is inevitably lighter than a comparable example in brass or bronze. Also it should be possible to mark the surface lightly (microscopically on the base!) of spelter with a finger nail, while bronze and brass are too hard for this to make a mark. Each example should be judged on its merits; quality of finish and definition is often of first importance and not necessarily the choice of material. Just because it is bronze it does not have to be a superb piece any more than one of spelter has to be poor — a well executed spelter case or figure can be superior to a poor bronze example.

When used for figures, statues, etc., bronze is frequently left to acquire its own patina with age, but when used for cases or mounts it was normal to fire-gilt or mercurial gild. The latter was introduced about 1785 and produced a beautiful finish, but due to health hazards to workers coming in contact with the mercury vapours given off during the process it has been banned. 'Mercury Madness', otherwise known as 'Hatters' Shakes', was the inevitable death of those working under these conditions. Electro-gilding was introduced by Elkington in 1836, coming into more general use by 1840. 'Ormolu' is the term used to describe cases or mounts cast in brass or bronze and gilt. It is possible to have cases regilded, but care has to be taken at each stage to ensure that the interstices are well drained and that no residual fluid remains. It would be criminal to use a cheap gold lacquer on any of these cases.

Spelter pieces are also generally electro-gilded although some of the extremely cheap examples were lacquered to resemble bronze or gilt.

Other price-influencing factors are commented on in the copy accompanying the illustrations.

In most instances the movements in these cases are the round French movements as found in the black marble-cased clocks, exceptions being the inevitable American copies. Care must be exercised in purchasing a case with the intention of adding or replacing the movement with one from another clock. Although conceding that the majority of the French movements were of standard sizes, the diameters of the bezels and dials were not so constant and it is remarkable how elusive the exact substitute can be!

Many of the clocks were originally part of a garniture and protected from the dust and atmosphere by a glass shade. Although it is obviously desirable to have examples in their original state it would be foolish to refrain from making a purchase if these were missing. It is more important to ascertain that side ornaments being sold with the clock started life with it and have not been added since to enhance the price.

Figure 309

This large clock and matching stand was manufactured by Coalport about 1837 and sold by Sparks of Worcester, a notable retailer and decorator of porcelain. Although the face and sides both feature the celest ground with encrusted flowers and painted decoration in a small reserve, the reverse of the clock has been left white with a number of finely painted floral studies. To a collector of Coalport a highly desirable piece, but although attractive to a horologist the price would deter his awakened interest in fine porcelains.

£850+

Courtesy of Parsons, Welch & Cowell

Figure 310

Two good examples of French clock garnitures of gilt-metal and porcelain.

The example shown above is a nineteenth century French clock with an eight-day striking movement. That shown below has a turquoise-blue dial painted with trophies as are the other porcelain insets on the clock and candelabra. Striking is on a gong. It is always wise when a clock and side ornaments are of such complexity of design to be meticulous in checking for damage, as the overall effect is so overwhelming that important points can be missed and the wrong valuation placed on the item in question. Missing hands or, in this example, the linked chains would be of nuisance value but not irretrievable, whereas broken porcelain panels or decapitated figures would be a disaster. It fetched £200 in 1976 and would have reached a higher price if the porcelain panels had been the product of one of the important factories.

£400 — £600

Courtesy of Sotheby's Belgravia

Figure 311

An uncommon late nineteenth century French bronze, porcelain and champlevé enamel clock garniture, with a two train movement striking on a gong. The backplate is stamped with the maker's name 'G. Mignon' and the information that they had won a Médaille d'Argent at one of the Exhibitions. Its value would lie in the high quality of the work and decoration put into the case and side ornaments and it is not surprising that it fetched £2,800 in a London saleroom in 1976. These clocks can rocket in price on occasion. A similar clock on its own sold in early 1977 for £780.

£1,500+ is a more realistic price although these ornamental decorative pieces have kept their prices well and even maintained a steady increase.

Figure 312

It is not surprising that this extremely decorative cloisonné and onyx garniture sold early in 1976 for £820, although it is basically the same movement as that shown in Figure 335 in a more attractive case. This is a good example of how the case influences the price and not the merits of the movement. However, if the movement had been American the price would have been slightly lower, but nothing of any great significance.

£1,200+

Figure 313

16ins. high. A further example of an exceptionally high quality clock garniture. The clock is of gilt-metal and alabaster with matching candelabra. This particular example provides several excellent points to note, examine and assess before purchasing a similar clock. The two gilt columns are not upright — is it a simple matter of tightening and straightening the supporting rods that run through their centres or have these become rusted and difficult to restore? The metal swag on the left-hand alabaster column is missing, what can be done to replace it? It is often possible to find replacement finials, swags, feet, etc., of a conventional pattern in an architectural ironmongers — one well-known London example being J.D. Beardmore & Co. of Percy Street, London, W.1. It is not known, however, how standard a pattern this particular swag is. Obviously one could be specially cast but even if sufficiently fortunate to find someone prepared to do this it could prove expensive (£20 to £25) and it may be necessary to have all the parts regilded in order that the colour of the replacement and original pieces match. This particular problem has two satisfactory solutions. Either sacrifice both the swags from the candelabra and manufacture one for the clock (the resulting pin holes could be covered by small brass rosettes) or replace all four swags by a new set. As it is not an insurmountable problem it should not deter from any intentions to purchase but may influence the price one is willing to pay.

£400 — £600

Courtesy of Sotheby's Belgravia

Figure 314

13½ins. high. The case of this mantel clock is silvered and gilt-bronze with a brass dial inscribed with the retailer's name (Elkington & Co. Silversmiths, Liverpool). This is an exceptionally good quality case both in material (bronze) and in design and has the added advantage of having a movement by an interesting maker. C.A. Richard et Cie were founded in Paris in 1848 and opened a branch in London in 1857. They were makers of good quality clocks including carriage clocks for which they received an Honourable Mention in the Paris Exhibition of 1889. The movement of this example of their work is stamped with their trade mark — the letters R C with two snakes entwined round a staff between them.

£550+

Figure 315

16½ins. high. This example of a French clock is in an ebonised and red tortoiseshell 'case' of Egyptian style. Many clocks from the early part of the nineteenth century have designs influenced by the rise of interest in Egyptology following the Napoleonic campaigns in Egypt. This later specimen is from the end of the century. The pendulum is a mock compensated one. The white enamel dial is recessed and signed John Hall, Paris. This is not an indication that John Hall was a Parisian clockmaker but that he had the clock made in Paris. Striking is on a count wheel. The value of this clock would again lie in its decorative qualities and it was sold in 1977 for £240.

£750+

Courtesy of Sotheby's Belgravia

Courtesy of Sotheby's, Chester

Figure 316

19ins. high. A gilt bronze mantel clock with the case cast in the form of Rheims Cathedral with polychrome panels simulating stained glass. The movement with outside count wheel, silk suspension is signed Richmond A. Paris. A miniature (8ins.) cast iron model was offered with the clock. Possibly not to everyone's taste but there is no disputing the fine quality of the castings — obviously a factor that most influences the price of these clocks. When sold in 1982 it realised a higher price than its estimated £500—£700 (with buyer's premium £825). These cast cases were more popular on the Continent than in this country. The fact that it has a silk suspension would date it as unlikely to be later than the mid-nineteenth century. A similar clock was sold at Sotheby's, Belgravia in 1979. It realised a hammer price of £580 (plus buyer's premium of 10%).

£800+

Figure 317

14½ins. high. A further example of a metal cased mantel clock. This specimen has a brass case with two matching ornaments. The movement was stamped with the name of Japy Frères et Cie. The value of this garniture is similar to the clock shown in Figure 314. Whereas it could be said that a bronze case was superior to a brass one, this clock has the advantage of being part of a set. It fetched £260 in 1977.

£300 — £400

Courtesy of Sotheby's Belgravia

Figure 318

17½ins. high. Although a nineteenth century copy of a Louis XV case this is a handsome clock in a brass and brown tortoiseshell boulle case. Boulle is a type of inlay frequently found in French furniture as well as smaller items such as boxes, clock cases, etc. The process was first used by André Charles Boulle (1642-1732) and involves the gluing of a thin layer of tortoiseshell to one of brass, pewter or silver and then pasting a paper pattern for the marquetry over the top. This is then cut out. To avoid wastage of materials the layers of brass and tortoiseshell are then separated and two different effects can be achieved by using the brass on the tortoiseshell background and vice versa. The resulting veneer is then firmly glued to the carcase of the piece being decorated. Various hues can be given to the tortoiseshell by placing coloured foil beneath it. The mounts in this example are of gilt-metal. They are added both for decorative appeal and in order to protect the fragile edges and corners of the inlay. The repair of boulle work is not to be undertaken lightly and, bearing in mind that before any restoration can be undertaken time is spent in removing the mounts, it can be a costly repair to have carried out professionally. Although, theoretically, it can be refixed by the application of heat, glue and pressure, it is often found in practice that the shell has lost much of its natural oil and become extremely fragile. Missing mounts would have to be specially cast.

Price for one this size £750+
one 12ins. high about £400+

Colour Plate 29

Ship automata clock manufactured by Vitascope Industries on the Isle of Man (see Figure 445 for details of the mechanism).

Colour Plate 30

Small shelf clock with automata in so far as the wheel of the waterwheel revolves while the bell in the tower swings. This is achieved by an arm attached to the pallet arbor having a projecting pin that engages with the fork of a second arm that is pivoted immediately above the bell in the tower thus 'ringing' the bell. The wheel is rotated by means of the balance wheel arbor. This is a mass produced factory made pin pallet clock dating from between the two World Wars and obviously of no great monetary value but is typical of its period and rather amusing.

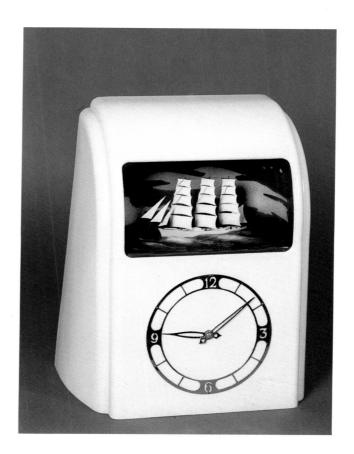

Colour Plate 31

A small four sided desk clock in burr walnut and ivory trim case that rotates to show timepiece, barometer, thermometer or calendar. A pleasing small clock with good quality workmanship throughout (Jaegar le Coultre movement).

£75+

Figure 319

The movement of this clock was examined by Lund & Blockley — a firm of whom little is recorded except that they carried on an extensive business in tower clocks, silverware and pocket watches in India and had possible connections with Lund of Barraud and Lund, London. They are listed as having traded in London between 1875 and 1881. As the movement has every appearance of being French they would have merely imported and checked it was in working order prior to casing. However, it is the case that is of great interest to anyone devoted to Victorian art pottery as it is a product from the Southall pottery of the Martin brothers. There is an excellent account of this pottery to be found in Victorian Art Pottery by E. Lloyd Thomas, but it is sufficient to say here that they manufactured jugs, vases, etc., between 1873 and 1930, although it is the work of Wallace Martin (1843-1923) with its grotesque birds and beasts that is particularly sought after. This example was sold 1973/4 for £280.

£600+

Courtesy of Riddett & Adams Smith

Figure 320

28ins. high. There is no doubt that the value of this piece lies purely in its excellent quality bronze figures and general aesthetic appeal. It could even be said that the clock dial is superfluous and added nothing to the composition of the group. It most certainly has an unusual pair of hands — the head of the snake pointing to the hours, while its tail indicates the minutes. One small casting has been lost from the right hand foot, but that could be recast. It sold for £840 in 1976.

£2,500+

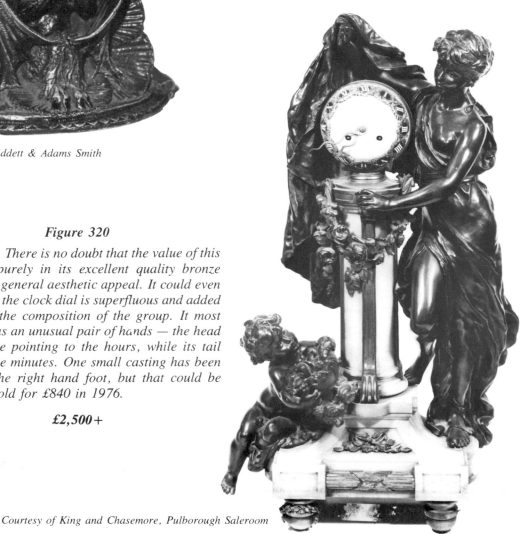

Courtesy of King and Chasemore, Pulborough Saleroom

Figure 321

This elegant timepiece has a white alabaster case, with gilt-bronze figures, feet and mouldings; however, the quality of the castings is poor and they lack definition as can be noted particularly in the hands and faces of the cherubs. It fetched £200 in 1977. It is interesting to note that as the dial is somewhat smaller than the total diameter of the movement, the arbor to the Brocot suspension has been placed outside and above the dial with a decorative escutcheon.

£500+

Courtesy of Sotheby's Belgravia

Figure 322

13ins. high. In view of the decorative dial and other porcelain insets, together with the high quality of the casting, this would be a desirable piece. Many of these clocks stand on decorative bases and are covered with glass shades; these have frequently become lost or damaged, which is unfortunate as they do protect the gilding from dust and atmospheric impurities. It fetched £400 in 1977.

£750+

Courtesy of Sotheby's Belgravia

Figure 323

14ins. high. An example of a pleasant compact French clock with porcelain dial and gilt-metal case. Note that the casting is not so sharp as that shown in Figure 322. It was possible to date the clock as c.1850 by identifying the maker of the movement (Popon à Paris).

£350+

Figure 324

1ft. 4ins. high. A nineteenth century French bronze ormulu mantel clock with the case in the form of an elephant supporting a howdah on its back surmounted by a cherub holding a bow. The two train movement is inscribed Gavelle, Paris. When sold in 1982 it realised £620. This price would have been influenced by the material and quality of the figures and not the movement. There are several variations on this design including copies of the example with case designed by Jacques Caffieri (1678-1755) the noted French bronzist. An example of his work can be seen in the Jones Collection in the Victoria and Albert Museum, South Kensington, London. In this instance the elephant is again surmounted by a howdah but with the figure of a Chinaman with umbrella.

£1,250+

Courtesy of Phillips Son & Neale

Figure 325

1ft. 8ins. The case of this clock is of ormulu and in the form of a prison entrance with a figure of a soldier pleading with a distraught young woman in the foreground. Further decoration takes the images of amorini, urns, flags and other military trophies. No doubt the theme of the French Revolution influenced the design especially in view of the dates 1755 and 1793 and initials AM surmounted by a crown engraved on the marble base are those of Marie Antoinette. The two train movement with countwheel strike is signed on the enamel dial Lepine A Paris. When sold in 1982 it realised £460. Not a high price in view of the quality and complexity of the case but then not everyone purchases a clock depicting such a sombre scene for decoration.

£750+

Figure 326

Not surprisingly this particularly high quality French ormulu mantel clock realised £2,250 when sold in 1982. The porcelain dial and plaque were colourful and well executed while the pierced base was decoratively modelled with amorini, a mask and scrolled swags and flowers. A seated figure of Minerva surmounted the case. The two train eight day movement with anchor escapement was unsigned.

£2,000+

Four-Glass Case Clocks

Although four-glass clocks are sometimes referred to as 'regulators' or 'library carriage clocks' neither term is correct. They are not to be confused with the large French mantel regulators of the early nineteenth century and their only resemblance to a carriage clock lies in the similarity of case style. There can be some superb examples with calendar dials, phases of the moon, etc., but those more commonly found have the French drum movements that were produced in such abundance from the mid-1800s and have been described in the chapter dealing with Marble-Cased Clocks. Naturally there are exceptions and these include copies manufactured in America by the Ansonia Clock Company of

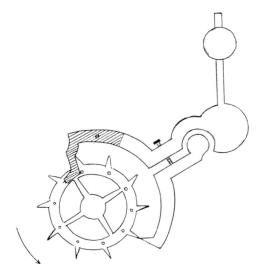

Figure 327b

Line drawing of the visible coup de perdu escapement attributed to Desfontaines in 1853 seen on the clock illustrated in Figure 327a.

Courtesy of Christie, Manson & Woods   **Figure 327a**

22ins. high. This is a fine French mantel clock — frequently referred to as a 'Mantel Regulator' which although a not strictly accurate term does, in this instance, reflect the quality and indicate correctly the accuracy of their timekeeping. This particular example has, apart from the perpetual calendar and barometer, a rather unusual and obscure visible escapement which could at first glance be taken to be that of Achille Brocot. It is, however, attributed to Desfontaines about 1853. Comparison of Figures 3a, 3b and 327b will demonstrate the difference. The movement had a rise-and-fall regulation, a gridiron pendulum and struck on a bell. The inscription on the dial reads 'Le Roy et Fils Palais Royal Galies Montpensier, 13 and 15 Paris 211 Regent Street, London'. The connection between Desfontaines and Le Roy is interesting. According to Charles Allix in his book Carriage Clocks, *on the 30th June, 1845, Charles-Louis Le Roy sold his business to an employee Casimir Halley Desfontaines on the condition that he continued to trade under the name 'Le Roy & Fils'. This request was complied with until 1889. 211 Regent Street was their London address between 1866 and 1875, thereby limiting the date of this particular clock to these years. Perhaps not a vital piece of information, but nevertheless an exercise in adding a little colour to the provenance of an excellent clock. It fetched £1,600 in 1977. A not unexpected price in view of the interesting escapement and perpetual calendar work.*

£2,000+

New York (their catalogue for 1914 has some eighteen pages of 'crystal regulators'). The examples in Figures 328 and 332-5 are, however, more typical and vary only in shape and decoration.

The enamel dials can be plain or decorated with swags of flowers, with others having a recessed centre and visible Brocot escapement (see page 20 for further details of this). The majority of the pendulums are mercury compensated although some examples appear with an Ellicott pendulum. The former was invented by George Graham (1673-1751) in 1721 and works on the principle that the mercury in the jar (or jars) expands upwards to compensate for any increase in the length of the pendulum rod caused by a rise in the temperature. The mercurial pendulum used in these clocks usually has two glass jars each containing an equal amount of mercury. This is for aesthetic reasons as well as ensuring a quicker response to any temperature change. The clocks illustrated in Figures 328, 332, 334 and 335 have mercurial pendulums. Difficulty can be experienced in obtaining replacement jars and it must be remembered that mercury is a dangerous and expensive commodity. It should be possible to obtain advice if not assistance from any of the restorers who specialise in barometers.

The Ellicott pendulum is an interesting mechanism. Upon any rise in ambient temperature the two outer brass rods expand and press downwards on two pivoted steel levers. This causes the other end of the levers to push upwards thus raising the bob. The pendulum was devised by John Ellicott (1706-72) and according to the records of the Royal Society, of which he was a Fellow, was shown to them in 1738. Figure 333 illustrates a clock based on the principle established by Ellicott.

Courtesy of Sotheby's, Pulborough, West Sussex

Figures 328a, 328b and 328c

14ins. high. A most unusual four glass chiming mantel clock, the circular white enamel dial having a subsidiary Silence/Westminster/Cambridge dial. The movement with lever escapement, going barrel for the time side and a fusee for strike bore no identifying marks.

When sold in 1983 it realised £1,100

£1,500+

328b

328c

Courtesy Sotheby Parke Bernet & Co.

Figure 329

17ins. high. A French nineteenth century perpetual calendar mantel clock in a glazed gilt metal and white marble case with enamel dials within a matt gilt surround engraved with acanthus scrolls. The dial is signed Chas Frodsham — Frodsham would have been as Clock Maker to The Queen, the retailer in this instance. The movement has spring barrels, bell strike, bimetallic compensated pendulum with a Brocot visible escapement. The matching calendar dial with year calendar ring marked for the equation of time has the centre with a moon aperture painted with stars and clouds and a separate day of the week and day of the month dial. The calendar work is geared to the striking barrel and mounted behind the single circular plate.

This is a high grade clock and in view of the fact that it was a presentation piece within the Royal Family — the inscription on the small brass plate on the base of the case reads "To Our Dear Alfred on the Day of His Confirmation from His Affectionate Parents Albert and Victoria R April 5 1860" — it was not surprising to see that the estimated price was £3,000 — £4,000 when it appeared on the market in 1982. The realised price including buyer's premium was £3,080.

£3,500+

Courtesy of Christie, Manson & Woods

Figure 330

14½ins. high. This example has been included in order to compare its evaluation with the preceding item. Externally the surround is not engraved and the case is entirely of brass and a slight inconsistency in the paintings of the clouds in the centre of the calendar dial! Movement stamped C.R. When sold in 1980 the hammer price was £1,500 (plus 10% buyer's premium).

£2,000+

Figure 331

17½ins. high. Again the comparison with previous clocks illustrated is not without interest. Basically this has the additional features of a centre seconds plus two thermometers — one Fahrenheit and the other Reamur — flanking the dial.

The small plaque in the centre of the surround is inscribed Wm Boore, 54 Strand, London. It has not been possible to trace this maker/retailer. When sold in 1978 this clock realised £1,700.

£2,000+

Courtesy of Phillips Son & Neale

Figures 332a, 332b and 332c

14ins. high. This is a particularly good specimen of a French four-glass mantel clock, with mercurial pendulum and was manufactured in Paris for Maple & Co. of London. The movement is stamped with the name of 'S. Marti et Cie'. The decorative bevelled glass side panels are etched with swags and geometrical borders, while the case is solid brass with attractive brass castings. The pied de biche *feet are especially pleasing (see Figure 332b). The white enamel dial is recessed with a visible Brocot escapement (see Figure 332c). The pallet stones appear to be agate. Hourly and half-hourly striking is on a bell although examples are common with striking on a gong. Generally gong striking is slightly less popular than bell striking. It is interesting to compare this with the example in Figure 335.*

£750+

Figure 333

9½ins. high. A further example of a French four-glass clock with a plain enamel dial, striking on a bell at the hour and half-hour, with the name 'Japy Frères et Cie' stamped on the backplate. Strictly speaking the pendulum is not an Ellicott compensated pendulum, although loosely based on his method of using the different coefficients of expansion of two metal rods within the pendulum bob. The greater expansion/contraction of one rod against the other operates a spring which raises/lowers the bob.

£200 — £300

Figure 334

15½ins. high. This example has the disadvantage of having a case of gilt-metal, not brass, otherwise it is an unusual shape for a four-glass case which is interesting. This also has a 'plain' visible Brocot escapement and would have been made in the late nineteenth century.

£250+

Courtesy of Sotheby's Belgravia

Courtesy of Sotheby's Belgravia

Figure 335

17½ins. high. Upon comparing this example with that shown in Figure 332 it is quickly noted that, apart from the lack of decoration to the case, this clock is less desirable for several other reasons. It is a larger clock and, therefore, not quite so versatile, the visible Brocot escapement is more functional than decorative, neither does there appear to be any maker's name on the movement. This would be a later specimen manufactured at the end of the nineteenth century. It was sold in 1977 for £210.

£200+

Figure 336

1ft. 9½ins. French four glass mantel clock surmounted by a rotating globe by Newton dated 1867 with horizontal chapter ring with red and white enamel numeral reserves. The striking movement had a visible Brocot escapement and mercury pendulum. The calendar dial showed day, date, month and phases of the moon section. The trade mark on the obverse of the calendar dial was a circle within which the letters E and S appeared either side of a bell. The letters Q and A appeared below in a lozenge shaped stamp. Tardy records Edouard Serin, Paris as having registered the bell trademark in 1876.

Although an additional feature the globe does not enhance the aesthetic appeal of this clock. The globe was original and not added at a later date. When examined, the connecting rods to the globe had been disengaged and some minimal restoration was required. When sold in 1983 this clock realised £1,550, slightly more than the estimated value.

£1,750+

Courtesy of Bonhams

Art Deco and Art Nouveau Cases

This is a further group of clocks that, although of no great interest to the horologist, would be of considerable importance to the collector of art deco or art nouveau pieces. Unless, by some remarkable coincidence, an interest in clocks was accompanied by some knowledge of the artists and their work, it would be advisable to seek a specialist dealer and rely upon his assistance. As can be seen by the example in Figure 337, prices can be extremely high. However, the inference should not be taken that all clocks in art deco or art nouveau cases are worth a considerable amount of money — this is only true of the pieces by known and important artists. Naturally their designs influenced the manufacturers of clock cases and this can be seen in contemporary examples. A movement housed in a glass case signed by Réné Lalique would be worth several hundreds of pounds, but one in a similar case by an unknown maker or designer would be valued accordingly with the quality of the movement coming to the fore. The clock in Figure 413, although it has an art deco-styled case, is highly valued because it has an extremely rare and virtually unique electric movement.

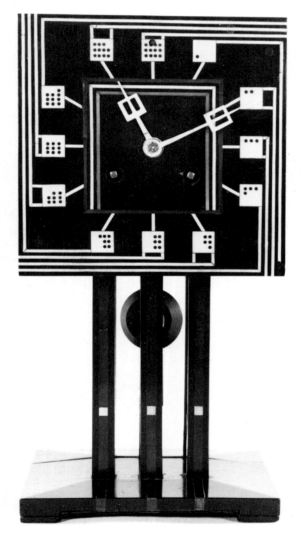

Figure 337

10ins. high. The sole reason for the high price of this clock (it cost over £7,000 at auction in 1976) is that is was designed for 'Derngate', Northampton, about 1917 by Charles Rennie Mackintosh (1868-1928) the leader of the Glasgow School of art nouveau artists. The case veneered with ebony and inlaid was made in the Isle of Man by German prisoners of war. It is aptly named 'The Domino Clock'.

£7,000 — £10,000

Courtesy of Sotheby's Belgravia

Figure 338

17ins. high. The information offered in a catalogue entry to the effect that the design on the case of this clock appears strongly influenced by Sir Edward Burne-Jones (1833-98), the painter and friend of William Morris, would greatly enhance the price of this item. The gilded and ebonised case is painted with scrolled leafwork and Gothic lettering, and is surmounted by stained ivory tiles finished with gilt metal minarets. The enamelled dial is signed 'W.A. Perry & Co. Birmingham' and 'Made in France'. The latter is possibly referring to the French movement. There is nothing of horological significance to affect the price of the clock. The auction price in 1977 was £320.

£500+

Figure 339

18ins. The design of the case of the clock shown here is typical of the flowing lines of the art nouveau style at the turn of the century. The figure is of a relatively high quality casting in gilding metal that has been electro-gilded. The unglazed dial is silvered with raised stylised numerals, and ornate hands. The movement is French with the maker's name 'Marti et Cie' stamped on the backplate together with the information that this maker received a Médaille d'Argent at the 1889 Exhibition, thereby indicating a date of manufacture after this date. Striking is on a gong. As has been stated previously, the American clock manufacturers were swift to copy designs from other countries especially France, so it is particularly interesting to note that in the 1906/1907 catalogue of the New Haven Clock Company of New Haven, Connecticut, a similar clock appears as part of a garniture. The figure is identical, but the dial of the clock is porcelain, with bevelled glass within a decorative bezel. The style of the hands is fleur de lys. *The caption to the American set reads:*

'LIBITINA SET
Clock, Height 17¼ inches; Width 9¼ inches. Vases, Height 14¾ inches. Four-inch Porcelain Dial, with Bevelled Glass. Ormolu Gold Plated Case. Eight-day, Half-hour Strike, Cathedral Gong Clock, List Price, $26.50. Vases, per Pair List Price $29.25. Set Complete, List Price $55.75'

If spelter, £100 — £200; if bronze £250 — £400

Courtesy of Sotheby's Belgravia

307

Courtesy of Sotheby's Belgravia

Figure 340

This clock was part of a garniture — the side ornaments being a pair of vases. The base of the clock is black marble with white inset pieces, while the figure is of spelter with ivorine hands and face. The resulting effect was rather attractive. The value of this clock would depend entirely upon its value as a piece of art deco, with the fact that it is a timepiece being of secondary importance. The movement was made in France (it had words to the effect of being Paris finished stamped on the backplate), but has a count wheel for the strike. Even for a French clock this is surprising on an example of this late date, and speculation is roused as to whether an old movement was utilised by the manufacturer. This would endorse the emphasis upon this being sold as a work of art rather than a clock.

£250+

Figure 341

19½ins. high. The case of this clock is of ebonised wood with china panels. The style is similar to the clock designed by Henry and Lewis F. Day (illustrated in The Aesthetic Movement *by E. Aslin). Other examples have been seen with panels reminiscent of blue and white delft tiles.*

£200+ the more porcelain the better

Courtesy of Sotheby's Belgravia

Courtesy of Phillips

Figure 342

There is, to the horologist, only one fascinating feature to this particular clock and this is the fact that it realised in a 1977 sale the astounding figure of £3,600! The movement has no particular merit, the value being entirely dependent upon the art nouveau figure gracing the case, which is by F. Preiss, whose work is extremely sought after.

Figure 343

The clock here and those on the following pages are a selection of the type of art nouveau clocks featured in the highly influential magazine, The Studio, *between 1893 and 1910. Many of these models were 'one-offs', and therefore the collector is unlikely to see the actual examples. Their importance, however, lies in the enormous influence such pieces had on contemporary designs.*

The prices of these clocks will depend largely on the materials used, but unless the case is by a well-known art nouveau name, the value is lower.

These pieces rarely have any feature of horological significance and are simply cases designed to fit into the appropriate setting.

The clock on the right was designed by Otto Prutscher, c.1901.

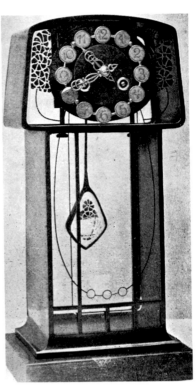

Two more clocks designed by Otto Prutscher of Vienna about 1905. He was then about twenty six, having won a Rothschild travelling scholarship. He studied in Paris and London and won a silver medal at Turin. Like most followers of the artistic movement he rebelled against the poorly made mass produced sameness of Victorian design and concentrated on making something different, even if it was more expensive.

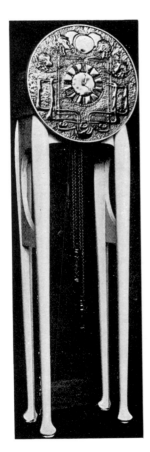
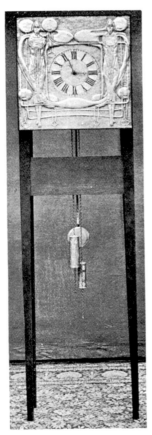

Two clocks, designed about 1897 and made by Margaret MacDonald (1865-1953) and Frances MacDonald (1874-1921), who were respectively the wives of Charles Rennie MacKintosh and Herbert MacNair.

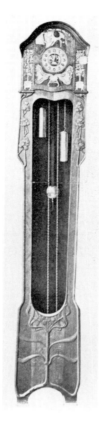

Left: A shelf clock made by the famous C.R. Ashbee and exhibited by the firm of Wylie & Lochead in the Glasgow Exhibition, 1901. Like so much of the art nouveau furniture the woodwork is solid oak and very simple, in contrast to the highly ornate carving of the standard Victorian mahogany furniture.

Right: A longcase clock exhibited in Glasgow by La Maison Moderne, but by contrast to Ashbee's clock, highly decorative. It has the typical curved lines in both the outline of the case and the decoration on the base. It was designed by F. Ringer.

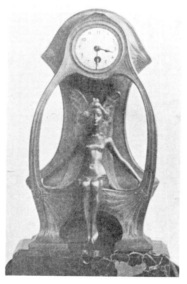

Two clocks by Albert Reimann, a young sculptor from Germany, and made about 1900. Exhibited at the Berlin Kunst Ausstellung, they were made in bronze, silver and majolica, and one can see the sculptor's approach in the designs which were then considered fairly advanced.

Clock designed by Otto Prutscher, c.1905.

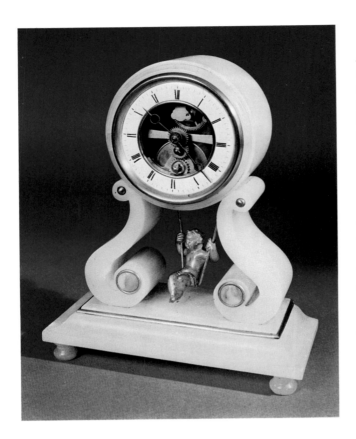

Colour Plate 32

An interesting and attractive clock with a cherub seated on a swing acting as the pendulum. A special form of anchor escapement, patented in 1862 by Farcot of Paris, is used (see Figure 275b). This particular example is housed in a white alabaster case, with some ormolu trim and a gilt-metal cherub.

£300+

Colour Plate 33

10ins. high. Although a relatively recent date (1955) this model known as 'The Kaiser' is of considerable interest. Made by J. Kaiser GMBH, Villinger, Schwartzwald, Germany the annular chapter ring has arabic numerals divided by the signs of the Zodiac. The painted sky and planets can be seen through the centre of this with a revolving moon sphere in the dial arch; another sphere representing the world rotates with the turn of the pendulum. According to Charles Terwilliger in the 9th edition of the 400 day Clock Repair Guide *The Kaiser Universe Clock as it was first named 'may go down in history as the best 400 day clock ever made'. There were in all probabilty only a few thousand manufactured.*

Complete with its dome £120 — £180

344a

344b

344c

Figures 344a, 344b and 344c

This clock cost £2. 2s. 0d. when sold by Liberty & Co. Ltd. in 1932 as part of their range of English Pewter Clocks. This large London store has always been associated with a distinctive style of goods whether they be their imported silks, textiles with the characteristic Liberty print, pottery and so on. Towards the end of the last century they were a major force in the developing interest in art nouveau with their merchandise much sought after here and abroad for the arts and crafts interiors that were so popular at that time. Noted designers of the period worked for them. The designs of Archibald Knox and Rex Silver were used in both their "Cymric" silver and "Tudric" pewter ware. It can hardly be claimed however that this particular clock case style is typical of art nouveau. It is a much later piece (art deco) but nevertheless still part of the Tudric range and as such commands a price far higher than it horologically merits. The name Tudric English Pewter Liberty's & Co. London together with the serial number are stamped on the base of the case.

The eight day movement is conventional but the winding and hand set arrangement is interesting. (see Figure 344b). The winding button is spring loaded and concentric with the arbor of the barrel wind.

The value would be £100 to a horologist but far more to a collector of Tudric ware!

Figure 345

8ins. Tudric pewter clock with copper bezel, roman numerals, and enamelled dial by Liberty & Co. Movement is typical round French movement with cylinder platform escapement. When sold in early 1983 the estimated price was £150 — £300 with it actually realising £180.

**To the horologist one price, to the collector another!
Arts and Craft movement**

Courtesy of Reeds Rains, Sale, Cheshire

Courtesy of Bonhams

Figure 346

Not surprisingly this clock appeared in a sale of Selected Silver and not a horological auction. The piece was made by Elkington's who undertook the majority of such commissions at the date of this item (1906). The train was of hallmarked silver and was modelled in great detail, the smaller train seen lower down on the plinth was called "La Portena". The figures represent Speed and Industry.

The plaques record in Spanish and English that the piece was given to David Simson by the members of the Buenos Aires Western Railway who had the pleasure of serving under him and that he was Chief Engineer between 1896 and 1906. The estimated value in 1981 was £4,000 — £7,000 and was obviously heavily influenced by its importance as a piece of typical Art Nouveau, the eminence of Elkington's, and its railroad associations. The movement was made by Swinden of Birmingham. The clock was subsequently purchased by the Birmingham City Museum and Art Gallery. (See 'A Horological Monument to Art Nouveau' by R. Good, Clocks, July 1985.)

CHAPTER XII

Alarm Clocks

It is recorded that it was an American, one Levi Hutchins of Concord, New Hampshire, who, needing assistance in awakening early each morning, devised in 1787 the alarm clock. Interesting though this version is, it is more accurate to state that timepieces with alarm devices are one of the oldest forms of clocks and date from the fourteen century. Man has always required more than his own biological clock and none more than the medieval monks with their need to have regular hours for prayer both day and night. Initially relying upon the manual ringing of bells, they eventually devised a simple clockwork mechanism wihich gradually over the centuries came to incorporate the telling of the hours and minutes as well as acting as a reminder of a pre-determined time. Inevitably their use spread outside the monastery walls and they became the 'toy' of the rich nobleman. It is rather amusing to consider how, during the centuries when the accuracy of clocks and watches was dubious, man preferred to depend upon these mechanical devices rather than his own acumen. Apparently gullibility for a novelty is not a recent human trait. Neither is man's ingenuity ever at a loss. The most bizarre solutions to waking a sleeping person were invented and often reached the stage of being patented; one of these being the brainchild of a German who evolved an ingenious device attached to a clock which at the appointed hour fired a cartridge which gave off a pungent gas causing the sleeper to sneeze until he awoke.

Centuries passed and what had been an amusing possession became a necessity. The industrial revolution of the 1800s, and later the advent of the railroads, with the need for railway time, made punctuality vital to the working man. The old night-watchman and his cry needed to be replaced. This urgent demand had to be satisfied by clocks that the workers could afford. Fortunately the technological advances that were forcing this social change were also capable of producing such a clock. The English clockmakers were extremely adverse to any of the mass-production methods introduced by their American and German competitors and so fell behind in the race for the home market. One report gives the figures for 1928 as:

75% imports from Germany
10% from France
4½% from Switzerland
4½% from United States of America
4½% from Italy

There had been a surprising post war development in the horological industry in Italy and by 1926 we were importing twice as many clocks from Italy as we were from the United States and at less than half the price asked by the latter manufacturers. Both Italy and France made startling inroads into the market for alarm and fancy metal timepieces. The English manufacturers chose to specialise in 8 day mantel and bracket clocks and the longcase. This stubbornness to accept new ideas struck the death knell for the English trade. Obviously not even the interruption of imports during the First World War nor the quick ominous massive importing programme reinstated within ten years of the end of that war made any impact on their thoughts and approach to the problem. It is astounding that a country that had led the world horologically could voice the following comments in the columns of the leading trade journal of the day — the *Horological Journal* for November, 1918:

> "*Dearth of Alarm Clocks* Is an alarm clock a necessity or is it a luxury? Whatever may be the opinion of people who rely on this little piece of mechanism to recall them to another day's doings those in authority over us have decreed that the alarm clock is something we do without. Only a few are coming in, from Japan. The result is that the alarm which before the war could be bought for 3s. 6d. or 4s. 6d. now sells at £1 or 25s. In the Woolwich area munition workers are unable to procure an alarm clock at any price and the jewellers are urging those who apply to make representation to their employers with the object of securing the importation of American clocks. If the clocks are not forthcoming from some quarter it is feared that there will be a good deal of time lost during the coming dark mornings. In some districts a 'caller-up' is employed — a system which works well where numerous workers living in a tenement require calling up at the same hour."

The English manufacturers did not learn from the experience and during the Second World War it became necessary to control the situation caused by lack of alarm clocks by fixing the price and making them only obtainable against a Buying Permit which was available in the first instance only to transport workers but by January 1944 to those whose work necessitated their rising between midnight and five in the morning. The problem regarding lack of timing devices for munitions was far more serious and the lesson was eventually learnt. In 1945 under the guidance of Sir Stafford Cripps (the President of the Board of Trade) successful efforts were made to resuscitate the horological industry of this country.

The American manufacturers, with their well established production methods using machinery and standardisation of parts, naturally included clocks with alarms in their range. The reprints of many of the firms' catalogues that have been published in recent years provide invaluable sources of information. Some of the American companies are listed here together with their dates and, in some instances, trade marks.

Ansonia

New Haven
USA

Seth Thomas

Ansonia Clock Company (1850-1929)
New Haven Clock Company (1853-1959)
Seth Thomas Clock Company (1853 to the present)
Waterbury Clock Company (1857-1944)
E. Ingraham Company (1857-1967)
William L. Gilbert Clock Company (1886-1964)
E.N. Welch Manufacturing Company (1864-1903)

Many of the wooden-cased shelf clocks made by these firms could, for extra cost, have an alarm included. Examples are shown in Figures 187 and 188.

The following illustrations give some idea of the diversity of designs even in the small metal-cased

316

drum alarm clock. The cases were nickel-finished, nickel-plated, oxidised copper with plain or hammered finish, brightly enamelled, plush (velour fabric finish), enamel and nickel with mother-of-pearl inlay, etc. Dials were either painted, white card or paper with inset dials for the seconds or alarm. The luminous dial appeared in America in the early 1880s. The Terry Clock Company of Pittsfield, Mass., claimed in their 1885 catalogue to be the exclusive manufacturers of 'Luminous Clocks' at that date (see Figure 360); patents for this process had been taken out in 1882 and 1883 respectively. The effect was achieved by mixing phosphorous with a small amount of radium. By the beginning of the 1900s several other manufacturers were advertising clocks with black card dials and luminous hands and numerals. The Ingersoll Company belonged to this latter group. This Company was founded in America in 1881 and in 1892 "startled the World with the 5s. pocket watch". In 1905 they opened a London office which eventually in 1930 became a separate entity under the name of Ingersoll Limited. The Second World War brought many problems and they were one of the firms taking advantage of the Government's post war offers of financial assistance to the horological industry. Mr. E.S. Daniells, the Chairman of the London office, originated the luminous watch dial immediately prior to the First World War, and it was but a logical step to use the process on the dials of alarm clocks when these were introduced to their range of products in 1918. The process (a mixture of radium and bromide) was given the name of 'Radiolite' and its use was continued for many years. An example of an Ingersoll alarm clock is shown in Figure 348.

The majority of the movements used would have been thirty-hour, with wire or lantern pinions. The exceptions tended to be movements housed in larger wooden cases that were intended as dual purpose clocks: decorative by day and utilitarian by night. These usually had an eight-day movement with the alarm train needing to be wound each night. Expressive and amusing names were used to indicate the character of the ring of the alarm — 'Bugaboo', 'Rattler', 'Drone', and 'Wasp' all appeared in the Waterbury Clock Company catalogue for 1908. Most firms offered three alternatives — an intermittent ring (the New Haven Clock Company in their catalogue of 1906 claims to have marketed the 'Tattoo Alarm' the 'Original Intermittent Alarm Clock'), a long continuous ring or a shorter standard ring. The bell or bells were fitted externally, and some manufacturers used the casing as the resonator.

Most manufacturers of early twentieth century alarm clocks appear to have been preoccupied with producing a sustained and extremely loud method of awakening their customers. The IWAKEALL (HAC manufacture) at 5/4d. or the "Phono — Novel and Startling Alarm Call — Rings continuously for 5 minutes" are but two further examples. Whether there was such a need for such 'attributes' or whether it made better marketing appeal is a matter for conjecture. Later examples certainly did not give their owners a heart attack each morning and eventually even the 'silent tick' came to be a virtue. Ingersoll for example marketed the Noiseless clock in early 1932. A new escapement was introduced, two escape wheels were used, one with internal and the other with external teeth with a single pallet pin engaging alternatively with the teeth on each wheel. In November of that year they were announcing improvements, they ceased to split the lever fork, the rubber round pin at the pillars was omitted and the two springs between which the pin oscillated was limited by two steel springs to prevent their being bent. It would be interesting to acquire a number of these clocks (cheaply!) in an attempt to find one of each type.

Although commencing factory methods at a later date than the Americans, the clock manufacturing area of the Black Forest in Germany made strenuous and highly successful efforts to compete. The making of clocks in this region had always been a cottage industry; the local populace making wooden weight-driven movements and cases when they could not find work on the land. The clocks were decorative, reliable and cheap and England had been a major importer. However, this market was temporarily lost to the Americans. In 1842 Chauncey Jerome sent his first consignment of factory-made clocks to England. The Germans could not compete at this period from the point of view of price or quantity. Eventually in 1861 Erhard Junghans of Schramberg decided to concentrate upon the clockmaking side of his business – the other being the manufacture of straw hats, etc. Together with his brother-in-law he formed a small company for manufacturing clocks by factory methods in place of the old unorganised methods.

Naturally there was a certain degree of local resentment at first but the venture prospered, and after his death his widow and sons carried on with the business. Various members of the family from both generations visited or worked in America primarily to observe first hand the 'American-style' and they both advised and arranged for the shipment of up-to-date equipment and machinery. As they so closely followed the methods and designs of the Americans, it is frequently only possible to ascertain a clock's origin by the trade mark it carries. The trade mark of the Junghans' Company is particularly helpful in so far as it changed several times through the years and it has been possible to ascertain the dates of these changes. It usually appears on the dial or stamped on the movement.

Another German firm — the Hamburg American Clock Company — formed in 1874 by Paul Landenberger and Philip Lang intially under the combined surnames, produced a similar range of clocks. These two men had worked with Junghans' widow but had decided to set up in business on their own after finding that she rigorously adhered to her husband's maxim that no outsider could participate financially in the company. They became worthy competitors of Junghans, but finally, after a few years of collaboration, merged with them in 1930. Their trade mark was a pair of crossed arrows. The American hold on the European market waned in the 1930s and the vast majority of alarm clocks imported into this country came from Germany.

The first Junghans' trade mark in 1877 was an eagle with outspread wings poised on a furled flag. Between 1882 and 1888 this was changed to a star with two eagles and a flag.

Trade mark after 1888

Left and right: trade marks after 1890.

Hamburg American Clock Company trade mark

Although the French did mass produce an alarm movement and included a plain metal drum case in their range (as can be seen from the advertisement shown in Figure 365a), these were not imported into this country in such vast quantities as those from Germany. Many of their earlier examples are of particular interest in so far as they have an unusual escapement with a short pendulum and spherical bob. A similar escapement was frequently used with a silk suspension on other early French clocks. Commonly referred to as 'tic-tac', the features of this escapement are:

Figures 349a and 349b

A further example from the Ingersoll range of dual purpose alarm clocks. Again the case is oak with the design of beading being of a stylised Jacobean pattern. This model was called the Puritan and had a plain white card dial, i.e. was not luminous. Reference to Figure 349b shows the method of fixing the movement into the case. Force and ensuing tight friction appears to be the answer! The movement was made in the United States and cased here. Ingersoll had an assembly works here in Clerkenwell. As well as serving as either a bedroom, kitchen or sitting room clock there was the added advantage with this type of clock that when necessary only a replacement was needed without the total cost of entire new clock.

£85+

DUO SUPERIOR Double Purpose Alarm Clock
Retail Price **15/-**

Many, many thousands of the "Duo" Double Purpose Alarm Clock have been sold since it was introduced a few months ago.

NOW! the "Duo Superior" Double Purpose Alarm Clock of taste and refinement at a price within reach of all.

Beautiful moulded case of Modern Design in a choice of several colours including Walnut and Mahogany. Smart $3\frac{3}{4}''$ Silvered Metal Dial with Fancy Hands. The Alarm Set is at the back, and from the front there is no indication that it is an alarm clock.

This ELEGANT CLOCK is a big seller; it RETAILS at the very low price of **15/-**

PLEASE WRITE FOR ILLUSTRATED LEAFLET.

Figure 350

Ingersoll 'Duo' clock being advertised and marketed during the late 1930s and 1940s. As can be seen from the description given in this advertisement the cases are now moulded and the movement is an integral part of the clock case.

£70+

Figures 351a, 351b, 351c and 351d

Oak cased alarm clock with exterior conical bells. The hammer and on/off lever appears on the top of the case between the bells. The trade mark on the small-alarm dial in 351d indicates that Friedrich Mauthe of Germany was the manufacturer. This trade mark was registered in 1902. See Figure 363 for further examples of clocks from this manufacturer. Not a particularly handsome clock but interesting as being of an early type.

Friedrich Mauthe was originally a merchant including among his merchandise clocks and clock parts. In 1870 he established a small clock factory in Schwenningen and by 1899 they were also manufacturing their own cases. Mauthe was one of the few firms to remain independent and did not amalgamate with competitors as times became more difficult. England was one of their major outlets and immediately before the second world war it is claimed that 60% of the total German- made clocks being imported by us came from the Mauthe factory, which finally ceased production in 1976.

£80+

Figure 351b

Movement of clock shown in Figure 351a. Note the folded winding keys. These are spring loaded in order to ensure their retaining folded or projecting position as required. This is an interesting transitional stage which would be logically followed by a fixed back to the case with the arbors lengthened in order for the winding keys, buttons, etc. to be manipulated without having to open the case.

Figure 351c
View of pin pallet.

Figure 351d
Mauthe Trademark (should have both wings and head!).

Figure 352a

12ins. high. It is relatively rare to find a wall-hanging alarm clock, and this is rather a handsome example in a solid wood case, that is hinged to a backboard upon which are mounted the two alarm bells. The bezel is sturdy and the glass bevelled — both indications of quality. The dial is of paper with an inset alarm dial. Paper and card dials were used extensively even in some comparatively high quality clocks. As cost of production was of prime importance, it is possible that this was an influencing factor. The crossed arrows on the dial indicate that the clock was manufactured by the Hamburg American Clock Company of Germany (1874-c.1930). The retailer's label is intact and is shown in Figure 352b. Both label and ability to trace the vendor enhance the price of this clock.

With visual appeal and solid case the expected price would be £90+

Figure 352b

Retailer's label inside the alarm clock in Figure 352a. It gives both his name (W.E. Watts of Nottingham) and instructions regarding care of the clock. It has been possible to trace the dates this man was in business (1874-1920).

Figure 352c

Interior view of 'Thunder Alarm' shown in Figure 352a. Note the thin stamped brass strip backplate, lack of cover to the spring, serviceable wheelwork, etc., all typical of a functional mass produced movement of this era. H.A.C. trademark of crossed arrows stamped on backplate below winding arbor.

Figures 353a, 353b, 353c

11½ins. high. This alarm clock is a magnificent example and, by the volume of noise achieved by it ringing both on bell and gong, speculation is aroused as to whether it was intended to awaken the living or the dead! The case is of walnut, with a cream celluloid dial (this needs close examination to detect that it was not ivory coloured porcelain), brass bushes to the winding holes and a recessed gilt metal centre with alarm disc. As indicated by the details on the label found pasted to the base of the case (Figure 353c), this particular model was marketed by Thomas Fattorini of Skipton, Yorkshire although a similar clock was sold by Fattorini & Sons of Bradford, Yorkshire. The two firms had a common ancestry springing from a watchmaker/jeweller, Antonio Fattorini, who settled in England after the Battle of Waterloo (1815).

Two separate business houses emerged — one in 1827 and the other 1831 — which to this day operate completely independently of the other: Thomas Fattorini (Skipton) Ltd., and Fattorini & Sons of Bradford. The latter among their other achievements designed and manufactured the Trophy Cup for the Football Association.

The patent for the clock illustrated in these photographs was taken out in 1901 by Joseph Arrigoni, a cousin of Thomas Fattorini. He was the retail shop manager at the Caroline Square premises and was also a qualified optician. The patent mentions an eight-day movement, but later a fourteen-day movement was introduced. This was offered to the public for 25s. or "Send 5/- Today . . . pay the balance by 4 subscriptions of 5/6d. per month".

About the same time Fattorini & Sons of Bradford were marketing a similar alarm called the "Automatic Caller". This had an oak case, eight-day movement, also rang on a bell and gong and also reset itself each night. The patent for this had been taken out in 1892 by T. Wood, but by 1897 the patent rights had been purchased by Fattorini & Sons. It was sold in 1902 for 26s. and the advertisement stressed that it was 'John Bull's Own Work' and 'English Manufacture'. The country of manufacture being a topical subject at this date as the previous year there had been legislation to ensure that clocks and watches marked as being made in this country must actually be made here and not just foreign movements assembled or cased in England and then sold as 'Made in England'. It had been noted on close scrutiny that the movement of the 'Bugler' Alarm Clock illustrated here (Figure 353b) bore the trade mark of the Hamburg American Clock Factory of Germany (crossed arrows). It would appear that although it was an English patent owned by the firm of Thomas Fattorini, the movements were made abroad and probably cased here. This was common practice to minimise the cost of import tax and transporting cases as well as movements from abroad. Fascinating device worth acquiring for sheer volume of sound.

£150+

353b

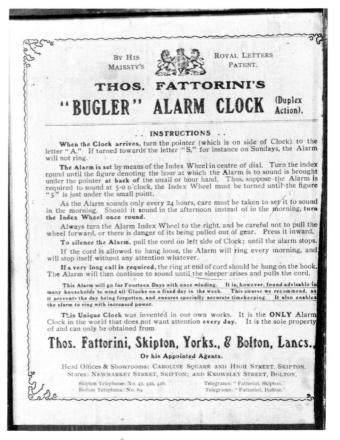

353c

Figure 354

Advertisement taken from A. Mayer & Son Ltd. 1906 catalogue showing two unusual alarm clocks. That on the left only needs winding once a week and should the alarm train run down through an excessive length of ring prior to being switched off a red spot is shown in the circle above the figure six. When the alarm is set the word Alarm appears in the aperture below the figure twelve. The prices given would be wholesale prices. Unfortunately there is no indication of manufacture except that it is foreign.

£70+

The 'XL' electric alarm shown on the right is similar in concept to that shown in the following illustration but if the text given here is to be believed was first marketed in 1895. The thought of using it as an alarm 'button' in case of burglary is certainly intriguing. When reading some of the early advertisements the impression is often gained that the advertising men of those days were most inventive as to the possible uses to which their products could be put.

£50+ as a curiosity

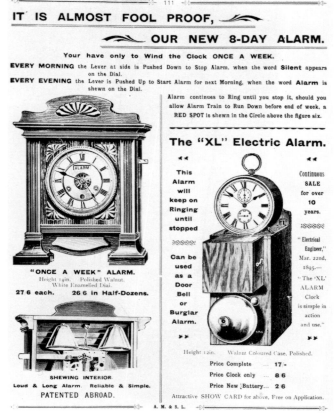

111

IT IS ALMOST FOOL PROOF,

OUR NEW 8-DAY ALARM.

Your have only to Wind the Clock ONCE A WEEK.

EVERY MORNING the Lever at side is Pushed Down to Stop Alarm, when the word **Silent** appears on the Dial.

EVERY EVENING the Lever is Pushed Up to Start Alarm for next Morning, when the word **Alarm** is shewn on the Dial.

Alarm continues to Ring until you stop it, should you allow Alarm Train to Run Down before end of week, a **RED SPOT** is shewn in the Circle above the figure six.

"ONCE A WEEK" ALARM.
Height 14in.   Polished Walnut.
White Enamelled Dial.
27 6 each.   26 6 in Half-Dozens.

SHEWING INTERIOR.
Loud & Long Alarm. Reliable & Simple.
PATENTED ABROAD.

The "XL" Electric Alarm.

This Alarm will keep on Ringing until stopped

Can be used as a Door Bell or Burglar Alarm.

Continuous SALE for over 10 years.

"Electrical Engineer," Mar. 22nd, 1895.—
"The 'XL' ALARM Clock is simple in action and use."

Height 12in.   Walnut Coloured Case, Polished.

Price Complete	...	**17 -**
Price Clock only	...	**8 6**
Price New Battery	...	**2 6**

Attractive SHOW CARD for above, Free on Application.

A. M. & S. L.

Figure 355

The patent for this clock was taken out by M. and S. Turton of Tiplow in 1888 but the patent rights were purchased from them by Fattorini & Sons of Bradford (not to be confused with Thomas Fattorini (Skipton) Ltd.). For further details of the Fattorinis see text accompanying Figure 353. According to the account of this clock in the Horological Journal *for December, 1888, it was not solely intended for use as a morning alarm. A handle at the back of the clock could be used to ring it at will — "a merchant can signal for one or all of his clerks, a teacher for his classes, or dismiss his pupils", etc. There was a double bell — one housed on the side of the box housing the electric battery and also the mechanical alarm which was within the casing of the clock. This was one of if not the first electric alarm clock.*

£50+

Figure 356

11½ins. high. A nicely produced, oval wooden base, with elegant mouldings and surround, make this a desirable and unusual alarm clock. The white card dial with inset seconds and alarm dial also has the Junghans eight point star with a capital 'J' trade mark denoting a date after 1900. The 3½ins. diameter alarm bell is mounted under the movement on the wooden base and has an internal hammer. Later it became more usual to use the case of the clock as a bell and eventually to house the bell inside the case.

£80+

An almost identical clock was illustrated in Junghans wholesale catalogue for 1903 where it appeared as the 'Brilliant' in black polished, walnut dull or old oak finish with a 3½ins. nickelled bell. The case was approximately 2¾ins. shorter than the example shown here. H. Samuels of Manchester also carried illustrations of this clock in their 1904 catalogue.

Figure 357

15½ins. high. Veneered softwood case with decorative beading and turned finials. Full glass door to front. White painted dial with skeletonised centre to show alarm setting dial. The decorative pendulum is a cheap imitation of the two jar mercurial pendulum often found on expensive clocks, the effect in this instance being achieved by two cylinders of polished steel. The movement is typical thirty-hour with striking of alarm on a bell mounted on the backboard of the case. Although very similar to many manufactured by the various American factories of this period, this was made by the Junghans factory in Schramberg, Germany. Evidence of this is found in the trade mark below the figure 12 of a five-pointed star with a capital 'J' in the centre. This dates the clock as prior to 1890. This particular clock was sold by T. Coombes, Watchmaker and Jeweller of 117 Walworth Road and 109 Westminster Bridge Road, London. According to Directories he was at these addresses after 1880. A nice piece of collaborating evidence. It is details like this that, when traced and recorded, can add the odd pound or two to the value of each item.

In good condition and working order
£120+

Figure 358a

7¾ins. high. This thirty-hour drum alarm, in a brass case, is supported by two side pillars which also act as standards for the two highly polished steel bells, firmly mounted on a wooden kidney shaped base. The decorative chapter ring is celluloid, with inset gilt metal centre — a similar style to those found on French movements. The bells are struck alternately with an external hammer. The Junghans' trade mark appears stamped on the movement as can be seen in Figure 357b and indicates a date of manufacture after 1890. It is noted that a similar clock appears in the catalogue of the American Company of William L. Gilbert dated 1901-1902. It is referred to as the 'Dewey Long Alarm', and although the external appearance (apart from finish on bezel and dial decoration) is identical, it would appear that the American clock was of a higher quality finish, with a more solid construction, an ivory porcelain dial and a movement with a seconds hand. No great

quality involved but unusual appearance has appeal.

This model was described in the 1908 Junghans' catalogue as the Nightingale "Brass case nickelled with nickelled trimmings and white dial or gold coloured with gilt trim and ivory coloured celluloid dial. Socle in walnut dull 3¼ inch bells". Overall height was given as 8½ins.

£75 +

Figure 358b

Interior view of the drum alarm illustrated in Figure 358a. This particular movement was introduced in 1875 and continued to be used unchanged for fifty years.

Figure 359

8ins. high. This is an intriguing clock with some curious features. The dial is of exceptional quality — ivory porcelain with a recessed centre and inset seconds and alarm dials — similar to those found on French movements in marble, four-glass cases, etc. The case is of lacquered brass with simulated jewels embossed around the bezel. Each boss is 'faceted' and painted red. There are no maker's marks either on the dial or stamped on the movement. The movement, although perfectly adequate, is not of the high standard of the dial and could have come from either a German or American factory. Reference has been seen in some American catalogues to models having a jewelled bezel for an extra charge. Rather splendid to wake up to this of a morning!

£95+

Figure 360

This illustration is taken from the 1885 Illustrated Catalogue of Clocks manufactured by the Terry Clock Company of Pittsfield, Massachusetts. At this date the Terry Clock Company (under the management of the grandsons of Eli Terry) claimed to be the sole manufacturer of luminous clocks (? in America): "If placed near the bedside at night, the time can be readily seen in the darkest night, without the aid of artificial light". Although not essential for the dials to be left in bright sunlight during the day it was necessary for them to be exposed to light for a certain length of time in order to "maintain the phosphorescent light". As stated by Chris Bailey in the historical section of his book, it is as well to remember that the phosphorescent properties of these clocks will have deteriorated with time and their dials will no longer be luminous. The two patent dates given on the dials are interesting. That of 1882 taken out by the executrix of William H. Balmain of Eversley, Ventnor, Isle of Wight, England, was for a self-luminous paint. It would appear that the first application for this patent had been filed in England in 1877, with others following in Italy, Spain, Austria in the ensuing years. The second date 1883 refers to the patent for "rendering paper uniformly luminous on its surface" that had been taken out by William Trotter, Jr. of Oyster Bay, New York, USA. Apparently there had been some technical difficulties in avoiding uneven patches of the luminous paint when applying it to the paper and card dials.

£40+ on the English market

Terry Clock Company. 41

METEOR ALARM.

ONE DAY TIME ALARM.   LUMINOUS DIAL.

THEY SHINE ALL NIGHT.

TIME VISIBLE IN THE DARK.

ONE-HALF SIZE.   4 INCH DIAL.

ONE DAY . LEVER . TIME . ALARM. $3.15.

This Clock is made in Hammered Metal, and is called Meteor Alarm Embossed Gilt. $3.15.

For illustration of finish see page 26.   Nickel finish 15 cents extra.

6

Courtesy of Chris H. Bailey

Figure 361

Square brass cased alarm clock surmounted with rectangular bells. Similar examples were being advertised by the manufacturers — Salvos Manufacturing Company Ltd. of 46 Duke Street, London E.C.3. in 1928. These clocks were of German manufacture, possibly they came from the factory of Thomas Haller.

£90+

Figure 362

Marvellous advertisement appearing in the Hamburg American Clock Company catalogue for 1908 which ably illustrates some of the models being marketed by them at this date.

MAUTHE

A **NEW** **Silent**
Alarm
Clock ...

... giving better service than
Clocks costing twice the price.

The handsome appearance of the New Mauthe
Alarm Clock interests the customer at once, and
its exclusive features make for an easy sale. No
tick is heard — the hands move round in perfect
silence. The alarm, which is of euphonious tone,
can be stopped by merely placing a finger on the
top handle. The case is heavily Chromium plated
encircled with a dark Blue, or Red enamelled Band.
Metal Silver Dial — luminous if required. Being
inexpensive, these Clocks will find a ready market.

SUPPLIED TO
WHOLESALERS ONLY

Announcement of

MAUTHE CLOCKS, Ltd.
2-3 Charterhouse Sq., E.C.1
Telephone · · · *Clerkenwell 0221*

MAUTHE

JAZ

ANGLIC Model

Selling Price: 12 6 Plain dial.   Luminous dial 15/-.

John Bull helps to make this New JAZ Model

THE new JAZ ANGLIC model, illustrated above, has been
produced at the request of a very large number of jewellers,
who wish to sell a guaranteed alarm of proved performance at
a competitive price.
This new JAZ model has been designed solely to meet the taste
of the British Public. The cases are made in England; final
assembly, testing and adjusting are thus carried out for this model
in this country. Thus, over 75% of the public purchase price
of this model remains in Britain. Notice particularly the large
clear dials — in both plain and luminous models. The dial is similar
in size to the JAZ Replic model. Height of clock 4¾ inches.

Selling Prices   Plain dial 12 6.   Luminous dial 15/-.
These models are the cheapest GUARANTEED Alarms in
moulded cases on the British market.

Finishes   The cases are available in the following finishes.
Walnut, Ebony, Mahogany and Mottled Green, in both
plain and luminous dials.

Sales Aids   Send to-day for illustrated folder giving details
of sales assistance, including showcards, leaflets for distribution
to the public, etc., to:—

JAZ Clock Co. Ltd., 6 Holborn Viaduct, London, E.C.1

Figure 363

*The Mauthe Silent Alarm first appeared in 1933 although this 1936 advertisement still called it "A new
Silent Alarm Clock". The silent action was achieved by modifying the usual form of pin pallet, utilising
a new method of mounting the escape wheel and manufacturing the escape wheel and lever from fibre.
'Euphonious tone'' replaced earlier claims of sufficient noise to awaken both the dead and the sleeping!
The cases were chromium plated encircled with a dark blue or red enamelled band. The dials were either
silvered or luminous. Note the trademark which is now of a sterner appearance than the original format
registered in 1902 and seen in Figure 351d.*

£45+

Figure 364

*This is a curious situation, the movement is French but the case was English made, no doubt to
circumvent import tariffs and sales resistance to foreign goods. The final assembly, testing and
adjustment and the making of the case meant that "over 75% of the public purchase price of this model
remains in Britain". The large clear dials (plain or luminous) were cited as being of great advantage.
It was also a good stable design that was not liable to be knocked off a bedside cabinet. The Jaz Silent
Alarm (not illustrated) was reviewed in 1936. The silent action being achieved by placing the escape
wheel and pallets so that long pallet pins were used. The impulse pin was also longer than usual.*

To the horologist £40+

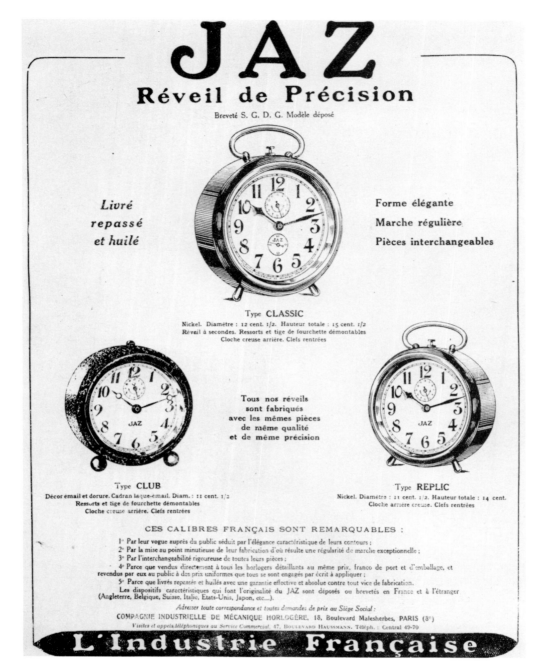

365a

Figures 365a and 365b

*This advertisement for alarm clocks manufactured in France by
Compagnie Industrielle de Mécanique Horlogère appeared in magazines,
journals, etc. around 1924. 'Jaz' the trade mark of this company was
registered in 1919 and continued to be used until 1941 when a parrot (as
in Figure 365b) was introduced.*

£30+

365b

Figures 366a, 366b, 366c

Three views of a French alarm clock. The front view shows a relatively conventional case style — square brass case, thick bevelled glass, pleasant designed clear dial, carrying handle, etc., with centre alarm hand. Turning the clock to look at the back, however, instantly revealed the novel feature of exposing the balance wheel as a decorative feature while protecting it by a small circle of glass. Fast/Slow regulation was by means of the small lever to the left of the balance. The third illustration is a closer view of the movement with the back of the case and the winding keys removed.

£80+   entirely due to visible balance wheel as novelty feature.

Figures 367a and 367b

Extremely well constructed and finished alarm clock manufactured by Zenith (see Figure 217 for further examples of small clocks from this noted Swiss factory). Aspects denoting quality are the exceptionally thick brass case, thick bevelled glass, well machined and finished movement. It is interesting to note that again cast parts have been used as with the other small Zenith Boudoir clocks.

£50+

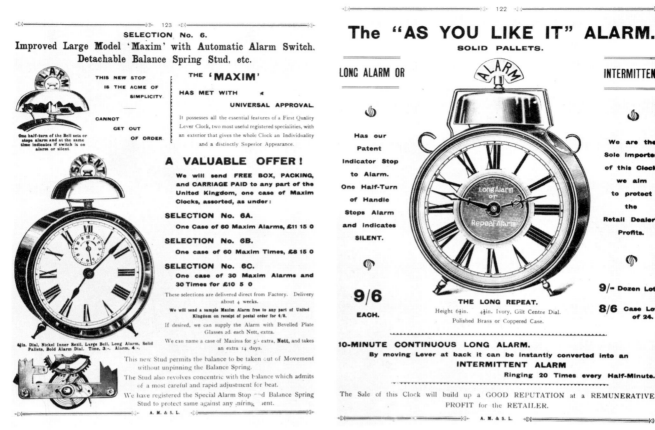

368a

368b

368c

368d

Figures 368a, 368b, 368c and 368d

Two imported alarm clocks marketed by A. Mayer & Son Ltd. in 1908 to the retail trade. It has not been possible to identify the actual manufacturer. That on the right with solid pallets, 10 minute continuous long alarm or by moving lever could be converted into an intermittent alarm that ran twenty times every half-minute. By turning the handle at the top of the case (see Figure 368c) the alarm was silenced with the appropriate word SILENT becoming visible. In the example on the left the whole bell needs one half turn in order to set or silence the alarm. For details of a stud permitting the balance to be removed without unpinning the balance spring as well as allowing for adjustment of the beat, see Figure 368d. Total height of this clock was 4½ins. with the nickel inner bezel, large bell, long alarm, solid pallets and bold alarm dial all being listed as the desirable features. Price of a timepiece 3/- and alarm 4/-.

£50+

Figure 369

5½ins. and 3¾ins. high. Two nickel-cased pin pallet alarm movements made by the Western Clock Company Limited of Peterborough, Canada, which was founded in 1885 under the name of United Clock Company in Illinois, USA. Unfortunately they were bankrupt within two years, but under new management recommenced business in 1895 as the Western Clock Manufacturing Company and by 1903 claimed to be making one million alarm clocks a year. They became the Western Clock Company prior to 1925 and in 1936 were again renamed, but this time adopting their trade mark 'Westclox'. This had been their formal trade mark since 1909. The name 'Big Ben' first appeared in 1910 and was retained as each new model appeared and so knowledge of the changes of name of the Company and appropriate dates are invaluable when dating these particular clocks. 'Baby Ben' is particularly neat and has been recognised as a collector's item for many years in America. They are not often seen on the market and this will be reflected in the purchasing price. The larger clock was made of parts manufactured in the USA but assembled at their works in Scotland. This is the de luxe version, on a stand and not on small legs.

£40+

Figures 370a and 370b

French brass cased alarm clock with good cylinder escapement. Note the neat positioning of the bell inside the case. The back cover is pierced and lined with silk to allow sound to be heard. Key wind, with a nice clear white enamel dial. It has not been possible to trace the trademark discernible to the left of the platform. Not all that exciting but a pleasant quality movement that well typifies one of the styles coming from the French manufacturers.

£80+

Figures 371a and 371b

This American table alarm clock was made by the Parker Clock Company, Meriden, Connecticut, USA. The Company is known to have been trading since 1890. The drum of the clock movement and base are made of brass. The bell is steel. The alarm is operated by an independent mechanism in the base (see Figure 371b). Alarms made by this Company tend to use the same clock movement but vary the style of base, number of bells, etc., to provide a variety of styles.

£120+

Figure 372

2½ins. high. Small French travelling alarm clock in a brass case, with a white enamel dial, black Roman numerals, steel trefoil hands and a pointer for the alarm.

This clock is not to be confused with 'the Alarm Clock with two Springs' patented by T. Maurel of Paris in 1868. This example has the alarm and time off a single barrel. It is possible as the style of hands and dial are so similar to the model by Maurel that this was a later copy. There are no maker's marks. Although collectable and an interesting example this clock would not fetch as much as one by T. Maurel.

£95+

Figure 373

View of back of movement of the small French travelling alarm shown in Figure 372. It has a tic-tac escapement (anchor encompassing two teeth), together with a short pendulum and spherical bob.

Figure 374a

2½ins. high. Rare small lacquered-brass case French alarm with stamped Arabic numerals and small centre pointer for setting the alarm. There is neither bell nor gong as the alarm hammer strikes the casing. This is purely a timing device as it is necessary to set it not at the hour one wishes to be awakened, but at the total number of hours between the time of setting and the time of awakening, e.g. at 10.0 p.m. the alarm is set to ring at 7.0 a.m. the next morning by moving the pointer to the number 9, as there are nine hours between 10.0 p.m. and 7.0 a.m. The movement is wound by turning the whole of the back cover anti-clockwise.

£150

Figure 374b

The movement has a tic-tac escapement (anchor encompassing two teeth) and an extremely unusual pendulum. One other known example has a similar escapement, but with a conventional spherical bob placed between the back and front plates. This second example had a partially legible label reading 'Maison Brevete a Paris, La rue Vivienne, horologerie boites à musique, reveils.'

Figure 374c

It has not been possible to trace the trade mark of a cockerel and letters A R, but it is likely to be that of Antoine Redier (1817-97) a prolific maker of alarms for the English market. It is interesting to comment that Japy Frères also used a cockerel as one of their marks, but of a different configuration. Figure 374c shows the cockerel trade mark inside the alarm in Figure 374a.

Figure 375

1½ins. high, 3⅛ins. diameter. The alarm mechanism shown in this illustration was patented in 1823 by William Gossage (1799-1877), an eminent industrial chemist. According to J. Fenwick Allen in his book Some Founders of the Chemical Industry, *Gossage, wishing to begin his studies at an exceptionally early hour, invented this device in order to ensure that his tutor was also awake! It was produced commercially with this particular example bearing the serial number '266', although it must be remembered that few manufacturers commenced numbering at '1' in order to make their output appear*

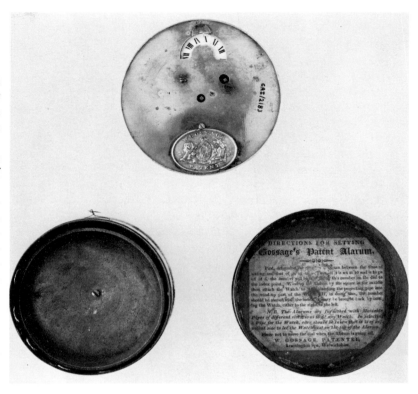

larger. This example has an outer protective box of japanned tin — black on the outside and red within. The alarm mechanism itself is housed in a silver plated case with pierced sides. This is on the left and viewed immediately from above in order to see the large bell that entirely fills the bottom half of the case. The top view shows the top plate under which is the movement. This drops into the lower half of the case and is secured by small pins at the side. This top plate is also of silver plate and carries a small plaque bearing the words 'W. Gossage — Patent' and the Royal Coat of Arms. The small dial visible through the crescent-shaped aperture is enamel on copper. The right hand view shows the instructions pasted to the bottom of the outer protective case. Before reading them it is necessary to understand that this is not an alarm clock, but an alarm mechanism for attaching to a watch. This was by no means a completely new idea as further examples of similar devices can be found. Clocks were still a comparative luxury whereas many men carried a watch.

"Instructions for Setting Gossage's Patent Alarm. First determine the number of hours between the time of setting and that of going off. (Thus, if it is set at 10 and is to go off at 4, the number would be 6.) Bring this number on the dial to the index point. Wind up the Alarm by the square in the middle then attach the Watch to it by inserting the projection pipe into the wind-up part of the Watch. If, in doing this, the number should be moved from the index, it may be brought back by turning the Watch, either to the right or the left.

N.B. The Alarms are furnished with Moveable Pipes of different sizes, so as to fit any Watch. In selecting a Pipe for the Watch, care should be taken that it is of sufficient size to let the Watch rest on the top of the Alarm.

Please not to move the dial when the Alarm is going off
W. GOSSAGE PATENTEE
Leamington Spa, Warwickshire"

Further details of another example can be found in an article 'Gossage's Patent Alarm' by Cedric Jagger that appeared in Antiquarian Horology *for December, 1959. Needless to say these alarms are choice collectors' pieces and have of recent years passed through the salesrooms for* **£300 — £400**

Figures 376a and 376b

Ansonia 8 day centre wound alarm with spring across full width of the movement. The patent dates on the back of the case (April 23rd 1878, October 21st 1884) (see Figure 376b) would have been American patents. The alarm set was by means of the small knurled knob on the right hand side of the case. The thin paper dial has the Ansonia trademark; these thin dials are inevitably discoloured.

£85+

Figure 376c

Close up of alarm mechanism housed under the bell surmounting the case. This illustration demonstrates how in order to facilitate production this virtually self sufficient conversion kit was simply added to a basic timepiece. Note lantern pinions and solid wheel.

Figure 377

3½ins. and 2½ins. high. These are two small German alarm clocks. That on the left is in a copper on brass case, with a white enamel dial and spade hands. The back of the case has knobs for winding and setting the hands or alarm, with a silent/alarm lever. The crossed arrow mark stamped on the movement indicates that it was made by the Hamburg American Clock Company of Germany (1874-c.1930). Upon comparing this with the example shown in Figure 122, also from their factory, the full range of their products is recognised. A small neat movement in a better than usual quality case and dial, so this clock would be worth a few pounds more than the usual drum alarm.

The smaller clock on the right is also of German origin and was made by Muller Schlenker AG of Schwarzwald and is advertised by them in the Deutsche Uhrmacher Zeitung throughout the 1920s. There is a marked similarity between this clock and a small alarm made by Veglia (an Italian manufacturer). Almost certainly this was intentional and was done in an effort to effectively compete with rival manufacturers and to appear to be undercutting their prices. The genuine Veglia has the name on the dial. The case is oxidised metal, and the dial again of enamel, with luminous hands and numerals. The pointer for the alarm is blued steel. The alarm shuts off by lowering the carrying handle. The most interesting feature of this clock is that it is wound by winding both clockwise and anticlockwise. This is made possible by an opposing ratchet mechanism on both time and alarm barrels — the movement being in two tiers rather than in the usual side by side arrangement. This immediately makes it a collector's item and also illustrates why it is necessary to examine closely an article before dismissing it as being of a standard range.

£35   **£40+**

Figure 378

3½ins. high. Neat case — copper on brass — with clear white enamel dial. Arrow indicating direction for alarm hand to rotate is fixed to tail of alarm hand. Made by Hamburg American Clock Company c.1917.

£50

Figure 379

2½ins. high. This is a small Swiss made alarm clock the INVENTIC, in nickel case made from designs patented by Swiss manufacturing company Ed. Kummer SA, Bettlach in 1929. The alarm is silenced by folding down the handle. The chapter ring is luminous.

£45+

Figure 380a

The value of this timepiece is greatly enhanced by the fact that the movement is housed in a well finished bronze figure holding a bell and torch. The dial is stamped on his ample stomach, with two spade hands and a pointer for setting the alarm mechanism. Winding and hand setting are through the cloak at the back. The cloak is removable (two small screws hold it in place) to allow access to the movement. When the predetermined hour is reached the whole arm holding the bell swings thus ringing the bell that is suspended from the hand. £350+ as a decorative piece, without the additional merit of telling the time and sounding an alarm. A similar clock but with type metal (spelter) figure has been seen in the Time Museum, Rockford, Illinois, USA.

£900+

Figure 380b

Rear view of Figure 380a with cloak removed. This is a typical French tic-tac escapement of the late 1800s of simple construction, a single barrel arbor producing the necessary power for both the alarm and time mechanism. There are no identifying maker's marks.

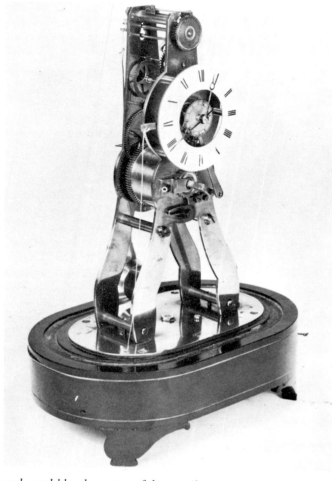

Figure 381

The small French alarm clock shown in this illustration has many interesting and desirable features. These clocks were introduced into this country at the Great Exhibition in 1851 by their maker Victor-Athanase Pierret (see pages 189 and 190 for further references to this maker). It is stated in the Jurors' Report that, "The Jury, however, agreed to mention some small alarm clocks by M. Pierret, of Paris, on account of their cheapness, and because alarms really are, for certain purposes, useful articles of household furniture." A rather grudging award when noting the fact that some 10,000 were sold during the Exhibition! Their price at this date was 25s. Many of them carry the oval stamp on their base plate advertising the fact that their maker had received an Honourable Mention, but it is not clear whether this was after the Great Exhibition or after the Paris Universal Exhibition of 1855 when he received a further Honourable Mention. This mark has frequently been mistaken for the name of the maker, M. Honourable! This example has the name 'R. Holt et Cie — A Paris' appearing on the frame and dial, and would be the name of the retailer or importer — it was common practice to imply French connections and this would indicate that the clock had been made in Paris.

The frames of this example are plain, but others have been seen with decorative engraving. The ebonised wooden base has brass stringing. The points to note are:

a) The alarm bell (a true bell shape and not the usual hemisphere), is housed in the base and is wound by pulling the small cord just visible on the right hand side of the base.

b) The alarm is set either manually by removing the dome or by pulling the cord on the left that protrudes through the small hole in the side of the base. This pulls the ratchet toothed alarm disc round via a pawl, tooth by tooth.

c) The tail of the hour hand acts as an index against the alarm disc to indicate the time of ringing.

d) The dial is porcelain, through the centre of which can be seen the alarm disc and motion work.

e) The ratchet wheel above the numeral twelve regulates a rise and fall mechanism for adjusting the length of the pendulum and thereby regulating the clock.

f) It is just possible to discern the silk suspension — a method much favoured by early French makers. It is often troublesome as temperature and other atmospheric changes readily affect it.

£450

382a

382b

382c

Figures 382a, 382b and 382c

6½ins. high. The style of this clock case — thin brass front, nickel-finished sides with glass panels — must have been extremely popular as at least two American and two German manufacturers used it for their movements. In 1879 Seth Thomas and Junghans used it to house either a musical alarm or strike movement, while the New Haven Clock Company model had either an ordinary alarm or strike movement. A catalogue of Adolph Scott of Birmingham, which appears to have been published c.1910, had the following offer which refers to these clocks as carriage clocks:

	BBC and other Makers	HAC Guaranteed Goods
1-day Time	5/4	6/4
1-day Alarm	6/8	7/-
1-day Strike	7/8	8/-
1-day Musical — 1 air Alarm	10/-	11/6
1-day Musical — 1 air Strike	10/6	13/-
1-day Musical — 2 air Alarm	11/-	13/-
1-day Musical — 2 air Strike	11/6	14/-

HAC refers to the Hamburg American Clock Company, but it is not certain who the other makers were.

The example illustrated here has a gilt front, nickel finish to the rest of the case and glass panels at the sides. The dial is of card with seconds and alarm dials. Although 'cheap and nasty' as regards finish of the case and movement these clocks are eagerly sought after and the present prices would be in excess of thirty times the original cost.

The movement (Figure 382b) is thirty-hour with the musical barrel and comb housed in the base as shown in Figure 382c. With a clock such as this, it is sensible to hear the music played through to determine whether it has one or two tunes and also in order to detect any broken or missing pins on the barrel or teeth on the comb. It is doubtful what repairs could be carried out as the original was so poorly made.

£125+

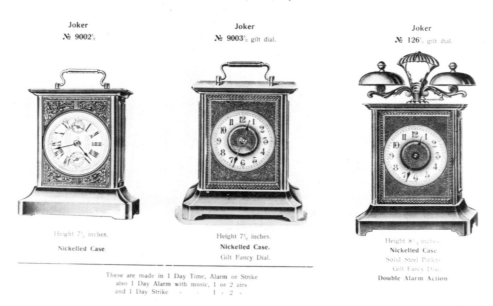

Figure 383

Three of the Hamburg American Clock Company models of the Joker alarm clocks being manufactured in 1908. It is worth remembering that nickel was a new material in the horological trade at that date and highly prized.

£90 — £120

Figure 384

These are rather late examples to be of great interest to collectors as yet (1939) but this advertisement was included to provide some information regarding styles of hands at that date and also the Haller trademark seen in the centre of the advertisement. The firm of Thomas Ernst Haller was founded in 1901. They amalgamated with Kienzle in 1929 but obviously the trademark and name Haller were retained long after this date.

£20+

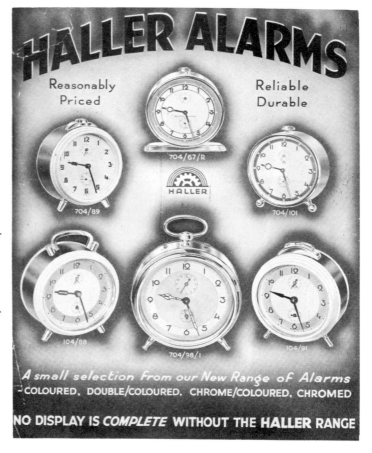

CHAPTER XIII

Electric Clocks

Although several stalwarts have persevered and ignored the initial derision of their fellow horologists, it is only in the past few years that examples of electrical horology have been more widely recognised as being worthy of study and collection. In 1970 the Electrical Horology Group was formed as a sub group of the Antiquarian Horological Society and, largely due to their work in this field of research, sufficient accurate information was available to form the basis upon which the exhibition held in 1977 at the Science Museum, London, entitled Electrifying Time, was built. 1977 was the chosen year as it coincided with the centenary of the death of Alexander Bain who is acknowledged as the 'Father of Electrical Timekeeping' in this country. Although as early as the sixth century BC electricity and magnetism were recognised phenomena, it was not until much later that they were fully understood and mastered. During the eighteenth and nineteenth centuries much research was carried out in England and on the Continent until finally in the mid-1800s sufficient technical advancements (magnets, reliable current sources, etc.) had been made for the practical application to be made possible. Steinhill (1801-70), a Professor of Munich University, built a master clock which sent out impulses to drive slave dials, but the pioneer in electrical timekeeping in this country was Alexander Bain (1811-77). During his boyhood in Caithness, Scotland, he had become intensely interested in the then extremely topical subject of applying electricity to telegraph communications, but soon began to realise that it had a potential use in timetelling. In 1841 Bain took out his first patent (Patent No. 8783), the contents of which were to provide the foundation stone for future electrical horology for nearly a century. Other patents soon followed. Most of the extant examples of his clocks are in museums (see Figure 385). To date, only the existence of two small models suitable for standing on a shelf have been recorded. Both are in private collections — the existence of the example appearing in Figure 386 only becoming known by virtue of the pre-exhibition publicity to the Electrifying Time exhibition 1977/8.

It now became possible to provide the distribution of a standard time over a much wider area. In the past an extremely localised standard time had been obtained from observatories where either a time ball was dropped or a gun fired at an appointed hour each day in order that the local inhabitants could set their timepieces. Now, using the wires of private telegraph companies, and later those of the Post Office, signals could be sent further afield either to activate the timeballs electrically or synchronise secondary dials. Examples of master clocks from this period can be found in various museums and their invention is attributable to such men as F.J. Ritchie of Edinburgh (Patent No. 2078 taken out in 1872); R.L. Jones of Chester (Patent No. 702 taken out in 1857), Lund of London (Patent No. 3924 taken out in 1876) to mention but a few of the English patentees.

Possibly the most important Continental pioneer in this field was Dr. Mattheus Hipp of Neuchatel (1813-93), as the principle of his toggle was used in modified forms in many of the later master clocks and battery electric clocks. This consists of a toggle or trailer attached to the pendulum, which passes quite freely over a notched block of steel until the swing falls below a predetermined arc, whereupon

Figure 385

This is generally acknowledged as being the most attractive of the master clocks attributable to Alexander Bain (1811-77). The carved case is exceptionally handsome, and the movement well finished with gold contacts and agate insulation. The silvered dial is signed 'Alex^r Bain's Patent Electric Clock' and carries the serial number 113, and is on view in the Time Measurement Gallery of the Science Museum, South Kensington, London. A museum piece and difficult to value. One in a conventional style longcase came on the market in June 1980 and through the auctioneer's erroneous catalogue entry implying it was incomplete, sold for the bargain price of £2,100. "Missing" parts had fallen into the bottom of the case and had not been lost during transit as implied!

£10,000 depending on case style

Crown Copyright

Figure 386

Until recently the existence of this small shelf clock by Alexander Bain was unknown, and its presence in the Electrifying Time Exhibition at the Science Museum, London, in 1977 roused great interest. The brass movement is mounted on an ebonised softwood base and covered by a glass shade in the manner of a conventional skeleton clock. The silvered chapter ring is signed 'A. Bain's Patent Electric Clock'. The contacts visible at the front of the movement and leading off through the base are operated at intervals of two hours. The reason for this is uncertain, which would imply that the clock was custom made for a specific purpose. It has subsequently been sold by private treaty and was last heard of in the USA. An exact replica, of value in its own right, is in the hands of a collector on the same continent.

Unique clock and therefore value can only be guessed at several thousands of pounds. If you see another one buy it at any price!

£10,000

the toggle is 'caught' in the notch, which depresses the block, closes the circuit, and the pendulum receives a fresh impulse from the electro-magnet. Hipp stated that he first conceived the idea of using what later became known as the 'Hipp Toggle' as early as 1834, but the first clocks using this did not appear until 1842. Examples of the original clocks are very rare in this country, the agent for them having been a colleague of Professor Wheatstone. They are more commonly found on the Continent and it is known that the firm of Peyer-Favarger & Company of Neuchatel and the Telegraphic Manufacturing Company, also of Neuchatel, were manufacturing his master clock after 1860. The backplates of the original clocks have some acknowledgement to the fact that they were manufactured to the specifications laid down by Hipp stamped upon them. This can be merely 'M. Hipp' or it can have the additional information 'Neuchatel — Swiss' and a serial number. Pure copies or variations of the Hipp toggle have always appealed to the model or precision engineer wishing to make an electric master clock. This is worth bearing in mind when assessing any unusual clock working on this principle.

The next step was to introduce systems of master clocks with subsidiary dials (slave dials) suitable for using in large factories, public buildings, etc., in order to provide a uniform time throughout the building. This had never been completely successful when using a group of mechanical clocks. It is from this group of electric clocks that the collector will be able, with varying success, to seek examples to acquire. Upon studying the patents taken out it quickly becomes evident that a great deal of ingenuity and enterprise was shown by the inventors of the day. Their main problems being to find a contact that did not wear or tarnish, thereby reducing the electrical efficiency, and a method of impulsing the pendulum but at the same time leaving it 'free'. Many of these designs never left the drawing board while many others died a natural death due to inherent faults that only became apparent after production had commenced. This is one field of collecting where the failures are often of more value than the successful examples.

Basically electric clocks fall into three main categories:
1. Those that are electrically impulsed. When the pendulum falls below a predetermined arc an electric circuit is made and the pendulum receives a further impulse.
2. Those that are electrically rewound. The same principle but instead of the pendulum receiving an impulse a small motor is rewound which powers the clock for a further period.
3. Those that are synchronous. These are the clocks that appeared so prolifically in the 1930s as small domestic shelf clocks running from the mains.

The following illustrations show a representative selection of master clocks by some makers. As can be seen from these illustrations they are housed in a variety of cases from the longcase style of the Bentley Earth Driven Clock (Figure 392) to the smaller wall hanging example made by the Silent Electric Company (Figure 397). Unfortunately it has not been possible to include every known type. A list of some of the known literature on the subject has been included in the bibliography, but in most instances these are contemporary references as, apart from the researches of some members of the Electrical/Horology Group of the Antiquarian Horological Society, little has been published in the last thirty or so years. Obviously some examples are more rare than others. It should be possible, however, to find a good working example manufactured by the Synchronome Company and these are highly desirable pieces for many reasons — one of which is the extremely high standard of workmanship put into their manufacture. The finish on the parts being commensurate with that used in instrument making rather than the more usual mass production look of most of the other clocks of this date.

The name of Frank Hope-Jones (1867-1950) is only second in importance to that of Alexander Bain when discussing the history of electrical horology. When reviewing one of his books in 1931, Professor Sir Charles V. Boys refers to him as ''The high priest'' and ''like St. Athanasius, his faith is clear and emphatic... St. Hope-Jones''. Possibly not all would subscribe to quite such extravagant praise today, as unfortunately in his enthusiasm, Hope-Jones swept all other contemporaries' thoughts to one side. However, it was he who, after some thirteen years of experimentation, patented in 1908 an electric master clock whose accuracy at that time was only matched by the most accurate astronomical regulators. Initially working with George Bennet Bowell (1875-1942) he formed the Synchronome Company at Birkenhead in 1895. This was the date of his first patent taken out in collaboration with Bowell. By 1908 the Company had moved to London and his partnership with Bowell had been dissolved. From this date onwards Hope-Jones fervently worked to make the Synchronome Master

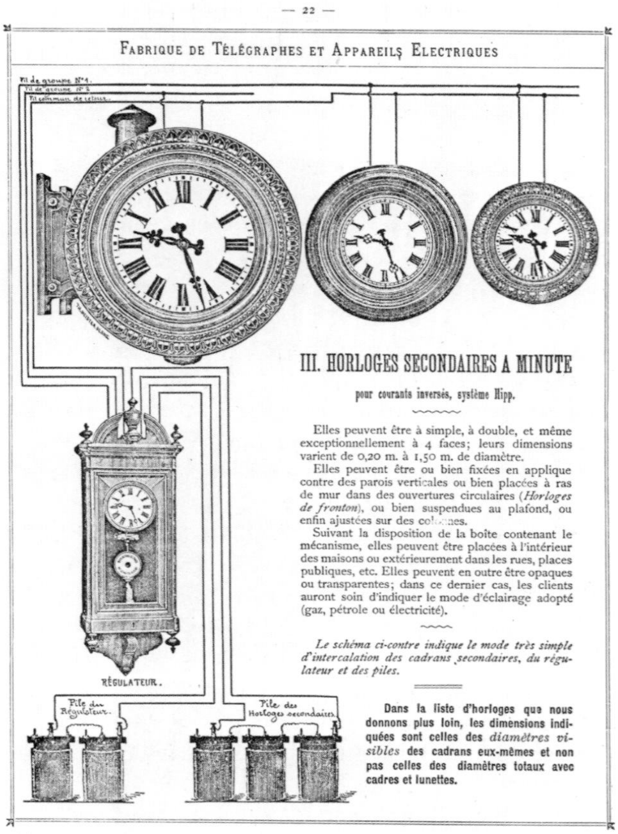

— 22 —

FABRIQUE DE TÉLÉGRAPHES ET APPAREILS ELECTRIQUES

Fil de groupe N°1.
Fil de groupe N°2
Fil commun de retour.

III. HORLOGES SECONDAIRES A MINUTE

pour courants inversés, système Hipp.

Elles peuvent être à simple, à double, et même exceptionnellement à 4 faces; leurs dimensions varient de 0,20 m. à 1,50 m. de diamètre.

Elles peuvent être ou bien fixées en applique contre des parois verticales ou bien placées à ras de mur dans des ouvertures circulaires (*Horloges de fronton*), ou bien suspendues au plafond, ou enfin ajustées sur des colonnes.

Suivant la disposition de la boîte contenant le mécanisme, elles peuvent être placées à l'intérieur des maisons ou extérieurement dans les rues, places publiques, etc. Elles peuvent en outre être opaques ou transparentes; dans ce dernier cas, les clients auront soin d'indiquer le mode d'éclairage adopté (gaz, pétrole ou électricité).

Le schéma ci-contre indique le mode très simple d'intercalation des cadrans secondaires, du régulateur et des piles.

RÉGULATEUR.

Pile du Régulateur.

Pile des Horloges secondaires.

Dans la liste d'horloges que nous donnons plus loin, les dimensions indiquées sont celles des *diamètres visibles* des cadrans eux-mêmes et non pas celles des diamètres totaux avec cadres et lunettes.

Figure 387

Page taken from Peyer, Favarger et Cie c.1890 illustrating Hipp installation.

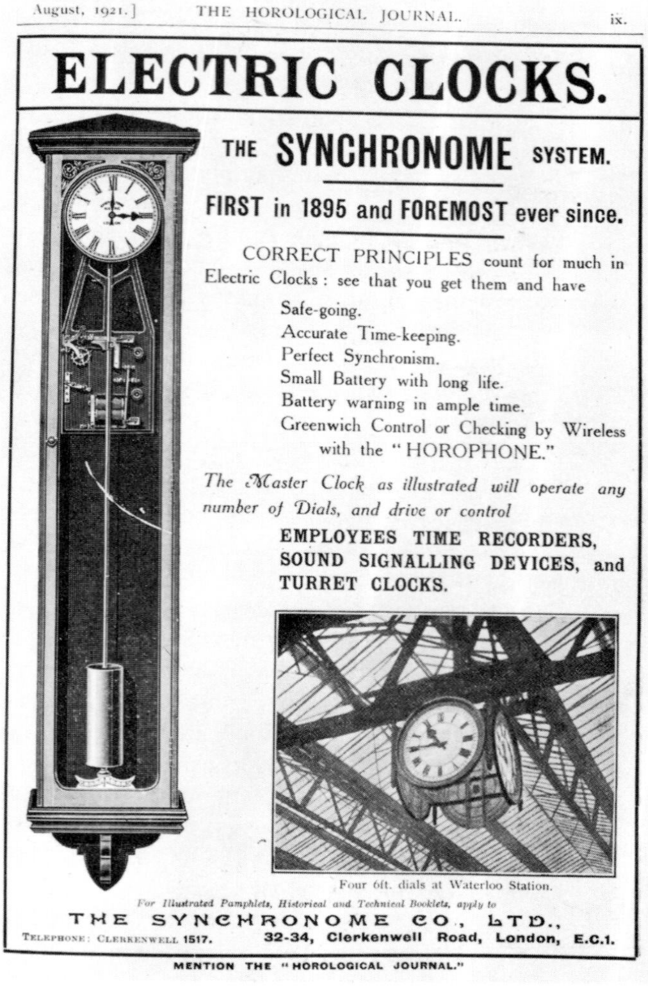

Figure 388

The advertisement by the Synchronome Co. Ltd., appeared in the Horological Journal *in 1920, and showed their earliest case style. In a contemporary catalogue they are offered 'In Polished Oak, Walnut or Mahogany Cases with Glass Fronts' at prices between £6 and £12.12s. depending upon whether they were of the ¾ seconds or seconds variety and whether their pendulum rod was of 'Invar' or 'Invar rod (certificated)'.*

Early examples in similar architectural pedimented mahogany cases £700. Later examples in ugly oak cases £550.

Clock the commercial success it deserved to be. Their installations included shops, government and public buildings, railway stations, etc., both here and abroad. William Hamilton Shortt, originally an engineer on the London and South Western Railway, became interested in precision timekeeping in 1906. Having met Hope-Jones in 1910, he continued his own line of research and patented his own clock in 1911. However, the early examples were not up to the standard Shortt had set himself, but by 1921 he had mastered the problems and the Shortt Free Pendulum Clock was patented (No. 187814). It was these clocks invented by Shortt (by then a Director of the Synchronome Company) and manufactured by the Company that became the standard timepiece for observatories the world over until superseded by the Atomic Clock. It is doubtful if it would be possible to find one of these clocks on the open market — their location is well documented — but there are many examples of the more standard models made to Hope-Jones' specifications to be found.

It is just worth bearing in mind that in the 1930s and early 1940s the Company made available to model engineers the castings, etc., of their clocks on the condition that the end product was for personal use and not for resale, so some examples on the market might not be from the commercial run of the Company. These examples would command a much lower price than those manufactured by the Synchronome Company. Often the cases were not to standard specifications and there would not be a serial number stamped on the base of the Retard/Advance plate on the left-hand side of the movement. Hope-Jones was the most prolific writer on electric clocks at this time. Although his books were basically tracing the evolution of the Synchronome Clock, and therefore omitted many other interesting contemporary examples, they are invaluable reference books for present day collectors. With one exception — *Electric Clocks* — they have the added advantage of having been reprinted which means that those who are not too worried about owning early editions can have a cheaper working copy. The titles of these reprints are: *Electrical Timekeeping, Electric Clocks and Chimes* and *Electric Clocks and How to Make Them*. Both of the later books provided the details of construction for model engineers using the castings provided by the firm. It is no longer possible to obtain these parts in England.

It is possible to date approximately the movements made by this Company. Basically there were three models. The earlier models (approximately before 1930) had a smooth finish to the casting holding the movement and were without a damper for the gravity arm. After about 1930 a 'crackle' finish appeared on the casting and a small damper was added to the gravity arm. Both these points are illustrated in Figure 389, although the damper is somewhat obscured by the pendulum rod.

The third model is that shown in Figure 390, and was an experimental movement manufactured in 1922. Although it eliminated some of the minor faults of the standard design it introduced others and this, together with the fact that it needed expert technical skill to set up and maintain, led to its discontinuation. Only one hundred were produced.

Other types of master clocks can be found with varying degrees of success. The sources range from the large London salesrooms to firms specialising in demolition. Often they were still in use until placed on the market, and so generally speaking the faults are of a mechanical nature, with only minor adjustments to the pawl, count wheel, toggle, etc. The majority of these clocks run on three volts for the master clock and one and a half for each slave dial. It is as well to familiarise oneself with the working of the example in question before assuming it is one of the exceptions and thereby burning out the solenoids with too strong a current.

One of the earliest examples of an English 'domestic' electric clock was that first patented by Herbert Scott in 1902 (Patent No. 10271). Herbert Scott was a Yorkshireman, born in Bradford in 1865, and brother of Alfred Scott the designer of the Scott two-stroke, water cooled engine for motor cycles. Initially Scott had problems finding anyone interested in manufacturing and marketing his clock. Eventually he managed to interest the Every-Ready Electric Specialities Company who were at that time promoting a number of novelty clocks, especially those using batteries. The most intriguing feature of this clock is that the pendulum has the unusual characteristic of swinging back to front instead of the more normal side to side. This, together with the fact that the later examples have the back of the case mirrored, can be most disconcerting. Examples of the two styles — early and late — can be seen in Figures 403 and 404. These clocks were not a great commercial success and production ceased by 1912. The figure of five hundred has been suggested as the quantity produced.

One of the most commercially successful clocks was that manufactured by the Eureka Clock Company

Limited between 1909 and 1914. The inventor of this clock was one Timothy Bernard Powers but the patent (No. 14614) taken out in 1906 was taken out jointly with the Kutnow brothers of New York, USA, the latter being a firm of manufacturing chemists with premises in the Clerkenwell area. Whether it was this tenuous association with horology that led them to set up a small clock factory in 1909 at 361 City Road, London, to commence manufacturing and marketing these clocks is not known. The three years between taking out this English patent and actually commencing production were spent in perfecting the design, negotiating the forming of the Company, acquiring the premises and some preliminary advertising. The Kutnow brothers utilised their expertise in the art of advertising to good purpose and even approached Edward VII, for an audience. They possibly hoped that he would accept one as a gift thereby setting the seal of approval on their product. There is no record of this occurring. By the time production had ceased in 1914 some 10,000 clocks had been manufactured. Clocks bearing

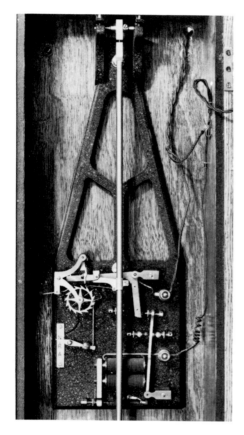

STANDARD PRICE LIST.

These Prices operative on and after 1st MARCH, 1947.

MASTER CLOCKS.

	£ s. d.	P. T. £ s. d.
Standard Master Clock, half-minute impulse with 7″ silvered engraved dial	34 0 9	8 10 3
Standard Master Clock, without dial	29 14 0	7 8 6
Standard Master Clock, without dial or case	28 6 6	7 1 7
Master Clock with seconds switch	53 11 0	13 7 9

The following OBSERVATORIES

have adopted the

SYNCHRONOME FREE PENDULUM

to assist them in their determination of time :—

Greenwich (3)	Lorenzo Marques
Edinburgh (2)	Loomis
National Physical Laboratory	Lick
Sydney	Copenhagen
Adelaide	Warsaw
Melbourne	Batavia
Singapore	Helwan
Cape of Good Hope (2)	Tokio (2)
Nairobi	Kyoto

Figure 389

This is a movement from a Synchronome Master Clock manufactured after about 1930 (assessed by crackle finish to casting and presence of damper to the gravity arm). The serial number is 2388. This appears across the bottom of the plate on the left indicating the rating of the clock. Unfortunately the screw holding this in place makes reading these numbers difficult. It is known that the two clocks with the serial numbers 2386 and 2387 were dispatched in October, 1937, so it has been possible to arrive at a more specific date for this particular clock of probably the November or December of 1937.

Although a later model the price would be enhanced for a collector by the documentation allowing for accurate dating.

£500+

Figure 390

This illustration shows an experimental movement made by the Synchronome Company in 1922. Only one hundred were manufactured and it was discontinued as it needed expert technical skill to set up and maintain the clock.

Through limited production £1,200

Figure 391

This is the movement used by the Synchronome Company Ltd. in their slave dials. The principle was extremely simple and was generally adopted by other manufacturers.

Slave dials are valued according to the design and quality of the case. They can be extremely handsome.

£50+

Figure 392

This master clock has the unusual distinction of having been designed for use with an earth battery (as had the earlier clocks of Alexander Bain). The energy driving the clock is obtained from a zinc carbon couple buried 3-4ft. deep and 1ft. apart in moist soil, which is said to provide a potential of approximately 1 volt at the clock terminals. The example in the Museum and Art Gallery of Leicester ran for forty years without attention on its original installation before being serviced in 1950. A more conventional modern battery can, however, be substituted! The patent for this clock was taken out by Percival Arthur Bentley in 1910 with the Bentley Manufacturing Company being formed shortly afterwards in order to commence their production. The factory was situated at Forest Gate, Clarendon Park Road, Leicester. Unfortunately the outbreak of the First World War halted the manufacture of this clock in September, 1914, but the firm moved to the Queens Road Factory, also in Leicester and undertook heavy commitments to help the war effort. The manufacture of clocks was not reintroduced and the firm was eventually absorbed into the Clore Group. This particular example is in a handsome, well made mahogany case, bevelled glass to the door, and a silvered skeletonised dial and subsidiary seconds dial. One other example is known in a superb shaped mahogany case, while others have been seen identical to that in the illustration but in oak. Two small plates appear below the contact mechanism one carrying the words 'Earth Driven Electrical Clock No. . . . ' while the other states 'Bentley's Mf. Co. Leicester, England' together with the patent numbers and dates (Patent 19044/10, 3236/12 and 8464/13). There are some minor variations in the movements but as these are rare clocks there is no merit in being fussy. In common with many manufacturers, numbering of the movements did not commence with No. 1 and although this example has the serial number of 193, it is doubtful if more than about seventy were ever manufactured. Further details on these clocks can be found in the article entitled 'The Earth Driven Clock' by Dr. F.G.A. Shenton which appeared in the December, 1972, issue of Antiquarian Horology.

Extremely rare clocks and although they do not play such an important role in the evolution of electrical horology they do seem to compete in scarcity and value with a Bain master clock if in a handsome case.

£7,000 — £8,500

a higher serial number than this are generally regarded as having been made from parts disposed of when the firm went into liquidation. Naturally the outbreak of the First World War meant that all their skilled workmen were needed for war work, but the firm had been in difficulties prior to this. If the clocks had a fault it was that they were too well made and, therefore, too expensive. Their stockists included Harrods, Knightsbridge, London, and Asprey Ltd. of Bond Street, London.

A few of the many case styles can be seen in Figures 405, 406 and 410. These range from those consisting of a base to house the battery and a glass dome to cover the movements, through to wall models and Sheraton balloon cases. Quality and style of case have some bearing on the price of the clock. As the main attraction of the clock is the large oscillating vertically mounted balance wheel, examples where this is completely hidden from view tend to attract a lower price than those where it is fully visible. The two exceptions to this generalisation are those in the baloon style case, or that in Figure 408.

There are no dramatic variations in the design of the movements. The first models were 'short', i.e.

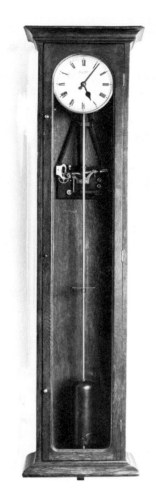

4ft. 2ins. high. A master clock manufactured by Gillett and Johnston Limited of Croydon, a firm world famous for their turret clocks and bell founding. The dial of this example is signed Gillett and Johnston The Croydon Bell Foundry Ltd, Croydon. Further details of the history of the company can be found on page 90. The patent for these clocks (United Kingdom Patent No. 194407) was taken out in the name of C.F. Johnston in 1921 and production has continued to the present day although nowadays not necessarily with mechanical movements. The distinctive feature of these clocks is that the design of the armature eliminates noise and makes it more acceptable domestically. The cases can be in mahogany or various shades of oak. In general any master clock in a mahogany case realises a higher price than one in an oak case. The pendulum is of Invar with a small platform half-way down its length to hold any weights it is necessary to add for precision adjustment.

Mahogany case £850
Late oak case £600

Figure 393b

Movement of Gillett and Johnston master clock.

the balance wheel was mounted behind the dial. Upon realising that part of the appeal to the public was this large wheel the manufacturers rapidly progressed to a 'tall' movement, i.e. the balance wheel mounted below the dial. The short movements however continued to be manufactured to fit the cases requiring this type of movement. All the movements were machined to a high standard, and although some collectors place great store on whether there were two or three small ball bearings in the bearing housing this has no definite dating significance. It would appear that many of the minor variations in the methods of finish could be attributed to several finishers each with their own individual approach.

Although there may be further examples of the clock shown in Figure 413 the location of only a handful is known at present. It was invented by Mr. Thomas Murday in 1908 and this particular example was presented to a member of the firm who manufactured the Murday Clocks — the Reason Manufacturing Company of Brighton — in gratitude for his assistance in a court case. Two years later, in 1910, Murday took out a further patent (No. 1326) in which he replaced the pendulum by a large balance wheel. See Figure 414. This is again an extremely rare clock. It is not known with any certainty as to how many were manufactured but as production commenced immediately prior to the First World War one would suppose that this, apart from any more inherent timekeeping faults would have had a serious effect upon both production and demand. A third fully encased model with balance wheel was also made in extremely small numbers.

Figure 415 shows the 'self wound' clock first patented by Frank Holden in 1909. The same movement can also be found mounted upon an oak base with a small compartment being provided for the battery and covered by a glass dome. The words 'Rebesi' and 'Regina' appeared on the dial, apparently as trade names. A later model with a large helical spring and horizontal oscillating balance appeared in 1923 — patents being taken out in England and France. It is not known where these clocks were manufactured but every indication points to it being France.

The most commonly found electric battery clock is the Bulle Clock. It was first patented in England during 1922. The initial research on this clock had been carried out by two Frenchmen, M. Moulin and M. Favre-Bulle and had reached the stage of a prototype being made immediately prior to the First World War. Both men had naturally become involved with war work and M. Moulin was killed during a military assault. Immediately after the war M. Favre-Bulle patented in his name and the widow of M. Moulin their original ideas. However, there were some technical difficulties still needing to be solved

Figure 394
Good mahogany cased slave dial.
£100

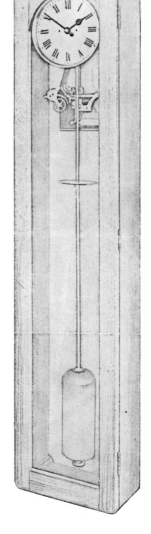

Figure 395
Further example of a late case.
£300

360

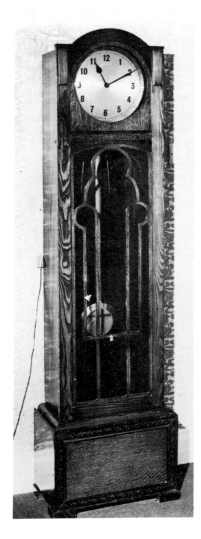

396c

Figure 396d

The side view of the movement of the clock in Figures 396a and 396b shows an arbor which can be displaced axially by a cam and lever arrangement to operate the hour striking train, and at the same time provision is made to increase the rate of striking of the hours by short circuiting a resistance placed in the motor circuit.

396a

396b

Figures 396a, 396b, 396c and 396d

At first glance this would appear to be a typical longcase clock of the 1920s or 1930s in an oak case and with a German movement. Although the latter is correct, the clock has some exceptionally interesting features including a single electrically rewound remontoire. It is rewound by an electric motor operating on 3 volts with a small step down transformer being provided so that this voltage can be obtained from the mains supply. The mainspring of the clock has a reserve capacity of fifteen hours, should for any reason the current be interrupted. The electric motor which is used to operate the quarter striking train also winds up the mainspring every quarter of an hour. Striking is on rod gongs. This clock was introduced and exhibited at the Polytechnic Institute in Northampton Square in London, during December, 1929, and was manufactured under Patent No. 319240 by Kienzle Uhrenfabriken, Schwenningen, Germany (Figure 396c shows their trade mark). Being introduced into this country during a recession, sales appear to have been limited and the design does not appear to have proved commercially viable. The widespread introduction of the mains-synchronous chiming movements about the same time would have sealed its fate. Movement of this clock is illustrated in Figure 396d.

The price would be influenced by whether case was of oak or mahogany and curiously in order to keep this in "period" oak would be more desirable £850+

and eventually, in 1922, Favre-Bulle patented in his name and the widow of M. Moulin their original ideas. However, there were some technical difficulties still needing to be solved and eventually, in 1922, Favre-Bulle took out a further patent alone. Within two years, 1924, the clocks were being marketed in this country by the British Horo-Electric Co., Ltd., whose brass plates can still be found on some of the early models. To date it has not been possible to trace any manufactory in this country although there may have been some assembling and casing of movements here. The main factory was in France with outlets in various countries including Belgium and England. Production continued until the intervention of the Second World War and it is claimed that some third of a million movements were made and that these were housed in at least a hundred different case styles. Some of these can be seen in illustrations accompanying the Chapter. One of the reasons this clock was so successful was that, apart from it being a good timekeeper, Favre-Bulle realised the principal reason for the failure commercially of many of the early electric clocks was the inability of owners to find anyone interested in minor adjustment or repairs. Conventional horologists were prejudiced and hostile and in any case understood little of the principle upon which the clocks worked. Other manufacturers played into their hands by not issuing any repair manuals. Favre-Bulle provided both a repair manual for the retailer and a small leaflet for the purchaser in the appropriate language.

The case style comprising a circular base with the movement covered by a glass dome was to remain constant throughout production, only changing in size and choice of materials. The first models were on a circular mahogany base, 8½ins. in diameter, with the battery housed in a vertical brass cylinder also supporting the movement. Smaller versions of this were soon introduced with the necessary modifications to the pendulum and supporting pillar. Provision for the batteries for these later models was made in the base. The most significant changes came in the early 1930s when more radical alterations in the design and materials of the movement slowly crept in. The wooden bases or cases were replaced by the 'new' material Bakelite, and nickel plated parts took the place of brass. The pendulum

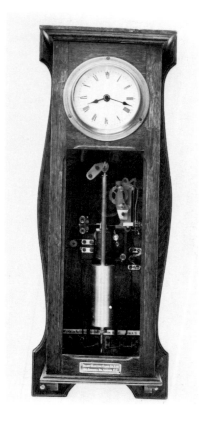

Figure 397

When G.B. Bowell and F. Hope-Jones parted company in 1897, Bowell continued with his researches into electrical timekeeping and took out several patents. However, in 1911, he took out, with H.T.W. Bowell, Patent No. 9287 relating to what was to become known as the Silent Electric Clock. A company was formed at 192 Goswell Road, London, and production of these clocks began. They managed to continue exporting clocks throughout the First World War and it is noted from their advertisements that a large floor standing model was sold to the Argentine Railways in 1919. Production ceased around 1925. The example shown in the illustration is the model used by the Post Office in England as well as many customers abroad. The case is oak, with a two-thirds glazed door and enamel dial in the remaining third. Other examples have been seen in a mahogany case. Note the small name label on the bottom of the door, in some instances the tradename 'Silectock' is shown. The name of the Company was derived from the fact that Bowell had successfully eliminated the usual noise emitting from the slave dials as they received their impulse from the master clock. A further example of a small neat master clock, domestically acceptable and technically interesting so a good collectors' item. Again a mahogany case is more desirable.

Oak £1,200
Mahogany £1,450

Figure 398

An advertisement for the Silent Electric
Clock Co. Ltd., which appeared in 1921,
showing a row of the small master clocks as
seen in Figure 397 and a selection of slave
dials as sent to the Siam State Railways.

Below: No. 1. Standard silent 'receiving'
mechanism as shown in the 1919 catalogue
of the Silent Electric Clock Co.

£40+

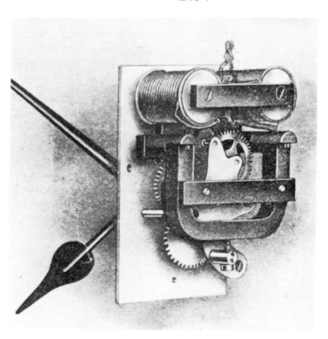

rod became a flat metal strip, a metal disc was used for the bob, and the introduction of cobalt steel meant that the use of a small magnet only ³/₁₆in. diameter and 2ins long could be used.

To anyone contemplating the purchase of any of these clocks, possibly a few words of caution would not be amiss. Beware of an apparently working example which is only doing so with the aid of too strong a battery. They should all run upon a 1½ volt battery; if one stronger than this is necessary it is because there is high resistance in the circuit due to oxidisation of the contacts or joints, or too great a variation in the magnetic field. Both can be tedious to trace and rectify. Quite often a non-working Bulle clock

Figure 399a

This is an example of a French Master clock of the type used in the Eiffel Tower from where signals were transmitted to the Paris Observatory. Unfortunately little has been published in this country concerning these clocks, but it would appear that they were a successful and widely used type of electric movement in France. Apparently invented by the Brillie brothers they were being marketed by 1910 first by them, but later their movements appeared in clocks manufactured by other companies (e.g. Magneta, Vaucauson, etc.). This particular example has the name 'L. Leroy & Cie, 7 Bould de la Madeleine, Paris' which can be traced to Leon and Louis Leroy who traded from this address between 1901 and 1938. The movement is mounted on a heavy white marble slab to provide extra stability; this is enclosed by a mahogany box with a fully glazed front. Other examples have been seen in a gilt brass four-glass case. Note the good quality enamel dial, seconds hand, spheroidal pendulum bob, horseshoe-shaped magnet and coil. These clocks should run from a 1½ volt battery, and the need for a stronger current indicates that the coil has, in all probability, been damaged by the earlier use of a strong battery. Although not rare in their country of origin, these clocks are not found in this country in any great number. This, together with the fact that they are aesthetically pleasing and of a compact size, would enhance their price.

£550+

399b

View of an identical movement but showing a fully glazed case with Electrique Brillie on the dial together with the name of the retailer.

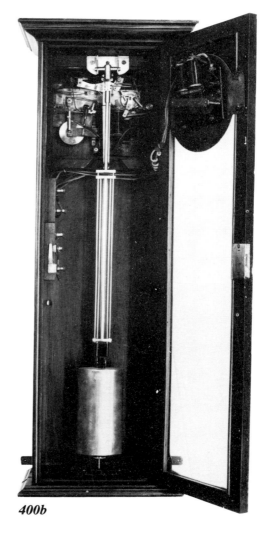

400a

400b

Figures 400a, 400b and 400c

Robert Mann Lowne, the inventor of the highly successful Lowne Electric Clock System, was born in 1840 and died in 1924, and during these eighty four years took out some eighteen patents on a variety of subjects. Further details can be found of these in the article entitled 'Robert Lowne and his Electric Clock System' by R.K. Shenton that appeared in Antiquarian Horology *for March, 1975. His first patent concerning electric clocks was abandoned but a more successful one (No. 25374) was filed in 1901. By 1903 the Lowne Electric Clock and Appliances Company Limited had been formed and in that year he obtained a contract from H.M. Royal Arsenal, Woolwich, for a master clock to which, connected in a series on a circuit nearly six and a half miles long, were forty six slave dials. It is understood that this installation ran for some thirty years before being replaced. Further patents were taken out for minor improvements. The Company continued after the death of Robert Mann Lowne under the directorship of his two sons, but competition in the 1920s and 1930s was extremely keen and the last Lowne master installation was completed about 1933 although a maintenance service was continued on past this date. They did manufacture a synchronous electric clock, and carried out war work for the air Ministry during the Second World War, but after this they gradually increased their production of anemometers and finally in 1960 changed the name of the company to Lowne Instruments.*

The example shown is typical of their master clocks. Although examples have been seen with a square painted dial on the clock it is more usual to find it serving merely as a transmitter with an accompanying

£1,900+

slave dial nearby. First impressions of the movement of these clocks cause speculation as to whether they ever worked, but as illustrated by the quoted installation at Woolwich they did and extremely successfully. The cases are of oak, although a few do appear in mahogany, the pendulum has been designed, as has all the movement, with an eye to serviceability and cost. The brass covering on the lead pendulum bob merely encloses that part of the bob which is readily visible. These are rare clocks and most certainly not for the amateur electrical horologist to become entangled in before reading the little information that appears on these clocks in the book Electrical Horology *by H.R. Langman and A. Ball, or* Science of Clocks and Watches *by A.L. Rawlings. These clocks should work quite satisfactorily from a 3 volt battery plus 1½ volts for each slave dial.*

Possibly only appreciated by a connoisseur of electrical horology who would realise its true collecting value. Perfect going examples are rare and therefore more expensive.

£1,000 if going example

Figure 400c

The illustration shows the movement of the Lowne master clock seen in Figure 400a.

needs nothing more than the battery connections reversed in order to activate it. A feature of its magnet is that it has consequent poles, i.e. both ends have the same polarity, while the centre has the opposite polarity. Usually the ends are South with the centre North, but examples have been seen in which this has been reversed. Another minor repair that might be necessary on the pre-1930 models concerns the suspension. The earlier models had silk suspensions which until quite recently it was possible to purchase, but the supply has now ceased. It should be quite simple to replace this with a small piece of lingerie ribbon (and not as has been seen a piece of sticky tape).

It should be possible to regulate any of these clocks to be quite reasonable timekeepers, although there is always the 'rogue' that will not bow to logic.

Replacement shades are no great problem when needed for the small Bulle or Holden, but any required for the larger clocks do create a problem and incur considerable expense.

Figures 401 and 402

The next two illustrations have been included as a cautionary point. The example in Figure 401a appeared in a sale at Christies on the 22nd June, 1977 and subsequently sold for £90. A remarkable amount in view of the fact that it was only a slave clock and not as must have been surmised a rare electric clock complete in its own right. Alternatively of course the name Paul Garnier that appears on the dial is associated with prestigious carriage clocks, etc. and it may be that the case and dial will appear next with a mechanical movement! Paul Jean Garnier was born in 1801 and when aged 46 took out a patent for an electric master clock system. Two years later in 1849 his electric clocks were exhibited at the Paris Exhibition and received a Gold Medal. The firm was taken over in 1937 by Leon Hatot.

£150+

401a

Figure 401b

View of clock shown in Figure 401a (mahogany bezel).

367

402a   *£120+*

Figure 402b

View of clock shown in Figure 402a ('beaded' spun brass case).

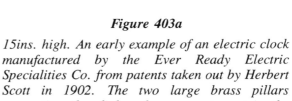

Figure 403a

15ins. high. An early example of an electric clock manufactured by the Ever Ready Electric Specialities Co. from patents taken out by Herbert Scott in 1902. The two large brass pillars supporting the dial and movement contain the batteries. This model on a lacquered wooden base under a glass shade was superseded by that shown in Figure 404. Original dome and base important.

£1,500+

Figure 403b

Back view of the Scott Electric Clock as shown in Figure 403a. The same movement was used in the later model as shown in Figure 404.

Figure 404

"The "Every Ready" Electrically Propelled Clock.
Specifications

Height: 17ins.	Width: 11ins.	Depth: 7½ins.

Base: Solid Mahogany, Oak or Walnut.

Cover: Front and Sides — best plate glass, bevel edge.

                     Back — Mirror of best plate glass.

Dial and Hands: Plain bold designs

Pendulum: Constructed of special alloy, which is practically insensitive to temperature changes.

Motion Work: Accurately machine cut on highest class precision tools.

Price: £5. 5s. 0d. complete Refills 4s. each"

This is the official description issued by the Every Ready Specialities Company of the clock manufactured by them from patents taken out by Herbert Scott in 1902.

The example in the illustration has a mahogany base and the name of one of the companies retailing the clock at one time — London Stereoscopic Co. 106 & 108 Regent St. London. — appears on the dial.

Other examples have been seen with a compartment in the base for housing the two batteries necessary for powering the clock.

Much of their value would depend upon undamaged case.

£1,850+

Figure 405

10½ins. high. The movement of this Eureka clock is of the 'short' variety, i.e. the balance wheel is behind the dial. The porcelain dial has been skeletonised in order not to hide completely the oscillating balance wheel. The base is gilt metal with the movement covered by a glass dome. The details of the manufacturer, etc., appear on the small brass plate immediately above the figure 6. See Figure 410b for further details. This is one if not the first model produced by the company.

Original dome vital

£650+

Figure 406

13ins. high. The movement of this Eureka clock is of the 'tall' variety, i.e. the balance wheel is visible below the porcelain dial. The seconds dial is an extremely unusual feature. The mahogany base houses the 1½ volt battery.

Without seconds £850
With seconds £950+

Figure 407

13ins. high. This is the rear view of a 'tall' movement on a rectangular thick brass base covered by a glass shade. Note the two balls clearly visible in the bearing housing behind the glass end plates. Impossible to find substitute shades if original is broken.

£900+

Figure 408

This is one of the most elegant case styles used by the Eureka Clock Co. Ltd. Although not an early model the shape of the case necessitates the use of a short movement. In this particular example the dial has been skeletonised but other models have been seen with a full porcelain dial. As the value of the former is slightly higher than the latter it is not unknown for examples that originally had a full dial to be skeletonised. Examples that started life skeletonised have the name plate of the company visible from the front, whereas those with full dials display this on the back of the movement. Note the three balls in the bearing housing behind the glass end plates.

One of the more elegant examples.

£850+ with examples with skeletonised dial being more desirable.

Figure 409

This is an example of a clock manufactured by the Eureka Clock Co. Ltd. in a mahogany case which, although pleasant, completely hides the movement. In this instance this would detract from the value of the clock. Note the '1,000 day electric clock' below the name of the Company — this is referring to the estimated duration before the battery needs changing.

Non-visible movement, so price is confined to £500+

Figure 410b

An enlarged view of the small brass plate found on many Eureka models with a rating star and scale on the lower section and the name of the company, address, patent number and date together with the serial number of the clock.

(For further reading on the Eureka see The Eureka Clock *by Dr. F.G.A. Shenton).*

Figure 410a

A Eureka movement housed in a wall clock is unusual and was probably intended for use in offices and factories rather than domestically.

Unusual example valuable to a collector.

£750+

Figure 411

One of the more unusual and rare case styles used by the Eureka Clock Company. Whether only a few were manufactured or whether some have been cannibalised to provide cases for mechanical movement is not known.

The movement in this example is of the 'short' variety with the Company name, patent number, etc. engraved on the dial. The centre of the dial is enclosed by a circle of thick bevelled glass.

Through rarity and originality of casing £1,200+

Figures 412a, 412b and 412c

Although not readily found in England, the Tiffany Never Wind electric clock is one of the more common American varieties. It would appear that the application for the patent relating to this clock was made by George Steele Tiffany of New Jersey, USA, in 1901. Two further patents were taken out in 1904, which was about the same time that these clocks began to be widely advertised. The earlier models were cased in a wood or brass case with a door back and front. The movements were impulsed on both the left and right turn of the torsion pendulum, the bob of which consisted of two balls (see Figure 412a). Later models produced after 1911 were on a brass circular base as illustrated in Figure 412b and covered by a glass dome. The movement was impulsed only on one turn of the torsion pendulum and the bob consisted of two inverted cups (see Figures 412b and 412c). The clocks were also made under licence and so the names on the dial can be those of the Cloister Clock Company, Tiffany Never Wind Company, Niagara Clock Company, etc. The example shown in 412b and 412c was made by the Cloister Clock Company of Buffalo, New York, U.S.A. Note the thin spun brass base and press brass frame and supports.

The Tiffany 'Neverwind' Clock, Description, Repairs and Adjustments by A. Amend NAWCC Bulletin Vol. XXIV No. 5, October 1982, pages 494-505.

£700+

Figure 413

19ins. high. An extremely rare example of an electric clock patented by Murday in 1908 and manufactured by the Reason Manufacturing Company Ltd., Brighton. The mahogany case has decorative brass work and ball-and-claw feet. Has original documentation of provenance as gift to employee.

£1,750 if movement enclosed, i.e. non visible although this particular example would realise more in view of documentation
£2,500 if movement fully visible

Figure 414

14ins. high. An example of the other clock manufactured by the Reason Manufacturing Company Ltd., Brighton, from a patent taken out by Murday in 1910. The mahogany base houses the 4½ volt battery, while a glass dome covers the movement. A few examples were made with clear glass skeletonised dials, but the majority of known examples have glass dials backed by white card. This is held in place by a brass retaining ring.

Example known to have realised £1,000+ in 1977, and by 1983 £2,000+.

£4,000+

Figure 415

This example of a clock patented by Frank Holden in 1909 is in a gilt metal four glass case with silvered dial. The alternative case style is shown in Figure 416. The movement in both models is identical. In this example the coil travels back and forth over a flat magnet (rectangular in cross section). This coil was easily damaged and provision was made for securing it in transit by a small screw through the pendulum bob that tightened into the centre pillar that supports the movement.

£1,500

Figure 416

View of the movement of a similar clock to that shown in Figure 415. This model has a square oak base (the two 1½ volt batteries are housed within), and circular brass ring to retain the glass shade in position. Not so aesthetically cased as previous example but nevertheless a collector's piece.

£1,000

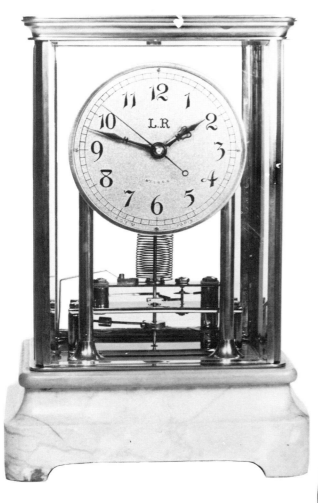

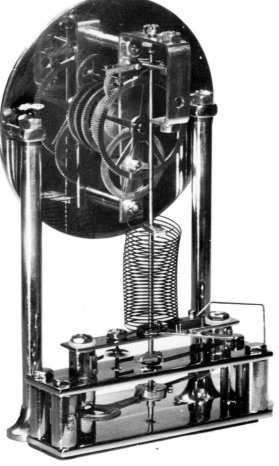

Figures 417a and 417b

This later concept was patented in 1923. Here the coil oscillates between the two poles of the magnet and acts as a balance under the control of the large helical spring. This is an extremely attractive movement to watch in motion. The base of the case is onyx. Very rare.

£1,500+ if in mint condition

Figures 418a and 418b

14ins. high. Example of an early Bulle movement on a circular mahogany base with the battery housed in a central brass pillar. The dial has been skeletonised to provide an additional visual feature.

In excess of £300; a missing dome would detract from this value.

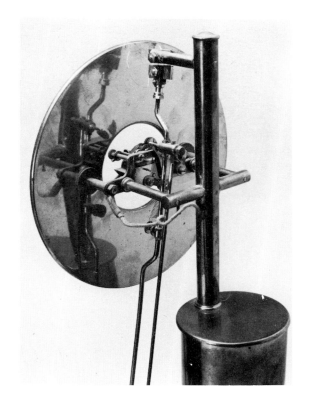

Figures 419a and 419b

10ins. high. An example of a smaller Bulle clock on a circular mahogany base, which holds the battery. A glass dome covers the movement. The dial is silvered. Note the clip which holds the pendulum steady during transit. It is rare to find an example which has retained this.

Mahogany base with brass finish £200+

Figure 420a

10ins. high. This is a Bulle clock made after 1930. This particular example was manufactured by Exide under the trade name of 'Tempex'. Note the Bakelite base, and disc bob.

Late example — mahogany base £150+
Bakelite base £75+
chrome finish

Figure 420b

View of the movement of the example shown in Figure 420a. All brass uprights, etc., have been replaced by metal plates. Note the smaller magnet.

Your Customers will look for these

"OLYMPIC" MODEL

Size 8″ × 7″ × 3″

Oak case, chromium plated bezel and movement, retailing at - - - - : - **50/-**

Same model without chromium bezel - **40/-**

MODEL No. 263

Size 7″ × 6″ × 3″

Jacobean Oak, chromium bezel, retailing at - - **45/-**

Be sure that your Bulle Clocks have the name

Figure 421

Two of the models exhibited at the 1934 Daily Mail Ideal Home Exhibition. The fully enclosed style shown in the bottom right hand corner is generally regarded as being one of the least desirable, i.e. no sight of the movement.

£100 — £130   **£100 — £130**

Figure 422

20ins. high. An exceptionally attractive cased Bulle clock. Brown and white marble top, bottom and columns decorated with brass reliefs and bronze lion battling with a snake enclose a conventional four-glass case. The movement is the standard type used in many of the wall clocks manufactured by this Company where the large central pillar has been replaced by a slender more aesthetic column and the battery is now housed in the base. The skeletonised dial is of gilt metal.

In the catalogue produced by the manufacturers in 1926 this example is designated: 'Modele S-Lion au Serpent, marbre skyros, sujet bronze cisele (signe Aubert). Hauteur 51cm. Largeur 32cm. Longueur du balancier 24cm. Prix unique: 1,800fr.'

Obviously originally one of the more expensive styles in the range produced. Comparative price is difficult as the whereabouts of only one other is known.

£1,750+

MODELE EE. — Grand œil de bœuf, type chemin de fer, tout en métal, décor vert antique, filets or ou laiton verni or, nouveau montage sur fond aluminium fondu.
Diam. total 41 cm, Diam. cadran 32 cm - Long. du balancier 24 cm
PRIX UNIQUE : 345 fr.

Figure 423

Four pages from the catalogue issued by the Bulle Clock Company in 1926. The rate of exchange at this date was 152 francs to £1.

Models EE and C: Rarity value would make these £150 — £200 plus. Models L and M: £100 — £250 depending on case.

Œil de bœuf lunette chêne naturel
MODELE C. — 9 pouces   MODÈLE C. — 12 pouces.
Diamètre total : 40 cm.   Diamètre total : 50 cm.
Diamètre du cadran : 24 cm.   Diamètre du cadran : 32 cm.
Longueur du balancier : 24 cm.   Longueur du balancier : 24 cm.
PRIX UNIQUE : 250 frs.   PRIX UNIQUE : 300 frs.

MODELE L. — Régulateur, 4 glaces bisautées, montants bronze doré, socle et chapiteau marbre vert de mer, griotte ou skyros. Article très riche.
Haut. 57,5 cm. — Larg. 24 cm. — Long. du balancier : 24 cm.
PRIX UNIQUE : 780 fr.

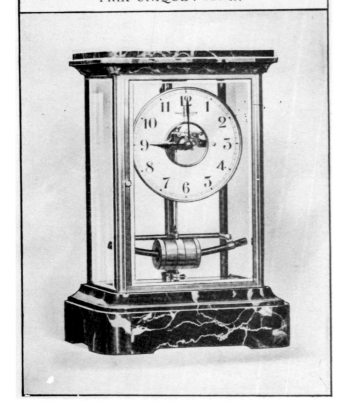

MODELE M. — Borne marbre, façon extra soignée, avec jonc bronze doré, skyros, portor, vert de mer, onyx du Maroc.
Haut. 40 cm. Larg. 26,5 cm. — Long. du balancier : 24 cm.
(La même en forme ogive)
PRIX UNIQUE : 1.430 fr.

Figures 424a, 424b and 424c

Although the Ato Clock Co. Ltd. was not registered in this country until 1928 to "carry out the business of manufacturing of and agents for clocks, electric and otherwise", Ato clocks had been manufactured for above five years by the Etablissement Leon Hatot in France to patents first taken out in 1923. Marius Laget the inventor had also been concerned with the Bulle clock. At first glance the Ato is remarkably similar to the Bulle clock but whereas in the Bulle the coil forms the pendulum bob, in the Ato it is the magnet that serves this role. The Bulle has a silk ribbon suspension while the Ato's suspension is made of double steel links. There were two basic movements — one with a half and the other with a quarter seconds pendulum. The quarter seconds movement being made in two styles, i.e. 'tall' or 'short'. This greatly facilitated the manufacture of a wide range of case styles. By 1929 the catalogues contained nearly fifty different models ranging from four glass cases through to fully encased movements in wood, Bakelite or even Lalique and the inevitable wall 'regulator'. The quality and material of case is often the major factor influencing the price of these clocks. For example Lalique glass is highly prized and so it is somewhat unfortunate for the collector of Ato clocks when one appears in such a case! Bakelite is also a collectable commodity in today's market place. Laget took out a French patent in 1953 for an Ato clock incorporating transistors. Clocks incorporating this design appeared on the market in 1955. They were of horological significance as this was the first time transistors had been applied to electrical timekeepers. Late examples of clocks using the Ato design were made under licence by other companies including the Hamburg American Clock Company.

£300+ depending on date and case style

Figures 425a and 425b

This is a French electrically rewound clock manufactured by Soc. Anon des Horloges Electric — Silentia à Besançon. An English patent (Patent No. 1039) was taken out in 1912. Both case and movement were manufactured and finished to a high standard — indeed the whole concept of the clock was advanced for such a timekeeper at this date. Little is known of the manufacturers but it would appear that only a small production run was made. As can be seen from the illustration of the movement in Figure 425b the case is far larger than necessitated by the actual movement and battery.

£1,500+

Figures 426a and 426b

A similar movement was exhibited in the Chronotome Exhibition held at Chaux de Fonds, Switzerland in May 1978. It was included as an example of the earliest successful use of an electrically maintained spring balance. At each vibration a contact controlled by the balance closes the circuit of a motor coil giving a sustaining impulse to the balance which carries a magnet. Patents were taken out in this country in 1893 by J. Cauderay with the clock being manufactured in France. The movement is held in place by three small brackets while two highly decorative silk covered coiled connecting wires go down to the two terminals in the base of the case.

Only a few known examples hence would command a good price of £1,500+

The battery electric clock shown here has a very interesting mechanism not only in the contact system employed but also in the use of a torsion pendulum in which the balance is approximately half way down a strained torsion wire. It has not been possible to trace details or dates of manufacturer and any help that readers could give would be much appreciated.

It is interesting to note that the name shown in the close up of the movement shown in Figure 427c is Telavox. The Telavox system was invented in 1927 by R.S. Wensley of the Westinghouse Electric and Manufacturing Company.

£200+

Figure 427b

View of the movement in situ in case. In this instance the case is far larger than the size of the movement necessitates and hence the manufacturers were not attempting to provide a small mantel clock.

Figure 427c
Close up of movement.

Figures 428a and 428b

This small battery clock was also marketed as a kit obtainable in Belgium ''L'Heure Electrique EXA pour le petit horloger electricien amateur'' between the two World Wars. The base is of Bakelite. An interesting curiosity and demonstrating the great interest in electric clocks shown at this time on the Continent, by the general public.

£250+

Figure 429a

Inlaid mahogany lancet style case, electrically rewound mantel clock with silvered dial signed Patent Moeller.

The original patents were taken out in 1899 by Max Hoeft for what was basically a mechanical clock with the spring being rewound by an electromagnet. Later patents were taken out in 1901 relating to methods of closing the circuits and the prevention of overwinding as well as the provision of powering a striking train in the same manner. In 1907 patents were granted for modification to the winding mechanism and for adapting conventional striking mechanisms for powering by electric current. These last four patents (Patent Nos. 10960 and 10961 in 1901 and Patent Nos. 9240 and 21623 in 1907) were all taken out by Max Moeller, Altona, Germany. Very few of these clocks have been seen and production dates or numbers are not known. It would be logical to assume that the outbreak of the First World War would have seriously interrupted if not caused cessation of manufacture. Those seen in this country appear in high quality cases with good well finished movements. Both movement and dials were mounted on a small seat board which slid out freely once the battery housed in the lower section of the case had been disengaged. A few examples reported from the United States of America however are of a completely different design and are in the style of a somewhat rococo skeleton clock on a base under a glass shade with pendulum and signed "Perpetual Moeller's Patent".

£1,750+

Figure 429b

View of the movement of Moeller self winding electric clock. Example would appear to comply with the details shown in the patent.

15ins. high. This well proportioned skeleton clock on a mahogany base under a glass shade has brass plates typical of those used in 1880/1890. It has, however, in place of the usual fusee movement an electrically rewound mechanism as patented by Chester Pond of New York, USA, in 1884. This system is commonly found in clocks marketed by the Self Winding Clock Company with movements manufactured by the Seth Thomas Clock Company of Connecticut, USA. It is both interesting and significant that in 1887 the manager responsible for the United Kingdom marketing of these clocks was Mr. Lund, who was later the founder of the Standard Time Company. It is not surprising that a few of these clocks were produced as a special line to demonstrate the 'new' electric clocks. These clocks proved to be very reliable with a remontoire hourly rewind which operated on 3 volts.

The name 'Wheatley Carlisle' that appears on the dial has been traced as referring to Wheatley & Sons, 65 English Street, Carlisle who exhibited at the Franco-British Exhibition of 1908.

Beware of amateur attempts to 'electrify' a standard skeleton movement. Only commercially produced examples have any value to a collector.

£2,000+

Figure 430b

Movement of clock shown in Figure 430a. Note the three-pole motor which rewinds a small remontoire mounted on the centre arbor of the movement, to which is attached a contact mechanism which operates once an hour.

*Reform movement manufactured by Schild &
Co. In this instance the trade name of the
clock is "Keynote". The case is of bird's-eye
maple with provision for housing the battery
in the base (foot) of the case. Note the quality
finish to the movement as seen in Figure
431b. The 'hand' is for fast/slow regulation
— the correct style originally used can be
seen in the materials for Reform clock,
Figure 433.*

**£150 depending upon case style and
condition**

Figure 431b
Reform electric clock movement.

Figure 432

1938 advertisement of the Reform clock as manu-factured by Schild & Co.

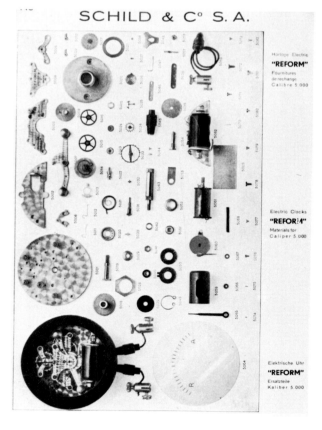

Figure 433

Page of materials for the Reform electric clock as it appeared in 'La Classification Horlogere' by Jobin — a Swiss material catalogue published c.1939.

Figures 434a and 434b

The movements of these small travelling battery electrically rewound clocks proved to be extremely reliable and found their way into a number of casing variations. This particular example is in a 4ins. high red leather case with compartment at the back for flat battery (4½ volts). Gilt finished dial has luminous numerals and hands. Signed PerPeTua — Electric. The movement (Figure 434b) would appear to be to the design patented in 1929 (Patent No. 126862) by A. Schild SA, Grencen, Switzerland, which appears to be a modified version of a patent taken out the previous year by Schild & Co., 137 Rue du Parc, La Chaux de Fonds, Switzerland. (See preceding illustration as manufactured by Schild under the name ''Reform''.) These movements were also marketed in Bakelite cases by the Riverside Manufacturing Company of Hammersmith, London, with similar examples being used extensively in the car manufacturing trade for dashboard clocks, the electric energy being supplied through the car battery. They were later developed for use in the delayed action mechanism of marine mines.

Example illustrated in mint condition   £95+

The Synchronous Electric Clock

The synchronous mains clock was basically an electric meter measuring the amount of current as it passed through and recording it on a dial. Alexander Bain in the 1840s had foreseen the possibility and Charles Wheatstone's impulse dial was actually a synchronous electric motor driven from the alternating current generated from his master clock. Renewed interest had been stimulated by Dr. S.Z. de Ferranti (1864-1930) demonstrating the advantages of using alternating current as opposed to direct current. Ferranti Ltd. of Hollinwood, Lancashire were at the time one of the leading manufacturers of electrical components.

Henry E. Warren an American engineer from Ashland, Massachusetts, had been interested for many years in the use of electricity in timekeeping. In 1912 he founded the Warren Clock Company to make and retail battery clocks but continued his researches into the possibility of a synchronous self starting clock using alternating current. His first successful model was demonstrated at Boston on the 23rd October, 1916. Patents were taken out and production commenced with the new range marketed under the name of 'Telechron' (Time from a Distance). However these clocks were not a practical proposition in this country until 1927, when the introduction of the National Grid assured a standard alternating current of 50 cycles.

Within a few years of their introduction claims were being made in the trade journals that the synchronous mains clock had completely revitalised the dying horological trade. In 1939 a well-known jeweller of the day reported that at least sixty to seventy-five percent of his total clock sales were of the 'new' electric clocks. A section devoted to these clocks was carried each month in the *Horological*

Figures 435a and 435b

The Synclock synchronous movement was one of the earliest clocks of this type to be marketed in this country. They were of the Warren type — gearing in oil filled bath, etc., and made by Everett, Edgcumbe & Co. Ltd. As the motors were of the self starting variety any interruptions in the electricity supply could occur without being immediately noticeable. To draw attention to the fact that the clock had actually stopped and restarted thus losing time, a small red indication disc appeared in the aperture immediately below the figure 12. Figure 435a shows a few of the case styles while Figure 435b shows a view of the encased movement. Case style and materials and also with/without seconds would influence value.

£60+

Journal. Advertisements emphasised their versatility and recommended "one for each room", authorative technical articles appeared in periodicals and as these clocks poured from the factories in their millions it did indeed look as if the spring driven mechanical clock would shortly be obsolete.

These clocks fell into two main categories — those that were self starting and those that were in need of manual assistance. Those that were self starting frequently had an indicator disc fitted in the dial to show when any interruption of electricity supply had occured. The merits and demerits of both types were fiercely contested with some people preferring their clock to stop completely rather than show the incorrect time. From the customers' viewpoint this feature and case style were the only deciding factors that influenced their purchase. The manufacturer, however, needed to overcome production problems and to cut manufacturing costs. The variations that occur even in one maker's product are therefore wide and to the enlightened few of great interest. Admittedly this cannot be called a purely horological interest but rather that of a horological social historian! By 1932 there were many models of synchronous clocks on the market, some utilising high speed motors with inductive starting, others had slow speed motors again with inductive starting. Examples with two-pole fields and two-pole rotors can be found while others had two open subdivided fields and multi-pole rotors. These clocks were part of a major step

Figure 437

Ferranti synchronous movement.

Figure 436

This is an example of a small synchronous mains electric clock manufactured by Hamilton Sangamo; this Corporation of Springfield Illinois, USA, according to Chris H. Bailey in his book Two Hundred Years of American Clocks and Watches, *"produced an expensive electrically wound clock about 1928 and did not survive the Depression". It is not known if this also applied to the subsidiary in this country.*

This particular model has a solid mahogany case, veneered and inset with circular movement, in a metal casing. The silvered dial has a central rotating disc which shows seconds. One of the problems with these early mains clocks was to convince the customer that they were working and to provide a way of detecting when variations in the current had caused the clock to stop. This was doubly important as many examples had to be manually restarted. The moving seconds dial and pointer in this instance providing the necessary visual assurance. Other makers left an aperture in the dial through which a moving line or series of dots could be seen while the clock was operating. Note the original flex and wooden plug.

*Interesting examples are bound to rise in price, as they form part of the history of electrical horology. As an early named example, would realise **£95+**, this would be a worthwhile acquisition.*

forward in the search for a cheaply produced reliable timepiece, as they required no winding and little maintenance. One of the disadvantages, however, was trailing wires — although some of the houses built at this date made provision for clocks running from the mains by means of a small two-pin socket in the wall immediately above the centre of the fireplace. The main problem was their complete dependency on the constancy of the electrical current and as this became more and more unreliable through first the war and then post-war difficulties they became troublesome to their user. Thus the way was admirably paved for the battery clock. It was self contained and could be moved from room to room as had the mechanical clock and the manufacturers could design strange and exotic case styles *ad nauseam*. The synchronous clock is becoming a collector's item for several reasons. As interior decorators turn to this period for inspiration and converts seek authentic pieces to match the rest of the decor; as the introduction of quartz crystal movements expands and mechanical clocks increase in value and, lastly, as the collecting of items using early plastics increases so will the desirability of these clocks. With regard to the last point, there is a rising interest in all of the early types of plastics, i.e. Bakelite, etc., and as many of the cases for these movements were made in the 'new' materials it will not be just the horologists that make selective purchases from this range of clocks. Purchasers do need to be selective as it is only early examples or those with any technical interest that deserve to be preserved for posterity.

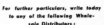

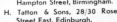

Figure 438

Selection of synchronous clocks being marketed by Ferranti Ltd. in 1933. The synchronous alarm was added to the range in 1935. They also made a model for fitting to the grille of the Ferranti radio. A number of manufacturers included such a style in their range as it must be remembered that reception was not continuous in the early days of radio, valves needed a preliminary warming up period and it was only too easy to miss a broadcast. See Figures 466 and 467 for examples of timing mechanisms especially for use with early radio sets.

£65+

397

Ferranti Synchronous Timepiece, in Bakelite Case.

Ferranti Synchronous Alarm, showing Dial and Case with 24-hour Setting Dial.

Ferranti Synchronous Timepiece, in Hammered Pewter Case.

Ferranti Synchronous Alarm, view of movement.

Ferranti Synchronous Timepiece, cased in Rare Woods.

£65 — £100

Figure 439c

Diagram of the Goblin movement.

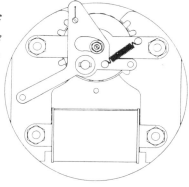

Figures 439a, 439b and 439c

The British Vacuum Cleaner and Engineering Company were the manufacturers of the well known Goblin vacuum cleaner. In late 1937 they decided to enter the synchronous clock market and produced the movement seen in Figure 439c. They were at this time working from premises in London but by November 1939, had moved to Leatherhead in Surrey to what was then described as 'Britain's Newest Clock Factory'. At this time they were also manufacturing under licence the Magneta clock and master clocks for use in the Post Office telephone exchanges. Figure 439b shows a range of the cases produced in 1939, i.e. glass, mahogany or walnut or Jacobean Oak with prices between 60/- and 71/6. Note the trademark — A Goblin against a Union Jack flag with the words around the perimeter "Made in England by the Originators of Vacuum Cleaning GOBLIN". Capricorn Model (introduced 1937) embodied a barometer, thermometer and clock (see Figure 439a). Combinations such as this or alternatively the clock as part of a lamp stand or wireless were not uncommon. In the latter case of course it saved the necessity of further plug etc., as the lamp or wireless would already be attached to the electricity circuit of the house.

The Goblin manufacturers also made a alarm mechanism which boiled the water, made the tea, lit the lamp and awakened the sleeper. This came to be known as the Teasmade. In fact this name has come to be used extensively even when referring to a similar mechanism manufactured by another company!

The Company had ceased production of their clocks by the 1950s.

£70+

399

The Temco mains electric clock was made by the Telephone Manufacturing Co. Ltd., Hollingsworth Works, London S.E.21. with sales being conducted by T.M.C. — Harwell (Sales) Ltd., Britannia House, 233 Shaftesbury Avenue, London W.C.2. Cases were of chrome, and chrome and Bakelite (see Figure 440a). Advertised features of their clocks were high torque easy starting motors, self lubricating phosphor-bronze bearings, and carry-over device to overcome momentary interruption of current.

Figure 440a

Chrome and black bakelite cased example of synchronous clock made by the Telephone Manufacturing Co. Note arrow showing through aperture below figure 12 to indicate continuous functioning (or non-functioning!) of movement. A necessary feature in a period when the electricity supply was not without interruption.

£75+

Figure 440b

View of casing to movement with pressed lettering. Note directions for manual start.

Figure 440c

Movement in use in 1934.

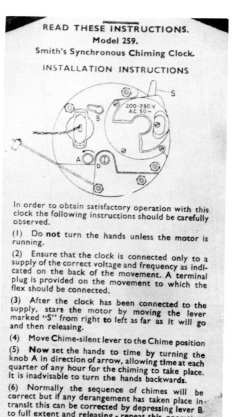

Figures 441a, 441b and 441c

With present furnishing trends these square cases are considered to be more desirable than the more curvaceous styles. The movement is a synchronous chiming movement by Smith's. Figure 441b clearly shows the movement and rod gongs, while Figure 441c is a copy of the Installation Instructions pasted on the door of the case. The movement is of a particularly compact design. The striking mechanism is driven at the appropriate times by being moved into gear with the main driving mechanism. This concept was the subject of patents taken out in 1931 and 1932.

£50+

Figure 442

This is a further example of an early mains electric clock, but this model was manufactured and marketed by Smith's English Clocks Ltd. of Cricklewood, London. It has a veneered case, silvered anodised dial with an aperture showing seconds placed above the 6 position to confirm that the clock is running. The trade mark of this Company being the letters 'SEC' although the dial on this example is inscribed

SMITH ELECTRIC

£30+

Figure 443

Two views of the Callboy Alarm clock marketed by Smith's English Clocks Ltd. in 1934. The case was of Bakelite. The movement consisted of a self starting synchronous motor, time train with the additional features of a contact making device, a switch, an alarm sounding coil and a buzzer tongue.

Figure 444

Japanese synchronous clock in Bakelite case. Similar clocks were being marketed in this country in 1937. The Japanese clock and watchmaking industry had resulted from the need to utilise the factories, machines and skilled technicians that had been employed in the ordnance business during the First World War. By 1923 they were producing 201,218 clocks, while in 1929 1,232,269 standing clocks and 506,504 hanging clocks are recorded. Naturally the industry was affected by the depression but with the crisis of the Manuchurian Incident in 1933/1 the demand for armaments increased and the factory largely returned to its previous occupation. In 1937 the watch factories were still preoccupied with the manufacture of time fuses, range finders, etc., and from 1934 little detailed information was released concerning the horological industry in Japan. The figure given at this date for the synchronous electric clocks was an aggregate of 51,000 with an estimated figure in 1937 being 300,000. The electricity supply in Japan was far from constant and raised problems but apparently they introduced a model with an auxiliary spring which automatically bridged periods of interruption. South Africa was one of their main markets for these clocks, with Manchukuo, China, British India, Australia and Kwantung Leased Territory providing their remaining outlets. Few were exported to this country.

Novelty value £40+

Figure 445

The first patent covering the design of this clock was taken out in 1941 by J.S. Thatcher with later amendments covered by patents taken out in 1944 and 1947 by J.F. Summersgill. The manufacturers were the limited company of Vitascope Industries on the Isle of Man. Examples seen to date have had green, cream or dark brown Bakelite cases. The manual start 250v-50 cycle synchronous motor drives both the hands and also the oval drum operating the automata in the upper third of the case. A further drum with lamp and sheet of dyed gelatine rotates to give the effect of the changing light of night and day. Sunrise, noon, sunset, etc., are most realistically simulated! An eye catching novelty and thus attracts a significant value as well as general interest.

£250+

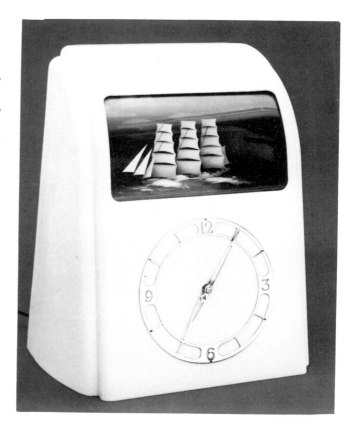

Figure 446

Mystery clock by Smith's Electric Clocks with a synchronous movement (200-250 volts — 50 cycles). Patents for this design were taken out in 1934. There are actually three plates of glass, the front plate has stencilled silver/black numerals, the rear plate is drilled for the drum holding the motion work while the third plate sandwiched between the other two is drilled centrally and shaped at the bottom to allow it to be rocked gently by a small roller arm connected to the synchronous motor housed in the base of the clock. This motion is sufficient to activate the motion work. The case/frame has a nickel finish. These clocks are always highly sought after and realise high prices.

£450+

Figure 447

The Golden House mystery clock was made by an American firm, Jefferson Electric Co., Bellwood, Illinois, USA. This company was still in business in 1972. As their products were intended for the American market they run on 110-125v-60 cycles. Handset is by turning the minute hand clockwise until it reaches the required figure while the correct hour is achieved by revolving the counterbalance on the back of the clock until the hour hand is at the desired position. Some repair details can be found in Best of Coleman — Clockmaker, *1979, pp.47-48. This includes an exploded view of the whole movement/mechanism. This example of a mystery clock with a synchronous movement also has three plates of glass but the middle pane has a 'ring gear' (metal toothed rim) attached in a similar manner to the earlier fully mechanical examples.*

£250

CHAPTER XIV

Time Recorders

Night-watchman Clocks

John Whitehurst (1713-88) is generally accepted as being the inventor of the 'tell tale' or night-watchman clock, although patents confirming this have not been traced to date. Born in Congleton, Cheshire, the son of a clock and watchmaker, he commenced business for himself at Derby about 1736, but moved to London upon receiving the appointment Stamper for the Money-weights. His firm continued after his death under the management of his nephew and later his great nephew. Examples of the tell tale clocks manufactured by the Whitehursts which were usually in extremely plain and functional oak or deal cases and were either floor standing or hung on a wall, had a rotating twenty-four hour dial with a set of spikes set round the periphery — one for each quarter of an hour. Upon arriving at the clock the watchman would pull a lever which pushed one of the spikes in. Hence if any pin was found protruding in the morning it was known that the watchman had neglected his duty for some reason.

As it became essential to prove either for personal satisfaction or insurance purposes that the

Figure 448

3½ins. An illustration of one type of night-watchman's portable tell tale. This particular example weights two pounds and would have been carried in a leather case and strap over his shoulder. The case is of lacquered brass and is stamped on the back 'Thos. Armstrong & Bro. Manchester'. It was claimed that this model patented by Hahn in 1888 was an extremely reliable model that could not be tampered with and falsified by the watchman — an important point. The movement is a good quality lever watch movement that runs for fifty hours.

£100+

watchman was actually patrolling a building thoroughly and regularly other methods were introduced. One of these being a portable tell tale upon which he could record his position at a certain time. Certain points on the watchman's round were selected as 'stations' and the watchman was appointed a definite time for visiting them. At each station there was an individually cut recording key firmly secured to a metal box or similar small receptacle. The watchman, upon arriving at the station, would take the key, insert it into the keyhole in the tell tale and give it one turn, thus recording upon a paper disc or tape the time and his position. The next morning upon it being opened it was possible to see where and when he had checked in during his patrol. The detectors would be made to suit individual requirements as to number of stations and time span between each recorded visit. Several examples appear in the following illustrations.

It is important when purchasing one of these portable tell tales to ascertain that it has its original set of keys, and if possible the leather pouch and carrying strap. Although still of interest without it is more a collector's item if complete.

Figure 449

A further example of a type of watchman's tell tale clock being marketed just prior to the First World War. As the print out is on tape the number of points of recording (stations) can be unlimited.

£95+

WATCHMAN'S TELL-TALE CLOCKS.

AUTO FIXED TELL-TALE No. SY/11.

The prefatory remarks on the previous page do not apply entirely to the above Model. It is a clock for one station only, and is screwed to the one point it is desired that the watchman should visit. Its advantage over the portable form lies in its low price and the simplicity of its operation, which consists merely of pressing a button—no keys being necessary. This pierces a paper dial (see illustration above), which shows the time the point is visited.

Figure 450

This is a clock intended to be used at a fixed point and not carried by the watchman. The makers claimed that its advantage was its low price and the simplicity of its operation, which merely consisted of pressing a button, with no keys being necessary. This pierced the paper dial shown on the right of the illustration to record the time that particular point or station was visited. This model was being manufactured just before the First World War.

£50+

Figure 451

A noteworthy example of a Watchman's portable Time Recorder complete with six station keys and heavy leather carrying case. The movement has lever escapement with jewelled pallet and escape arbor. A good complete example.

£80+

Time Clocks

The Industrial Revolution with the rapid increase of factory employment also led to the need for clocks upon which to record the time of arrival and departure of the employees. Names associated with these early 'Clocking in Clocks' are Bundy, Dey, and Rochester.

The Bundy Key Recorder was invented by W.L. Bundy in the USA in 1885, with production commencing in 1890 by the Bundy Manufacturing Company. It first appeared here in 1893. A key was issued to each employee who, upon arrival at the factory, inserted it into the keyhole at the front of the clock, whereupon the engraved number on his key was recorded, together with the precise time, on to a continuous roll of paper — the entries being in chronological order of the employees' time of arrival and not their numerical order. The Dey Dial Recorder was similar in appearance to that shown in Figure 453 and was invented by Alexander Dey of Scotland in 1888, and placed on the market by him and Mr. John Dey of New York in 1892. In this instance the recordings appeared in their numerical order instead of chronological order and thus obviated a great deal of work in abstracting and noting an individual's timesheet.

Alexander Dey was one of His Majesty's Inspectors of Schools in Scotland. The Dey clocks were marketed in this country (Lancashire only) from 1896 by the Howard Bros. but later they bought the patent rights outside Canada and the USA. In 1908 the two brothers founded Dey Time Register Ltd.,

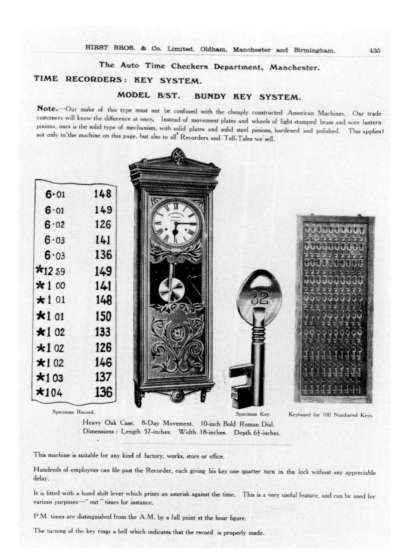

Figure 452

Illustration from a catalogue of time clocks being marketed just before the First World War. The name on dial reads Auto Time Checkers Co., Corporation St., Manchester.

£150+

and amalgamated with the Bundy Company in 1914 to form the International Time Recording Co. Ltd.

Later examples introduced the use of a card for each employee which he inserted so it could be stamped with the time of arrival or departure. This is the system generally in use today. The Rochester Recorder was the first of this type. One English firm manufacturing these clocks was Gledhill-Brooks Ltd. The two founders being Mr. G.H. Gledhill who had founded a cash register business in Halifax, and Mr. F. Brooks who made the original time recorder. The story runs that a dispute broke out at Martins Cloth Mill in Huddersfield where Frank Brooks as a young man worked as a weaver. The work force accused the timekeeper of favouritism when recording the times of arrival and departure of the workers. The method of checking was by means of their throwing a check (personalised by name or number) through an open window of the timekeeper's office that was officially closed at 6 a.m. He was accused of opening it for personal friends and closing it for those with whom he was not on good terms! During discussion with the manager Brooks pointed out that it would be far better to have an impartial mechanical method of recording the times of arrival, etc. He was told to make one! In 1889 he supplied the first check clock to Martins. It had been made for Brooks by a local Swiss watchmaker Ulricht Feichter. Brooks himself had sold watches to workmates and so great had grown the demand for repairs, etc., that he actually employed a few other watchmakers to carry out this work for him. Thus when his popularity at the Mill waned as his check clock was installed he was left without work for these men. He turned his attention to his check clocks and in 1896 formed the Brook Time Checking Clock Co. Ltd. to market the Paragon Check Clock. The movements were made for him by Haycock of Ashbourne,

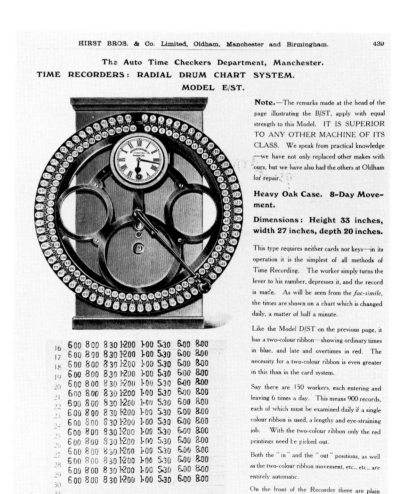

Specimen Record.

Figure 453

Further illustration from a catalogue of time clocks being marketed just before the First World War. The external appearance of this is similar to that manufactured originally by the Dey family in New York. Interesting collector's item but rather large.

£150+

The Auto Time Checkers Department, Manchester.

TIME RECORDERS: CARD SYSTEM.

MODEL K/ST, SEMI AUTOMATIC.   ONE COLOUR RIBBON CARD MACHINES.

MODEL D/ST, FULLY AUTOMATIC.   TWO COLOUR RIBBON CARD MACHINES.

Note.—The quality of their materials—the finest procurable—their superiority of construction, their improvements and advantages, combine to make these the highest developments in Time Recording mechanisms ever brought out.

Card Rack for 100 Employees.

Heavy Oak Case, 8-Day Movement, 10-in. Dial.
Dimensions : Length 57 inches, width 21 inches, depth 13½ inches.

The Model K/ST is semi automatic, that is, the "in" and "out" positions are changed by hand, only, however, by the authorised person, by whom it is locked into position. But the card positions from the A.M. to the P.M., and *vice versa*, as well as from one day to the next are changed automatically.

The D/ST is fully automatic, and carries a two-colour ribbon, printing ordinary times in blue and late and overtimes in red. The tremendous saving of time in making up the wages sheet is obvious—a glance at the card, no matter how hurried, will detect the red figures and, of course, only these need attention. It never errs in this. Absolute reliance can be placed upon it.

One of the advantages of the D/ST machine, possessed by no other machine on the market, is the detachable lever for shifting the card sheath to "in" or "out" when a workman is leaving at a legitimately irregular hour. But after he has stamped his card the sheath RETURNS AUTOMATICALLY TO ITS ORIGINAL POSITION. The detachable lever, of course, is in charge of a responsible person and is only handed to the workmen when properly entitled to use it.

On the front of the Recorder there are plain indicators showing the day, A.M. or P.M., and which "in" or "out" position is in operation, and, on the two-colour machine, an indicator shows the colour of the ribbon.

The record is made by dropping the card into the sheath and pressing the lever. This rings a bell, indicating that the record has been properly made.

BOTH MACHINES ARE IDEAL FOR JOB COSTING.

For Specimen Card Records, see page 441.

Figure 454

Illustration from a catalogue of time clocks being marketed just before the First World War. This is more advanced model, i.e. semi or fully automatic or one or two ribbons.

£140+

Derbyshire, with the castings for the mechanisms and the gear wheels being made by Edward Hollingworth. Gledhill and Brooks joined forces in 1912 in order to make recorders. The movements were again manufactured for them by Haycock but as demand grew it became feasible for them to be produced in their own factory.

Later examples introduced the use of a card for each employee which he inserted so it could be stamped with the time of arrival or departure. This is the system generally in use today.

Although spring-driven clocking in clocks were extensively used, with many still in use today, they were superseded by models manufactured with electrically rewound movements or synchronous electric motors working from the mains.

One other type is the time stamp which prints the date and time — these have a multitude of uses including timing length of stay in car parks, recording the arrival of incoming mail, etc. Their advantage over a hand stamp and ink pad is that they are proof against fraudulent use, as only the supervisor having the key can open the case and adjust the mechanism. Most of these have high grade lever movements and are spring-driven.

Most people looking at a Time Recorder for the first time are surprised by its quality of manufacture. It must be remembered that in order to serve their purpose satisfactorily it is a waste of time devising a system to entrap a devious night watchman or employee if the basic timekeeper cannot keep excellent time itself. They are exceptionally well made and possibly only through lack of domestic appeal have not realised the prices deserved by their quality.

Figure 455a

This illustration shows a time stamp manufactured by Blick Time Recorders Ltd. during the 1930s. Earlier models had movements manufactured by the Seth Thomas Company, USA, but by this date they were being made in England. As well as this model (The Universal) the Company also produced a range of time recorders, key recorders, radial recorders, etc.

The example shown here is in a sturdy metal box with a black crackle finish and plain silvered top. The silvered dial is signed 'Blick Time Recorders, Limited, 188 Gray's Inn Road, London, W.C.1.' The card to be stamped is placed under the handle which is then pressed down by means of a lever. Small and compact which enhances price.

£100+

Figure 455b

View of the fully opened Blick time stamp shown in Figure 455a.

Figure 455c

Movement of the Blick time stamp shown in Figure 455a. Note the solid construction — thick plates — and platform lever escapement.

Figure 456a

13ins. x 13ins. x 9ins. This is an extremely compact time recorder made by the British Time Recorder Co. Ltd. of 149 Farringdon Road, London. The case is of plywood faced with oak veneer, with a painted dial behind a glass 'window'. This illustration shows the clock as it is normally in use, case closed and locked securely from meddling fingers. It would stand on a small shelf, but be screwed to the wall by fishplates.

£100+

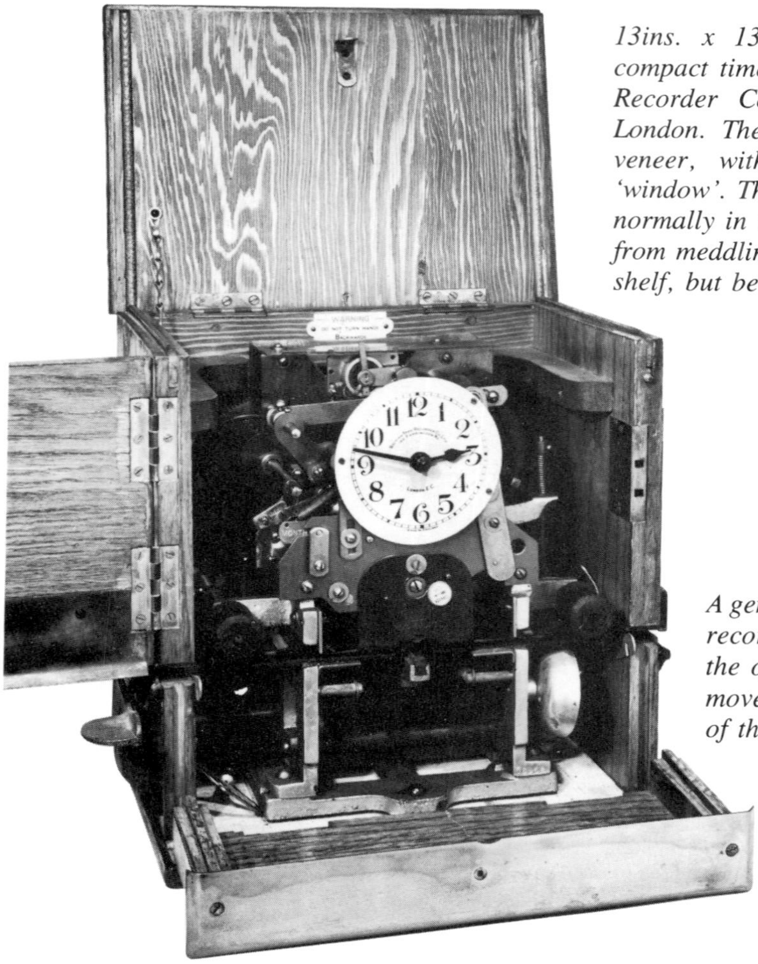

Figure 456b

A general view of the movement of the time recorder shown in Figure 456a provides the opportunity to note the quality of the movement, lever escapement, mechanism of the inking tape, etc.

Pigeon Clocks

It was not only their work people began to time but also their pastimes and sports! This was the inevitable result of many activities becoming commercialised either by centres offering facilities or by the large prize monies offered to the winners.

Although a popular sport in many countries, the breeding and racing of pigeons is considered to be the national sport in Belgium. The interest shown by the Prince of Wales (later Edward VII) and the founding of a Royal Loft at Sandringham in 1886 with birds given by the King of the Belgians provided a sufficient *raison d'etre* for an increase in the number of fanciers in this country. Next to the pigeons themselves, the most important piece of equipment when racing is a reliable timepiece which can also record the precise time of the bird's return. The velocity of the winning bird is calculated by knowing the distance covered by the bird between its point of release and its place of return (i.e. loft) and the exact time this took. The names of Alexander and William Henry Turner are often quoted in connection with the first pigeon clocks (Patent No. 1886 taken out in 1903), but earlier patents by Kitson and Bulmer (Patent No. 3451 taken out in 1891) and S. Gibson (Patent No., 18996 taken out in 1892) have been noted. As most of the clocks were covered by patents taken out in the country of origin (mostly Germany or Switzerland) and as most pigeon fanciers are only interested in their clock's accuracy, little has been recorded regarding the early examples. It is, however, among these older clocks, now considered obsolete for pigeon racing purposes, that the collector should find items of interest. Many variations occur. Each inventor was attempting to improve upon the accuracy of the recording of the timepieces produced by his predecessor and at the same time to make his version less vulnerable to fraudulent use. The conditions under which these races were judged were extremely stringent and in the past each country had its own set of rules according to the governing body. The sport is now run on an international basis with the result that designs have emerged made to mutually acceptable specifications.

As with the night-watchman tell tale, there were basically two types of pigeon clocks — those that recorded the time by pricking a hole in a paper dial or tape and those that printed it. Two examples of the former are to be seen in the following illustrations.

Courtesy of Christie's New York

Figure 457

Pigeon Racing Clock signed on the dial The "Homing Pigeon" Clock Turner's Patent with the case inscribed 'Hately's Patent'. A nice example of an early type. The estimated price in 1982 was between £150 to £250. If in good working condition it would have realised the higher price on the market in this country.

£400+

Photographs and details by Courtesy of F.A. Greensword

Figures 458a, 458b and 458c

An example of a Belgica Pigeon Racing Clock sold in England in 1927. When supplied by Jos. Dusesoi, Lederberg, Gand of Belgium it cost complete with its wooden case and basket £4.10s. Figure 458a shows the clock in situ in its case. Paper dials calibrated in hours, minutes and seconds were placed on brass carriers, the glass cover placed in position and the clock slipped into the wooden case. The metal stud which pushes against the case lock to prevent it being opened also starts the clock. When the pigeon returned the racing number was removed from its leg, placed in a small thimble and inserted into the numbered aperture. The thimble was locked into the aperture ring upon pressing the lever and the time recorded by means of metal pins puncturing the paper dial. Figure 458b shows a view of this ring. Figure 458c shows a more general view of the movement. It also has an 'anti-shake' device! The makers fitted a ratchet operated pointer which was activated by 'see-saw' weights on an arm fitted to the underside of the top plate. This is not connected to the clock movement. In the bottom of the wooden case there are also two metal pots containing loose ball bearings which gave an audible warning if someone shook the clock!

£175 as a collector's piece

Figure 459

View of the movement of a clock used to time racing pigeons, marketed and serviced in this country by The Automatic Timing Clock Co. Ltd. but manufactured to the Benzing Original patent (taken out in France in 1925 and described springs fitted to the pallets to prevent fraud by shaking the clock to accelerate timing). This particular example is in an oak case with a leather carrying handle, and was intended to be used to time twelve birds. The instructions for setting the clock appear on the lid.

£175+

Instructions

1. Wind Clock. Keyhole between 1 and 2.
2. Before Setting, move Brass Pin (as seen above 11) to the left, this allows the minutes to be completed when the Clock will stop.
3. Set Clock by the Minute Hand in front to time required and re-start by releasing Pin as mentioned above, to original position.
4. To set Day, pull Small Key on Square as seen between 3 and 4 and turn to No. 12 on day Wheel to line under 12 o'clock.
5. Dial Roll must be signed on the End, before connecting to Hook.
6. Place Ring-Plate on with No. 5 against Striking Piece.
7. The Small Glass MUST be signed on the INSIDE.
8. After Setting the Clock to the Master Timer, close the lid, draw the knob home to the LEFT. Place the Striking Key on the Square, then turn forward and Strike Time, this action moving the Puncture Hole and leaving the receptacle in position for the First Thimble.
9. To read variation at Locking, continue gentle forward action with Striking Key until Dial appears.
10. When Clock is Closed a puncture is made in Paper Roll, the Same occurs when Opening.
11. When fixing New Roll on, unscrew Adjusting Wheel covering the same.
12. Set Dolometer by turning the Pointer the way indicated on Dial.
13. Clock can be regulated by Screw as seen on side of movement.
14. To release Roll for reading, first put Key on Striking Arm and turn sufficient to bring Blank Part of Wheel as seen at base of same in contact with paper carrier, then release small finger click and draw paper from Roll.

When the bird returns to the loft the identifying ring is removed from its leg, placed in a small brass thimble which is then dropped in place through the lid of the clock into the ring dial, this is rotated by means of the large key at the back and the time of insertion is pricked on to the paper tape. This can then be noted through the small glazed window at the side of the case. The locked clock is taken to the Secretary of the local racing club when all the birds have returned to that loft, and opened under surveillance and the times, etc. compared with those of the other competitors.

Figure 460

5ins. high. An interesting example of a clock used to time racing pigeons patented in 1907 by W. McMillan and A.H. Osman which incorporated a lever escapement perfected and patented by W.G. Schoof in 1874. The outer case is brass with a glazed front, and the movement has a silvered dial. A paper disc would have been attached immediately below the dial, and as the thimble containing the appropriate pigeon ring was inserted through the aperture at the bottom of the case two pricks would be made in the edge of the disc.

Early neat example   £300+

Other Time Recorders

It was inevitable that both genuine and stimulated demand for timers for specific uses would be exploited by the manufacturers, and examples can be found intended for use in the photographer's dark room, in the kitchen, in laboratories or factories timing certain processes, or in the world of sport timing races, games, etc. This area does provide a veritable goldmine of collectors' items!

Figure 461

By the number of patents taken out at the time, it is known that clocks especially used for timing billiard games were extensively used in the 1890s. This example was being marketed immediately before the First World War. One invented by Thomas and Atkinson in 1897 was intended to have an arm or rod placed over the table to prevent play, the removal of which started a timing device. Others have been noted in which the insertion of a coin enabled the table to be used for a specified length of time, after which some form of obstruction made play impossible, i.e. the light illuminating the table was switched off!!

£125+

416

Figure 462

An amusing example of a rather elaborate eggtimer! By the case style (cheap brass case with steel alarm bell) it can be deducted that it was manufactured sometime prior to the Second World War. The movement is a conventional pin pallet.

£25+

Figure 463

The metal case of the time stamp manufactured by the Stromberg Electric Company of Chicago is matt black, with a small one inch monitoring dial on the side. It operates off 110 volts and was used in conjunction with a master clock and secondary dial installations, by being included in the slave dial circuit. The stamping mechanism is housed in the spring loaded arm. The patent for this was taken out in November, 1909.

£75+

Figure 464

Further example of a timer marketed for a specific purpose. Added interest to this particular example is the fact that it has its original box and instruction leaflet.

£45 +

Figure 465

5 ¼ ins. Eastman Photographic Timer made for Eastman Kodak Co., Rochester, New York, USA, pre-1930. The centre knob sets the minute hand, or when pulled to maximum position will set the seconds hand. The dial records up to 60 minutes with the centre hand registering 0-60 seconds. The stop/start lever is at the 6 o'clock position on the bezel with a light spring activating the edge of the balance wheel. The movement is of a cheap pin pallet variety. No great merit apart from being an example of a particular type of timer and also carrying a famous name in the world of photography.

£45 +

Figure 466

"Assuming the apparatus is to be used in the reception of radio broadcast it will be placed in series with the filament circuit of the thermionic valves. If a listener desires only to hear a particular item or items of a previously published broadcast timetable, he or she will plug into a socket or sockets any desired number of said wander plugs appropriate to advertised time of rendition". This was the suggested use by the manufacturers of this timing device. Housed in a small mahogany case with a copper anodised surround to the enamel dial it is an attractive small piece. The dial carries the name 'Electone' together with the Patent No. 250001 (which was applied for in March, 1925) and the name Fredk. J. Gordon & Co. Ltd., 92 Charlotte Street, London, W.1. This is the name of the manufacturer; the patentee was Graham Cotterell of Wanstead, London.

£125

Figure 467

9ins. high. The Axuel Radio Programme Selector, in brown crackle finish Bakelite of foreign manufacture. The outer dial with five minute markings for programming and with the inner dial for timekeeping. The keywound 30 hour spring wound movement has a pin pallet lever escapement and lantern pinions. Terminals on the rear of the case are for connection to the radio. An insulated arm behind the dial moves with the hour wheel to complete a circuit as it comes into contact with the setting peg in the programming dial. The features of this device bear a strong resemblance to those covered by two patents taken out in 1925 by K. Kopatschek. Obviously there was a good market for these timers in the early days of wireless transmission.

£150+

468a

468b

Figures 468a and 468b

6ins. high. The origins of the game of chess are lost in obscurity but in 1846 important publications and handbooks appearing in this country stimulated a revival here. Associations were formed and tournaments, both local and international, arranged. Rules and regulations had to be laid down in order to standardise competitions and one of the innovations was the limitation of the time allowed for each player to make his move. As the use of sand-glasses for measuring intervals of time whether for private devotions, lessons, auctions, or sermons was widespread at this time it was natural to extend their use to timing chess games. They were, however, soon superseded by clockwork timers. These took the form of double clocks which measured the total elapsed time each player took while playing. The example in this illustration is in an oak case with white enamel dials; the movement was manufactured by the Hamburg American Clock Company of Germany as confirmed by the presence of their trade mark (crossed arrows) on the dial. Note the small flags at the top of the dial which fall when the total time allowed has elapsed. Only one clock is operating at a given time. As a player makes his move he depresses the knob at the top of the case nearest him which, by means of a system of levers, etc., seen in Figure 468b, automatically raises the other knob thus releasing the balance wheel of his opponent's clock and stopping his own. When his opponent makes his move he presses the knob nearest to him and reverses the process. By looking at the two dials it is possible to see at a glance the length of time each player has consumed during the match.

£75 depending upon quality of case

Figure 469

This clock has the name Fattorini & Sons, Bradford, raised on the upper surface of the cast iron base. The cases of the actual clock movements are nickel plated and the dials are printed card. Each clock has a small spring driven pendulum movement with a very long main spring in going barrels, the barrel forming the winding key. The clocks are mounted on a platform having central cone pivots, this allows the higher clock to run while the lower one is stopped by its pendulum being out of beat and touching the side of the case. A quick depression of the higher end of the platform reverses the procedure. The pendulum bobs are friction fit for adjustment and the hands are set by conventional key from the rear of the movements. This would appear to be an early example of a chess clock and it is difficult to comment over the actual manufacturer. Fattorini & Sons although stating in advertisements that they made clocks, etc., probably imported examples made to their specification rather than actually being the owners of factories (see text accompanying Figure 353 for further comments and details of the company).

£150+ as obviously an early example

Figure 470

6ins. high. The Chronoscope as manufactured by the Hamburg American Clock Company in 1911-12. A downwards push on the top of the dial set the clock in motion and set the hands. A second push brought the hand back to the starting position. The manufacturers listed its possible uses as a telephone time checker, egg boiler, for tea tasting or for use in a photographer's darkroom.

£30+

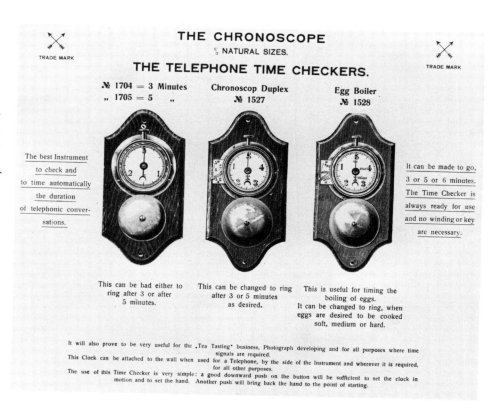

471a

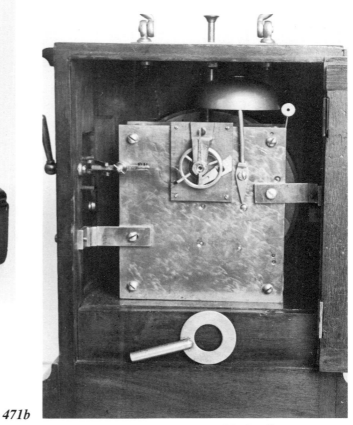

471b

Photographs and details by courtesy of D. Rundle

Figures 471a and 471b

This is a gas referee's clock and part of the apparatus required for the calorific testing of gas. About 1922 all statutory gas undertakings were required to supply gas at a declared calorific value. The stop clock timed the rate of the calorimeter gas meter whose dial was also divided into 100 divisions. The rate of the meter was adjusted to keep pace with the stop clock. A bell sounded on the minute to avoid the tester having to keep glancing at the dial to establish the time. The decimal minute dial also became popular with gas engineers for the purpose of calculating gas flow. This particular example was owned by the Glastonbury & District Gas Company but had been supplied to them by Alexander Wright Co. Ltd. of Westminster. There are several interesting features. The barrel wheel meshes with a brass pinion — a practice commonly found in gas meter index trains — the spring barrel has stop work allowing one turn only between up and down; the massive vertical platform escapement is of the English lever pattern with the start/stop acting on the balance rim in an extremely positive action. A well made unusual piece that might pass unidentified to those unfamiliar with the gas industry.

£200+ as collector's item rather than useful timepiece

Figure 472

*1937 advertisement showing three
of the time switches made by
Automatic Light Controlling Co.
Ltd. of Bournemouth, Hampshire.*

£25+

Dr. Thuger of Norwich is attributed to being the man who first thought of using a clock mechanism to regulate switching street lighting on and off at the appropriate hours. Its practical application was not fully realised, however, until it was further developed by John Genney of Bournemouth during the late 1890s. The Gunfire controllers seen in Figure 472 evolved from these early beginnings. The Horstman Gear Co. were another firm closely linked with a large range of gas controllers and time switches that were marketed under the name of Newbridge. They produced the first controller having a 'solar dial' that switched the lights on and off at different times according to the seasonal changes. Their factories were just outside Bath — further details of the Company can be found on page 182 where one of their mechanical domestic clocks is shown.

As street lighting became more widespread the need grew for mechanical automatic methods of control and many companies appeared including the Gas Meter Co. of London who manufactured the 'London' Controller, Venner Time Switches and the Reason Manufacturing Company Ltd. of Brighton who are possibly better known by horologists for their electric clocks (see Figures 413 and 414).

Venner Time Switches was founded in 1906, became a private company in 1911 and was made a public company in 1930, with premises initially at Horseferry Road, Westminster, and then in 1933 they moved to Shannon Corner, Raynes Park. As well as time switches for use with street lighting they and indeed most of the other companies engaged in making these switches also made time switches to light telephone boxes, hen houses, billiard tables, late shop window lighting and the automatic switching out of lights on staircases in blocks of flats, etc. To identify the function of an odd switch purchased from a market stall can be quite a feat in itself.

473a

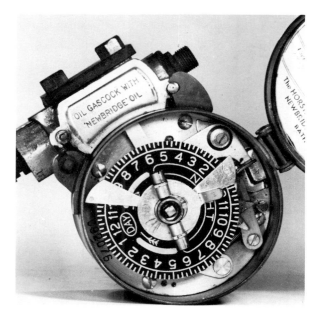

473b

Figures 473a and 473b

3⅜ x 3ins. Gas controller manufactured by Horstmann Gear Co. Ltd. for use with automatic street lighting. The movement is a standard single train Type 3 with a duration time of 15 days. Some models were modified to run 21 or 40 days. The extra run was obtained by the addition of secondary mainsprings. According to the label pasted inside the lid this particular example was cased on the 21st March, 1949 and timed three months later but is typical of the gascocks marketed from a much earlier date. The club foot lever platform escapement is stamped EH English Made ABEC which indicates that it was made by All British Escapements Company (part of Smith's Industries) of Cricklewood. Manufacture of these escapements continued until the end of the 1960s.

£25+ *from shop premises, less if purchased on demolition site!*

Figure 473c

The key of the gascock illustrated in Figure 473b is interesting in itself in so far as it acts as an advertising medium and also can be used as a spanner to tighten nuts on various parts of the casing, etc., of the installation.

APPENDIX I

Pages from a Trade Catalogue of Watch and Clock Materials c.1916 (Hirst Bros. & Co. — Wide Awake Catalogue).

For those who obtain part of their pleasure in 'doing up' some of the more robust examples it is hoped that the necessary details regarding parts and styles can be gleaned from the pages in this Appendix showing the materials offered to the clockmaker at the beginning of this century. It is still possible to obtain some of the more standard parts — spring, weights, spandrels, etc., from material dealers today. Many of these are willing to supply catalogues and conduct business through the post. List of addresses as well as display advertisements appear in current copies of the *Horological Journal*, *Antiquarian Horology* or *Clocks*. Failing that, local trade directories at the local library can be consulted.

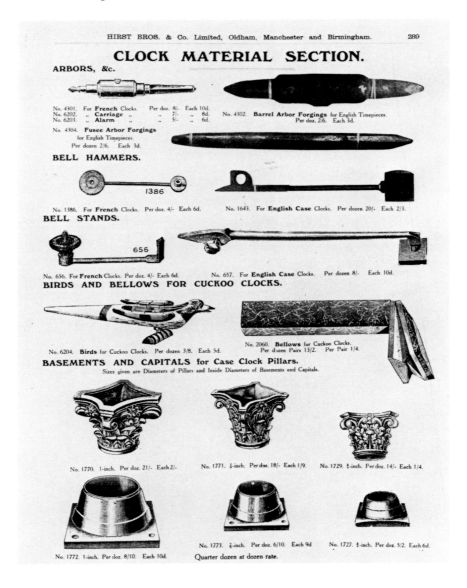

HIRST BROS. & Co. Limited, Oldham, Manchester and Birmingham.

BEZELS, Various.

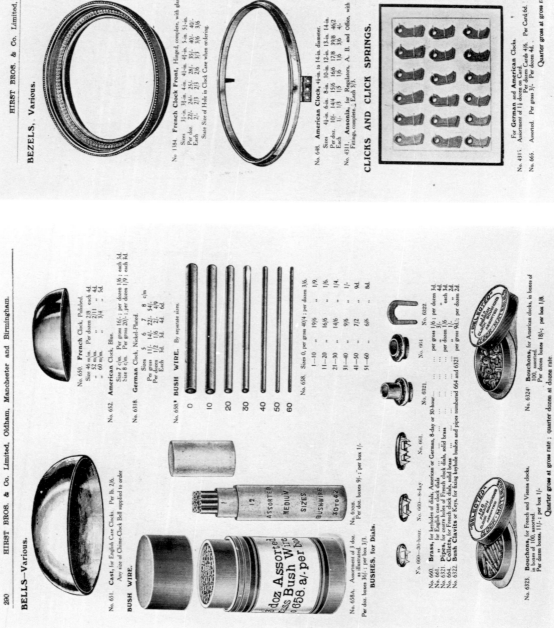

No. 1385. **French Clock Back.** Hinged, complete.

Sizes	3½-in.	3¾-in.	4-in.	4½-in.	4¾-in.	5-in.	5¼-in.
Per doz.	16/-	18/-	19/-	25/-	30/-		
Each	1/6	1/8	1/9	2/3	2/9		

State Size of Hole in Clock Case when ordering.

No. 1384. **French Clock Front.** Hinged, complete, with glass.

Sizes	3½-in.	3¾-in.	4-in.	4½-in.	4¾-in.	5-in.	5¼-in.
Per doz.	22/-	24/-	25/-	28/-	35/-	40/-	40/-
Each	2/-	2/3	2/3	2/6	3/3	3/6	3/6

State Size of Hole in Clock Case when ordering.

No. 648. **American Clock,** 4½-in. to 14-in. diameter.

Sizes	4½-in.	6-in.	8-in.	10-in.	12-in.	13-in.	14-in.
Per doz.	10/-	14/4	15/6	16/6	17/8	39/8	46/2
Each	1/-	1/3	1/5	1/6	1/6	3/6	4/-

No. 4311. **Ansonia,** for Regulators, A, B, and Office, with Fittings, complete. Each 5/3.

No. 1117. **Weight Regulator.** Bezels only. Size 8-in. Per dozen 8/4. Each 9d.

2053

668 B

668

667

668A

1107

No. 2053. For inside English Clock Fuzes, Brass. Per dozen 1/9.
No. 668A. For French Clocks. Per gross 10/9. Per doz. 1/3. Each 2d.
No. 668B. " Weight Regulators. 9/-. " 1/3. " 1d.
No. 1107. " Weight Regulators. Per gross 10/6. Per doz. 1/-. Each ¾d.
No. 667. French Clocks, with Spring combined. 1/3. " 2d.
No. 668. " " Per gross 9/-. Per doz. 1/3. " 2d.

No. 1725. For English Fuzes. Per gross 7/6. Per doz. 9d.

No. 669. For Bee Clocks. Per doz. 4d.

CLICKS AND CLICK SPRINGS.

For German and American Clocks.
No. 4311. Assortment of 1½ dozen on Card. Per Card 6d.
No. 665. Assorted. Per dozen Cards 4/6.
Per doz. 3/-. Per box 4d.

Quarter gross at gross rate.

HIRST BROS. & Co. Limited, Oldham, Manchester and Birmingham.

BELLS—Various.

No. 650. **French Clock,** Polished.
Size 46 c/m. Per dozen 2/8 each 4d.
" 52 " 2/11 " 4d.
" 60 " 3/4 " 5d.

No. 651. **Cast,** for English Case Clocks. Per lb. 2/6.
Any size of Chime-Clock Bell supplied to order

No. 652. **American Clock,** Blue.
Size 7 c/m. Per dozen 16/-; per dozen 1/6; each 3d.
Size 8 c/m. Per dozen 20/-; per dozen 1/9; each 3d.

BUSH WIRE.

No. 6318. **German Clock,** Nickel-Plated.

Sizes	5	6	7	8 c/m.
Per gross	11/-	14/-	22/-	54/-
Per dozen	1/2	1/6	2/-	6d.
Each	3d.	3d.	4d.	6d.

No. 658 **BUSH WIRE.** By separate sizes.

Sizes 0, per gross 40/4; per dozen 3/6.

1—10	"	19/6	" 1/9.
11—20	"	16/6	" 1/6.
21—30	"	14/6	" 1/4.
31—40	"	9/6	" 1/-.
41—50	"	7/2	" 9d.
51—60	"	6/6	" 8d.

No. 658a. Assortment of 3 doz. as illustrated.
Per doz. boxes 36/-; per box 3/3.

BUSHES, for Dials.

No. 6322

No. 664

No. 6321

per gross 1/6; per gross 3d.
per gross 3/-; " 4d.
" 5/-; each 3d.
per gross 9½/-; per dozen 2d.

No. 660. **Brass,** for keyholes of dials, American or German, 8-day or 30-hour.
No. 661. " " for English case clock dials.
No. 6321. **Pipes,** for centre holes of French clock dials, solid brass.
No. 664. **Collets,** for French clock dials, solid brass.
No. 6322. **Bush Clavits** or **Keys,** for fixing keyhole bushes and pipes numbered 664 and 6321.

No. 6323. **Bouchons,** for French and Vienna clocks, in boxes of 100, assorted.
Per dozen boxes 11/-; per box 1/-.

No. 6324. **Bouchons,** for American clocks, in boxes of 100, assorted.
Per dozen boxes 18/-; per box 1/8.

Quarter gross at gross rate; quarter dozen at dozen rate.

BLOCKS FOR MOVEMENTS, BARRELS, &c.

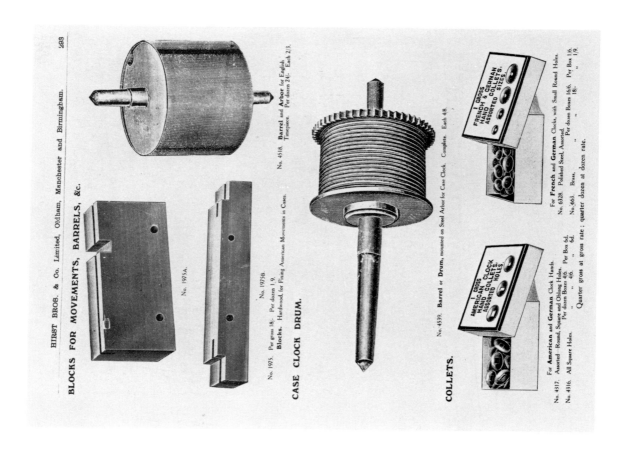

No. 4518. **Barrel and Arbor** for English Timepiece. Per dozen 24/-. Each 2/3.

No. 1975A.

No. 1975B.

No. 1975. Per gross 18/-. Per dozen 1/9. **Blocks.** Hardwood, for Fixing American Movements in Cases.

CASE CLOCK DRUM.

No. 4539. **Barrel or Drum**, mounted on Steel Arbor for Case Clock. Complete. Each 4/8.

COLLETS.

For **French** and **German** Clocks, with Small Round Holes.
No. 6328. Polished Steel, Assorted. Per Box 1/6.
No. 665. Brass. Per dozen Boxes 16/6. Per Box 1/6.
18/-. " 1/9.

1 GROSS FRENCH & GERMAN HAND ASSORTED COLLETS SIZES.

Quarter gross at gross rate ; quarter dozen at dozen rate.

For **American** and **German** Clock Hands.
No. 4317. Assorted - Round, Square and Oblong Holes. Per dozen Boxes 4/6. Per Box 6d.
No. 4316. All Square Holes. " 4/6. " 6d.

1 GROSS AMERICAN CLOCK HAND COLLETS ASSORTED HOLES.

CLICKS AND CLICK SPRINGS Continued.

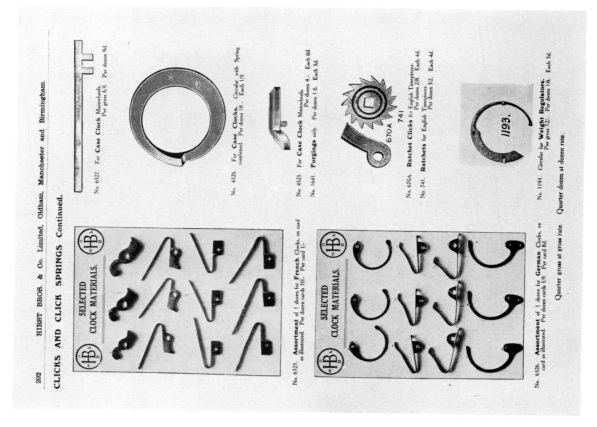

No. 6327. For **Case Clock** Mainwheels. Per gross 6/9. Per dozen 9d.

No. 4326. For **Case Clocks.** Circular with Spring combined. Per dozen 18/- Each 1/9.

No. 4325. For **Case Clock** Mainwheels. Per dozen 4/- Each 6d.

No. 1641. **Forgings** only. Per dozen 1/6. Each 3d.

670A

741

No. 670A. **Ratchet Clicks** for English Timepieces. Per dozen 2/8. Each 4d.

No. 741. **Ratchets** for English Timepieces. Per dozen 3/2. Each 4d.

1193

No. 1193. Circular for **Weight Regulators.** Per gross 12/- Per dozen 1/6. Each 1d.

Quarter dozen at dozen rate.

SELECTED CLOCK MATERIALS.

No. 6325. **Assortment** of 1 dozen for **French** Clocks, on card as illustrated. Per dozen cards 10/-. Per card 1/-.

SELECTED CLOCK MATERIALS.

No. 6326. **Assortment** of 1 dozen for **German** Clocks, on card as illustrated. Per dozen cards 5/9. Per card 8d.

Quarter gross at gross rate.

HIRST BROS. & Co. Limited, Oldham, Manchester and Birmingham.

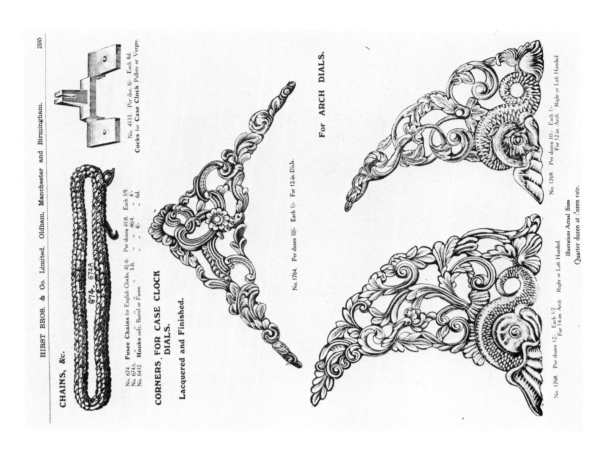

CHAINS, &c.

674. 674A.

No. 674. **Fusee Chains** for English Clocks. 4½-ft. Per dozen 41/8. Each 3/9.
No. 674A. " " " 5-ft. " 46/4. " 4/-
No. 6432. **Hooks** only, Barrel or Fusee. " 4/- " 6d.

CORNERS FOR CASE CLOCK DIALS.

Lacquered and Finished.

No. 1764. Per dozen 10/-. Each 1/-. For 12-in. Dish.

For ARCH DIALS.

No. 1769. Per dozen 10/-. Each 1/-. For 12-in. Arch. Right or Left Handed.

No. 1768. Per dozen 12/-. Each 1/2. For 14-in. Arch. Right or Left Handed.

Illustrations Actual Sizes.

Quarter dozen at dozen rate.

No. 4533. Per doz. 6/-. Each 8d.
Cocks for **Case Clock** Pallets or Verge.

HIRST BROS. & Co. Limited, Oldham, Manchester and Birmingham.

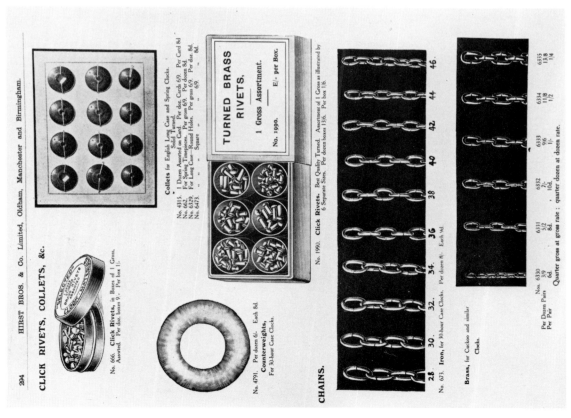

CLICK RIVETS, COLLETS, &c.

No. 666. **Click Rivets**, in Boxes of 1 Gross, Assorted. Per dox. boxes 9/-. Per box 1/-.

No. 4791. Per dozen 6/-. Each 8d.
Counterweights,
For 30-hour Case Clocks.

Collets for English Long Case and Spring Clocks.
Solid Turned.

No. 4315. 1 Dozen Assorted on Card. Per doz. Cards 6/9. Per Card 8d.
No. 662. For Spring Timepieces. Per gross 6/9. Per dozen 8d.
No. 6329. For Long Case—Round Holes. Per gross 6/9. Per doz. 8d.
No. 6473. " " " Square " " 6/9. " 8d.

TURNED BRASS
RIVETS.
1 Gross Assortment.

No. 1990. E/- per Box.

No. 1990. **Click Rivets.** Best Quality Turned. Assortment of 1 Gross as illustrated by 6 Separate Sizes. Per dozen boxes 13/6. Per box 1/6.

CHAINS.

28. 30. 32. 34. 36 38 40 42 44 46

	Nos.	6330	6331	6332	6333	6334	6335
Per Dozen Pairs	3/9		5/2	7/-	9/6	11/8	13/8
Per Pair	6d.		8d.	10d.	1/-	1/2	1/4

Quarter gross at gross rate; quarter dozen at dozen rate.

No. 673. **Iron,** for 30-hour Case Clocks. Per dozen 8/-. Each 9d.

Brass, for Cuckoo and similar Clocks.

HIRST BROS. & Co. Limited, Oldham, Manchester and Birmingham.

Illustrations Actual Sizes.

CORNERS, &c., FOR CASE CLOCK DIALS.

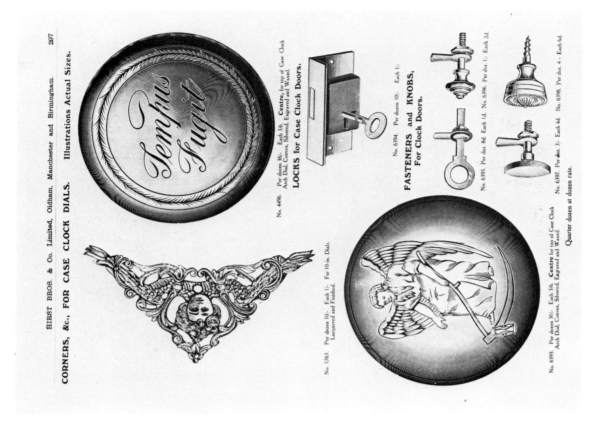

No. 4496. Per dozen 36/-. Each 3/6. **Centre**, for top of Case Clock Arch Dial, Convex, Silvered, Engraved and Waxed.

LOCKS for Case Clock Doors.

No. 6394. Per dozen 10/-. Each 1/-.

FASTENERS and KNOBS, For Clock Doors.

No. 6395. Per doz. 8d. Each 1d. No. 6396. Per doz. 1/-. Each 2d.

No. 6397. Per doz. 3/-. Each 4d. No. 6398. Per doz. 4/-. Each 6d.

Quarter dozen at dozen rate.

No. 1763. Per dozen 10/-. Each 1/-. For 10-in. Dials. Lacquered and Finished.

No. 6393. Per dozen 36/-. Each 3/6. **Centre** for top of Case Clock Arch Dial, Convex, Silvered, Engraved and Waxed.

HIRST BROS. & Co. Limited, Oldham, Manchester and Birmingham.

CORNERS FOR CASE CLOCK DIALS.

Lacquered and Finished.

Illustrations Actual Sizes.

No. 1765. Per dozen 12/-. each 1/3. For 13-in. and 14-in. Dials.

No. 1766. Per dozen 10/-. Each 1/-. For 12-in. Dials.

No. 1761. Per dozen 10/-. Each 1/-. For 12-in. Dials.

Quarter dozen at dozen rate.

DIALS.

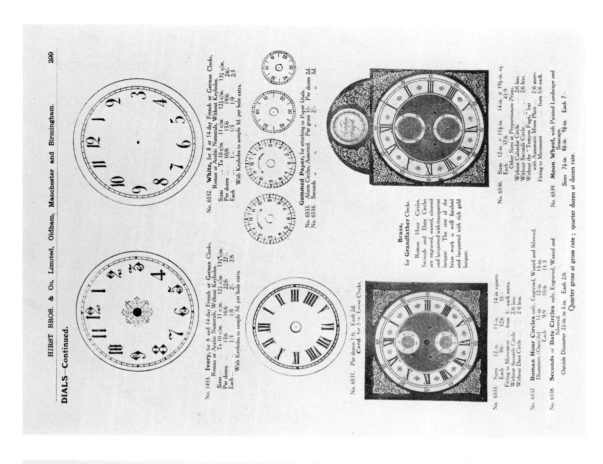

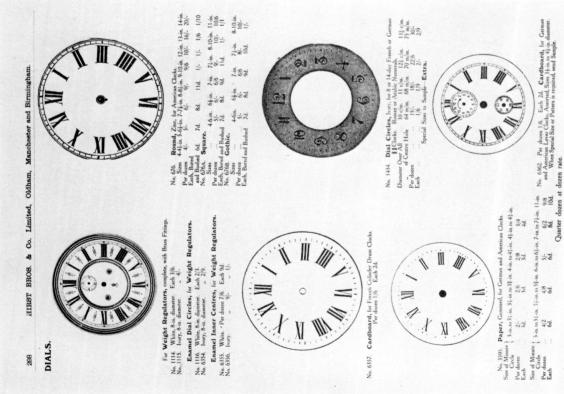

For **Weight Regulators,** complete, with Brass Fittings.

No. 1114. White, 8-in. diameter. Each 3/6.
No. 1115. Ivory, 8-in. diameter. „ 4/-.

Enamel Dial Circles, for **Weight Regulators.**

No. 1116. White, 8-in. diameter. Each 2/3.
No. 6354. Ivory, 8-in. diameter. „ 2/9.

Enamel Inner Centres, for **Weight Regulators.**

No. 6355. White. Per dozen 7/6. Each 9d.
No. 6356. Ivory. „ 9/- „ 1/-.

No. 676. **Round,** Zinc, for American Clocks.

Sizes	4-4½-in.	5-6½-in.	7-7¼-in.	8-8½-in.	9-10-in.	12-in.	13-in.	14-in.
Per dozen	4/-	5/-	6/-	9/-	9/6	10/-	16/-	20/-
Each, Bored and Bushed	6d.	7d.	8d.	11d.	1/-	1/-	1/6	1/10

No. 676A. **Square.**

Sizes ...	4-6-in.	6½-in.	7-in.	7½-in.	8-10-in.	11-in.
Per dozen ...	5/-	6/-	6/6	9/-	10/-	10/6
Each, Bored and Bushed	7d.	8d.	9d.	11d.	1/-	1/1

No. 676B. **Gothic.**

Sizes ...	4-6in.	6½-in.	7-in.	7½-in.	8-10-in.
Per dozen	5/-	6/-	6/6	8/-	10/-
Each, Bored and Bushed	7d.	8d.	9d.	10d.	1/-

No. 6357. **Cardboard,** for French Cylinder Drum Clocks.
Per dozen 1/6 Each 2d.

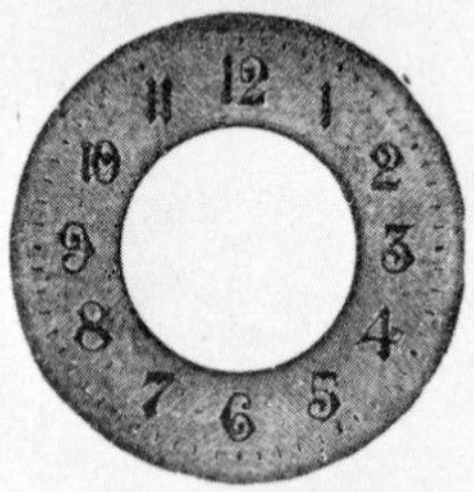

No. 1434. **Dial Circles,** Ivory, for 8 or 14-day French or German Clocks. Roman or Arabic Numerals.

Diameter Over All ...	10 c/m.	11 c/m.	12½ c/m.	13½ c/m.
„ of Centre Hole	64 m/m.	68 m/m.	70 m/m.	75 m/m.
Per dozen	15/-	18/-	22/6	30/-
Each	1/6	1/9	2/-	2/9

Special Sizes to Sample—**Extra.**

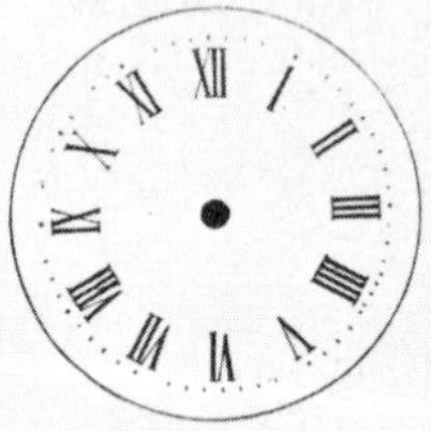

No. 3595. **Paper,** Gummed, for German and American Clocks.

Size of Minute Circle	3-in. to 3¼-in.	3½-in. to 3¾-in.	4-in. to 4¼-in.	4½-in. to 4¾-in.
Per dozen ...	2/-	2/6	2/8	3/4
Each ...	3d.	3d.	3d.	4d.

Size of Minute Circle	5-in. to 5¼-in.	5½-in. to 5¾-in.	6-in. to 6½-in.	7-in. to 7½-in.	11-in.
Per dozen ...	4/2	4/8	5/-	6/2	9/8
Each ...	6d.	6d.	6d.	8d.	10d.

No. 6362. Per dozen 1/6. Each 2d. **Cardboard,** for German and American Lever Clocks, Assorted, Sizes 3½-in. to 4½-in. diameter. When Special Size or Pattern is required, send Sample.

Quarter dozen at dozen rate.

DIALS—Continued.

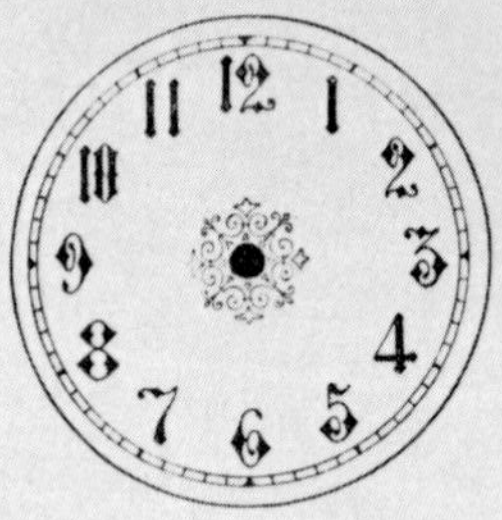

No. 1433. **Ivory,** for 8 and 14-day French or German Clocks, Roman or Arabic Numerals, Without Keyholes.

Sizes	To 10 c/m.	11 c/m.	12½ c/m.	13½ c/m.
Per dozen ...	13/6	16/6	22/6	27/-
Each	1/3	1/6	2/-	2/6

With Keyholes to sample 3d. per hole extra.

No. 6532. **White,** for 8 or 14-day French or German Clocks, Roman or Arabic Numerals, Without Keyholes.

Sizes ...	To 10 c/m.	11 c/m	12½ c/m.	13½ c/m.
Per dozen ...	10/6	13/6	19/6	24/-
Each ...	1/-	1/3	1/9	2/3

With Keyholes to sample 3d. per hole extra.

No. 6531. Per dozen 1/6. Each 2d.
Card, for 2-in. Lever Clocks.

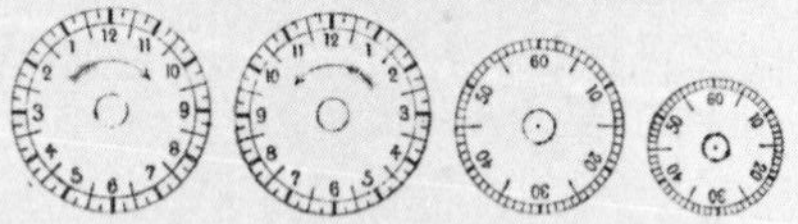

Gummed Paper, for attaching to Paper Dials.
No. 6533. Alarm Circles, Assorted. Per gross 1/- Per dozen 2d.
No. 6534. Seconds „ „ 2/- „ 3d.

Brass,
for **Grandfather** Clocks.

Roman Hour Circles, Seconds and Date Circles are engraved, waxed, silvered and lacquered with transparent lacquer. The rest of the brass work is well finished and lacquered with rich gold lacquer.

No. 6535. Sizes 12-in. 13-in. 14-in. square.
Each 30/- 32/6 35/-
Fitting to Movement ... from 5/- each extra.
Without Seconds Circle ... 2/6 less.
Without Date Circle ... ... 2/6 less.

No. 6536. Sizes 12-in. × 15¼-in. 14-in. × 19½-in. sq.
Each ... 37/6 43/9
Other Sizes at Proportionate Prices.
Without Calendar Circle ... ... 2/6 less.
Without Seconds Circle... ... ... 2/6 less.
Without the "Tempus Fugit," but with Automatic Moon Plate ... 7/6 more.
Fitting to Movement ... ... from 5/6 each.

No. 6537. **Roman Hour Circles** only, Engraved, Waxed and Silvered.
Diameters (Outside) 11-in. 12-in. 13-in.
Each 9/9 10/6 11/3

No. 6538. **Seconds** or **Date Circles** only, Engraved, Waxed and Silvered.
Outside Diameter 2½-in. to 3-in. Each 2/6.

No. 6539. **Moon Wheel,** with Painted Landscape and Seascape.
Sizes 7⅜-in. 8⅜-in. 9⅜-in. Each 7/-

Quarter gross at gross rate ; quarter dozen at dozen rate.

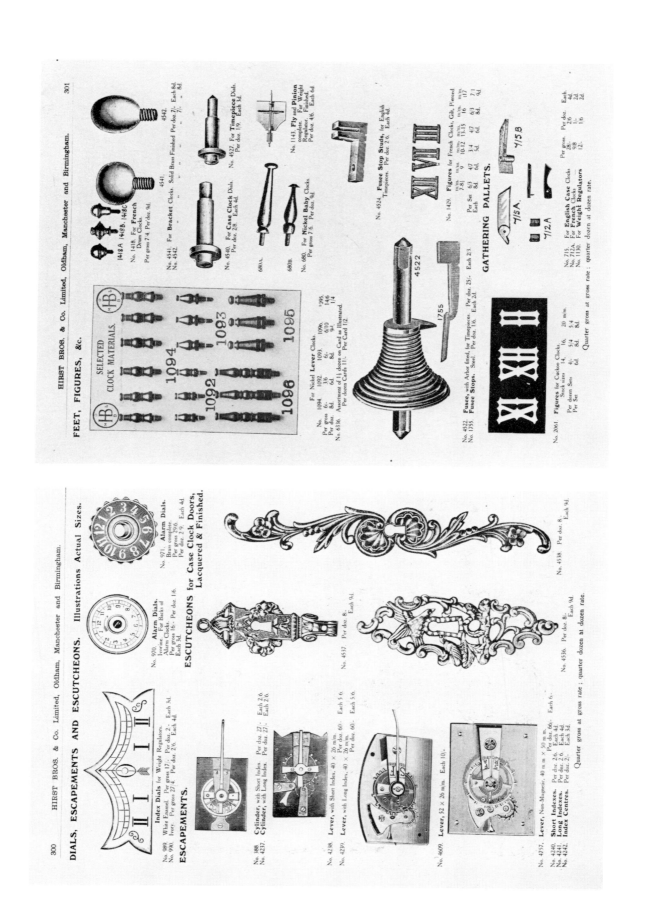

301

FEET, FIGURES, &c.

SELECTED CLOCK MATERIALS.

1094 1093 1095

1092 1096

1094 1092 1093 1096 1095
Per gross 6/- 3/6 6/- 6/10 1/6
Per doz. 8d. 6d. 9d. 9d. 1/4

No. 636. Assortment of 1¼ dozen on Card as illustrated.
Per dozen Cards 11/- Per Card 1/2

For Nickel Lever Clocks.

1418A /1448B. /1448C.
No. 1418. For French Drum Clocks.
Per gross 7/4 Per doz. 9d.

4541. 4542.
No. 4541. For Bracket Clocks. Solid Brass Finished Per doz. 7/- Each 8d.
No. 4542. .... 7/- .... 8d.

No. 4540. For Case Clock Dials. Each 4d.
No. 4540. For Case Clock Dials. Per doz. 2/8. Each 3d.

680A.
680B.
No. 680. For Nickel Baby Clocks. Per gross 7/6. Per doz. 9d.

4542.
No. 4527. For Timepiece Dials. Per doz. 1/9. Each 3d.

No. 1143. Fly and Pinion complete. For Weight Regulator. Finished.
Per doz. 4/6. Each 6d.

No. 4524. Fusee Stop Studs, for English Timepieces. Per doz. 2/6.

4522
1755
No. 4522. Fusee, with Arbor fitted, for Timepieces. Per doz. 25/- Each 2/-
No. 1755. Fusee Stops. Steel. Per doz. 1/6. Each 2d.

XI VII II

No. 1429. Figures for French Clocks, Gilt, Pierced.

m/m. 7-8½	m/m. 9	m/m. 10-12	m/m. 13-15	m/m. 16	m/m. 17
Per Set 6/3	9	3/4	4/7	6/3	7/1
Each 8d.		5d.	6d.	8d.	9d.

GATHERING PALLETS.

7/5B
7/5A.
7/2A

	Per gross	Per doz.	Each.
No. 715. For English Case Clocks	28/-	2/6	4d.
No. 712A. For French Clocks	9/8	1/-	2d.
No. 1130. For Weight Regulators	12/-	1/6	2d.

Quarter gross at gross rate; quarter dozen at dozen rate.

XI. XII. II

No. 2061. Figures for Cuckoo Clocks.

Stock sizes	14.	16.	20 m.m.
Per dozen Sets	4/-	5/4	5/4
Per Set	6d.	6d.	8d.

DIALS, ESCAPEMENTS AND ESCUTCHEONS. Illustrations Actual Sizes.

No. 969. White Enamel. Per doz 17/- Each 3d.
No. 990. Ivory. Per gross 27/4. Per doz. 2/6. Each 4d.
Index Dials for Weight Regulator.

No. 970. Alarm Dials. Ivorine. For Backs of Alarm Clocks.
Per gross 16/- Per doz. 1/6. Each 3d.

No. 971. Alarm Dials. Brass complete.
Per gross 29/6. Per doz. 2/9. Each 4d.

ESCUTCHEONS for Case Clock Doors, Lacquered & Finished.

No. 4538. Per doz. 8/- Each 9d.

No. 4537. Per doz. 8/- Each 9d.

No. 4536. Per doz. 8/- Each 9d.

ESCAPEMENTS.

No. 388. Cylinder, with Short Index. 40 × 26 m.m. Per doz. 27/- Each 2/6.
No. 4237. Cylinder, with Long Index. Per doz. 27/- Each 2/6.

No. 4238. Lever, with Short Index. 40 × 26 m.m. Per doz. 60/- Each 5/6.
No. 4239. Lever, with Long Index. 40 × 26 m.m. Per doz. 60/- Each 5/6.

No. 4609. Lever, 52 × 26 m.m. Each 10/-.

No. 4757. Lever, Non-Magnetic. 40 m.m × 50 m.m. Per doz. 66/- Each 6/-
No. 4240. Short Indexes. Per doz. 2/6. Each 4d.
No. 4241. Long Indexes. Per doz. 2/6. Each 4d.
No. 4242. Index Centres. Per doz. 2/- Each 3d.

Quarter gross at gross rate ; quarter dozen at dozen rate.

431

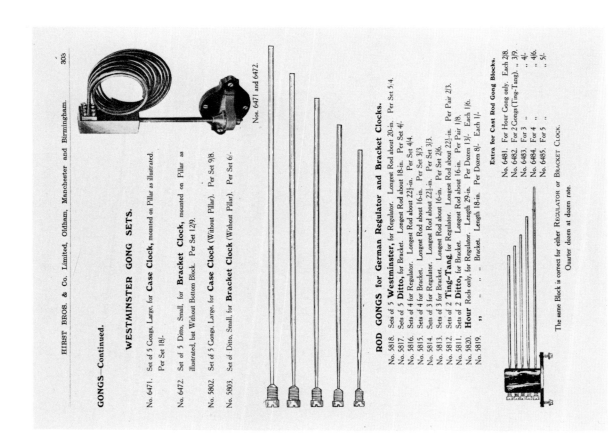

GONGS—Continued.

WESTMINSTER GONG SETS.

No. 6471. Set of 5 Gongs, Large, for **Case Clock**, mounted on Pillar as illustrated. Per Set 18/-

No. 6472. Set of 5 Ditto, Small, for **Bracket Clock**, mounted on Pillar as illustrated, but Without Bottom Block.   Per Set 12/9.

No. 5802. Set of 5 Gongs, Large, for **Case Clock** (Without Pillar).   Per Set 9/8.

No. 5803. Set of Ditto, Small, for **Bracket Clock** (Without Pillar).   Per Set 6/-

Nos. 6471 and 6472.

ROD GONGS for German Regulator and Bracket Clocks.

No. 5818. Sets of 5 **Westminster**, for Regulator. Longest Rod about 20-in.   Per Set 5/4.
No. 5817. Sets of 5 **Ditto**, for Bracket. Longest Rod about 18-in.   Per Set 4/-
No. 5816. Sets of 4 for Regulator. Longest Rod about 22½-in.   Per Set 4/4.
No. 5815. Sets of 4 for Bracket. Longest Rod about 16-in.   Per Set 3/3.
No. 5814. Sets of 3 for Regulator. Longest Rod about 22½-in.   Per Set 3/3.
No. 5813. Sets of 3 for Bracket. Longest Rod about 16-in.   Per Set 2/6.
No. 5812. Sets of 2 **Ting-Tang**, for Regulator. Longest Rod about 22½-in.   Per Pair 2/3.
No. 5811. Sets of 2 **Ditto**, for Bracket. Longest Rod about 16-in.   Per Pair 1/8.
No. 5820. **Hour** Rods only, for Regulator.   Length 29-in.   Per Dozen 13/-   Each 1/6.
No. 5819.   „   „   „   for Bracket.   Length 18-in.   Per Dozen 8/-   Each 1/-

Extra for Cast Rod Gong Blocks.

No. 6481. For Hour Gong only.   Each 2/8.
No. 6482. For 2 Gongs (Ting-Tang).   „   3/9.
No. 6483. For 3   „   „   „   4/-
No. 6484. For 4   „   „   „   4/6.
No. 6485. For 5   „   „   „   5/-

The same Block is correct for either REGULATOR or BRACKET CLOCK.
Quarter dozen at dozen rate.

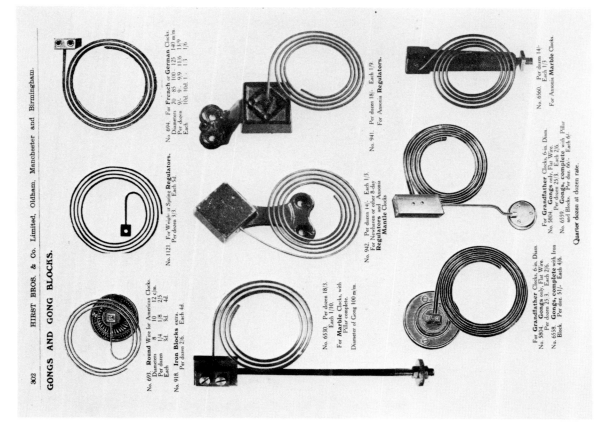

GONGS AND GONG BLOCKS.

No. 693. **Round Wire** for American Clocks.

Diameters	8	10	12 c/m.
Per dozen	1/4	1/8	2/5
Each	3d.	3d.	4d.

No. 918. **Iron Blocks** extra.
Per dozen 2/6.   Each 4d.

No. 694. For **French** or **German Clocks.**

Diameters	70	85	100	125	140 m/m.
Per dozen	9/-	9/-	9/9	11/6	13/9
Each	10d.	10d.	1/-	1/3	1/6

No. 1121. For Weight or Spring **Regulators.**
Per dozen 3/3.   Each 5d.

No. 6530.   Per dozen 18/3.   Each 1/10.
For **Marble Clocks**, with Pillar complete.
Diameter of Gong 100 m/m.

No. 941.   Per dozen 18/-   Each 1/9.
For Ansonia **Regulators.**

No. 942.   Per dozen 14/-   Each 1/3.
For Newhaven or other 8-day
Regulators and Ansonia
Mantle Clocks

No. 6560.   Per dozen 14/-
Each 1/3.
For Ansonia **Marble Clocks.**

For **Grandfather** Clocks, 6-in. Diam.
No. 5804. **Gongs** only, Flat Wire.
Per dozen 25/3.   Each 2/6.
No. 6559. **Gongs, complete** with Pillar
and Blocks.   Per doz. 66/-   Each 6/-

For **Grandfather** Clocks, 6-in. Diam.
No. 5804. **Gongs** only, Flat Wire.
Per dozen 25/3.   Each 2/6.
No. 6558. **Gongs, complete** with Iron
Block.   Per doz. 51/-   Each 4/6.

Quarter dozen at dozen rate.

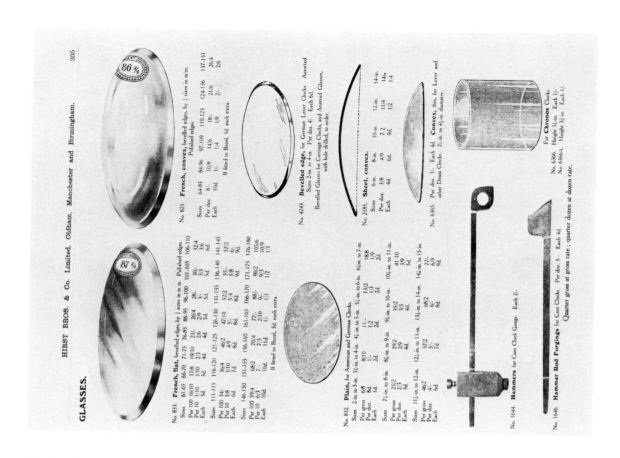

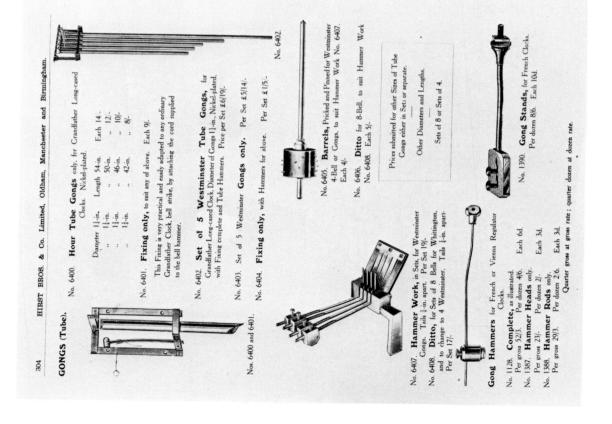

HIRST BROS. & Co. Limited, Oldham, Manchester and Birmingham.

GLASSES.

No. 833. **French, flat,** bevelled edges, by ¼ sizes in m.m. Polished edges.

Sizes	61-65	66-70	71-75	76-85	86-95	96-100	101-105	106-110
Per 100	16/10	17/8	19/10	23/-	26/4	28/-	30/-	32/4
Per 10	1/10	1/10	2/3	2/6	2/9	3/-	3/3	3/6
Each	3d.	3d.	4d.	4d.	5d.	5d.	5d.	6d.

Sizes	111-115	116-120	121-125	126-130	131-135	136-140	141-145
Per 100	34/-	36/4	46/2	47/10	55/-	58/-	57/2
Per 10	3/8	3/10	4/9	5/-	5/8	6/-	6/-
Each	7d.	7d.	8d.	8d.	8d.	9d.	9d.

Sizes	146-150	151-155	156-160	161-165	166-170	171-175	176-180
Per 100	59/4	68/2	68/2	77/-	88/-	90/2	105/6
Per 10	6/3	7/-	7/3	7/10	9/-	9/3	10/9
Each	10d.	10d.	11d.	1/-	1/-	1/2	1/3

If fitted to Bezel, 3d. each extra.

No. 825. **French, convex,** bevelled edges, by ¼ sizes in m.m. Polished edges.

Sizes	64-83	84-96	97-109	110-123	124-136	137-151
Per doz.	8/-	10/8	14/6	18/-	21/6	26/4
Each	10d.	1/4	1/9	2/-	2/3	2/6

If fitted to Bezel, 3d. each extra.

No. 4243. **Bevelled edge,** for German Lever Clocks. Assorted Sizes 2-in. to 4-in. Per doz. 4/-. Each 6d. Bevelled Glasses for Carriage Clocks, and Aneroid Glasses, with hole drilled, to order.

No. 2185. **Sheet, convex.**

Sizes	6-in.	8-in.	10-in.	12-in.	14-in.
Per doz.	3/8	4/9	7-2	11/4	14/-
Each	4d.	6d.	9d.	1/2	1/4

No. 6365. Per doz. 3/-. Each 4d. **Convex,** thin, for Lever and other Drum Clocks. 2½-in. to 4½-in. diameter.

For **Chronos** Clocks.
No. 6366. Height 3½-in. Each 1/-.
No. 6366A. Height 3½-in. Each 1/-.

No. 832. **Plain,** for American and German Clocks.

Sizes	2-in. to 3½-in.	3½-in. to 4-in.	4½-in. to 5-in.	5½-in. to 6-in.	6½in. to 7-in.
Per gross	6/8	8/10	11/-	13/2	18/8
Per doz.	8d.	1/-	1/2	1/3	1/9
Each	1d.	2d.	2d.	2d.	2d.

Sizes	7½-in. to 8-in.	8½-in. to 10-in.	10½-in. to 11-in.
Per gross	23/2	29/2	35/2
Per doz.	2/3	2/9	3/3
Each	4d.	4d.	4d.

Sizes	11½-in. to 12-in.	12½-in. to 13-in.	13½-in. to 14-in.	14½-in. to 15-in.
Per gross	41/10	57/2	68/2	77/-
Per doz.	3/9	5/-	6/-	6/9
Each	5d.	5d.	7d.	8d.	9d.

No. 1644. **Hammers** for Case Clock Gongs. Each 2/-.

No. 1646. **Hammer Rod Forgings** for Case Clocks. Per doz. 3/-. Each 4d.

Quarter gross at gross rate; quarter dozen at dozen rate.

HIRST BROS. & Co. Limited, Oldham, Manchester and Birmingham.

GONGS (Tube).

No. 6400. **Hour Tube Gongs** only, for Grandfather Long-cased Clocks. Nickel-plated.

Diameter	Length	Each
1½-in.	54-in.	14/-
1½-in.	50-in.	12/-
1¼-in.	46-in.	10/-
1⅛-in.	42-in.	8/-

No. 6401. **Fixing only,** to suit any of above. Each 9/-.

This Fixing is very practical and easily adapted to any ordinary Grandfather Clock, bell strike, by attaching the cord supplied to the bell hammer.

No. 6402. **Set of 5 Westminster Tube Gongs,** for Grandfather Long-cased Clock, Diameter of Gongs 1½-in. Nickel-plated, with Fixing complete and Tube Hammers. Price per Set £6/19/-.

No. 6403. Set of 5 Westminster Gongs only. Per Set £5/14/-.

No. 6404. **Fixing only,** with Hammers for above. Per Set £1/5/-.

No. 6402.

No. 6405. **Barrels,** Pricked and Pinned for Westminster 4-Bell or Gongs, to suit Hammer Work No. 6407. Each 4/-.

No. 6406. **Ditto** for 8-Bell, to suit Hammer Work No. 6408. Each 5/-.

Prices submitted for other Sizes of Tube Gongs either in Sets or separate.

Other Diameters and Lengths.

Sets of 8 or Sets of 4.

No. 6407. **Hammer Work,** in Sets, for Westminster Gongs. Tails ½-in. apart. Per Set 19/-.

No. 6408. **Ditto,** for Sets of 8 Bells for Whittington, and to change to 4 Westminster. Tails ½-in. apart. Per Set 17/-.

Gong Hammers

for French or Vienna Regulator Clocks.

No. 1128. **Complete,** as illustrated. Per gross 52/3. Per dozen 4/6. Each 6d.

No. 1387. **Hammer Heads** only. Per gross 23/-. Per dozen 2/-. Each 3d.

No. 1388. **Hammer Rods** only. Per gross 29/3. Per dozen 2/6. Each 3d.

No. 1390. **Gong Stands,** for French Clocks. Per dozen 8/6. Each 10d.

Quarter gross at gross rate; quarter dozen at dozen rate.

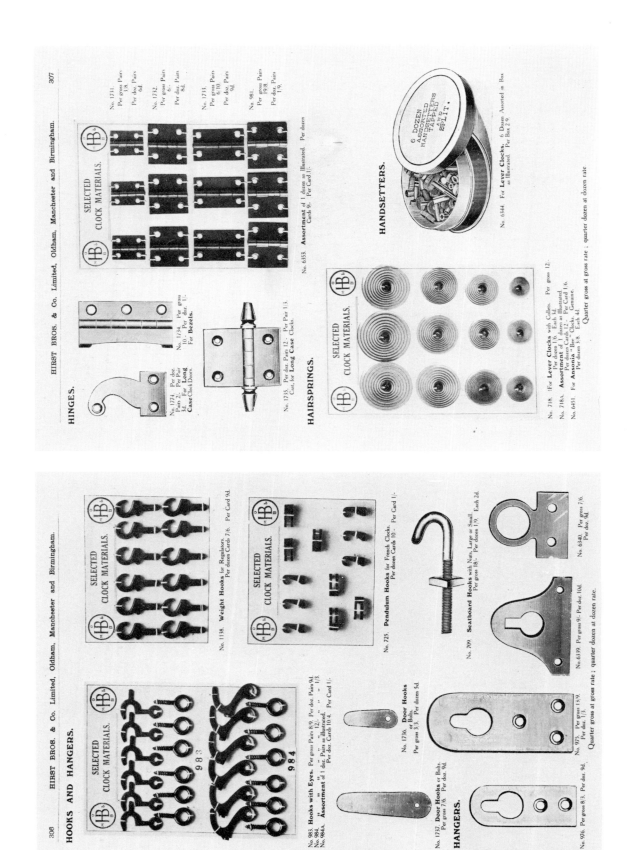

HINGES.

SELECTED CLOCK MATERIALS.

No. 6353. Assortment of 1 dozen as Illustrated. Per dozen Card 9/. Per Card 1/-.

No. 1731. Per gross Pairs 3/8. Per doz. Pairs 6d.
No. 1732. Per gross Pairs 6/-. Per doz. Pairs 8d.
No. 1733. Per gross Pairs 6/10. Per doz. Pairs 9d.
No. 981. Per gross Pairs 19/8. Per doz. Pairs 1/9.

No. 1774. Per doz. Pairs 2/-. Per Pair 3d. For Long Case Clock Doors.

No. 1734. Per doz. Pairs 12/-. Per gross 10/-. Per Pair 1/-. For Bezels.

No. 1735. Per doz. Pairs 2/-. Per Pair 1/3. Cast. for Long Case Clocks.

HANDSETTERS.

6 DOZEN ASSORTED HANDSETTERS TAPPED AND SPLIT.

No. 6344. For Lever Clocks. 6 Dozen Assorted in Box as Illustrated. Per Box 2/9.

HAIRSPRINGS.

SELECTED CLOCK MATERIALS.

No. 718. 1For Lever Clocks with Collets. Per gross 12/-. Per dozen 1/6. Each 1d.
No. 718A. Assortment of 1 dozen as Illustrated. Per dozen Card 12/-. Per Card 1/6.
No. 6451. For Ansonia "Bee" Clocks. Genuine. Per dozen 3/8. Each 4d.

Quarter gross at gross rate at dozen rate.

HOOKS AND HANGERS.

SELECTED CLOCK MATERIALS.

No. 1158. Weight Hooks for Regulators. Per dozen Cards 7/6. Per Card 9d.

SELECTED CLOCK MATERIALS.

No. 725. Pendulum Hooks for French Clocks. Per dozen Cards 10/-. Per Card 1/-.

No. 209. Seatboard Hooks with Nut, Large or Small. Per gross 18/-. Per dozen 1/9. Each 2d.

No. 6340. Per gross 7/6. Per doz. 9d.

No. 6339. Per gross 9/-. Per doz. 10d.

SELECTED CLOCK MATERIALS.

No. 983. Hooks with Eyes. Per gross Pairs 8/9. Per doz. Pair 9d.
No. 984. " " " 12/-. " " 1/3.
No. 984A. Assortment of 1 doz. Pairs as illustrated. Per doz. Cards 10/4. Per Card 1/-.

No. 1736. Door Hooks or Bolts. Per gross 3/3. Per dozen 5d.

No. 1737. Door Hooks or Bolts. Per gross 7/6. Per doz. 9d.

HANGERS.

No. 976. Per gross 8/3. Per doz. 9d.

No. 975. Per gross 15/9. Per doz. 1/3.

Quarter gross at gross rate; quarter dozen at dozen rate.

HIRST BROS. & Co. Limited, Oldham, Manchester and Birmingham.

CLOCK HANDS—Various.

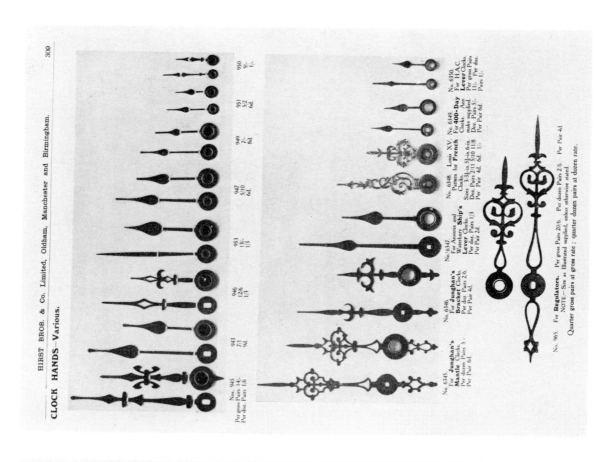

HIRST BROS. & Co. Limited, Oldham, Manchester and Birmingham.

HANDSETTERS,
HAMMERSPRINGS AND
SECONDS HANDS.

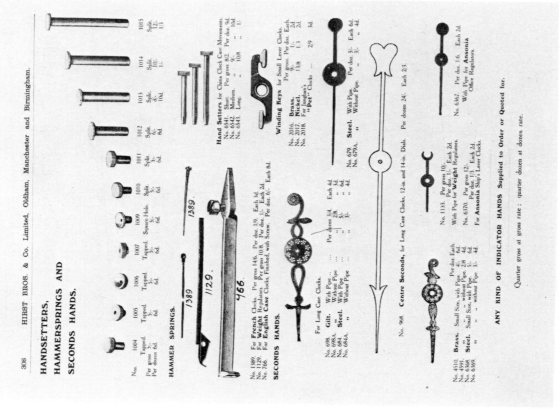

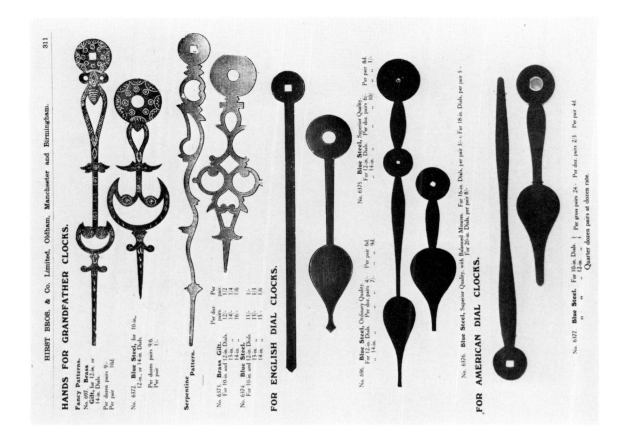

HIRST BROS. & Co., Limited, Oldham, Manchester and Birmingham.

HANDS FOR GRANDFATHER CLOCKS.

Fancy Patterns.
No. 697. **Brass Gilt**, for 12-in. or 14-in. Dials.
Per dozen pairs 9/-
Per pair 10d.

No. 6372. **Blue Steel**, for 10-in., 12-in., or 14-in. Dials.
Per dozen pairs 9/6.
Per pair 1/-

Serpentine Pattern.

	Per doz. pairs	Per pair.
No. 6371. **Brass Gilt**, For 10-in. and 12-in. Dials.	12/-	1/2
13-in. "	14/-	1/4
14-in. "	16/-	1/6
No. 6374. **Blue Steel**, For 10-in. and 12-in. Dials.	11/-	1/-
13-in. "	13/-	1/3
14-in. "	15/-	1/6

FOR ENGLISH DIAL CLOCKS.

No. 690. **Blue Steel**, Ordinary Quality. For 12-in. Dials. Per pair 4/-. Per pair 6d.
14-in. " Per pair 7/-. Per pair 9d.

No. 6375. **Blue Steel**, Superior Quality. For 12-in. Dials. Per doz. pairs 6/-. Per pair 8d.
14-in. " Per doz. pairs 10/-. Per pair 1/-

FOR AMERICAN DIAL CLOCKS.

No. 6376. **Blue Steel**, Superior Quality, with Balanced Minutes. For 16-in. Dials, per pair 3/-. For 18-in. Dials, per pair 5/-. For 20-in. Dials, per pair 8/-.

No. 6377. **Blue Steel**. For 10-in. Dials. } Per gross pairs 24/-. Per doz. pairs 2/3. Per pair 4d.
12-in. " Quarter dozen pairs at dozen rate.

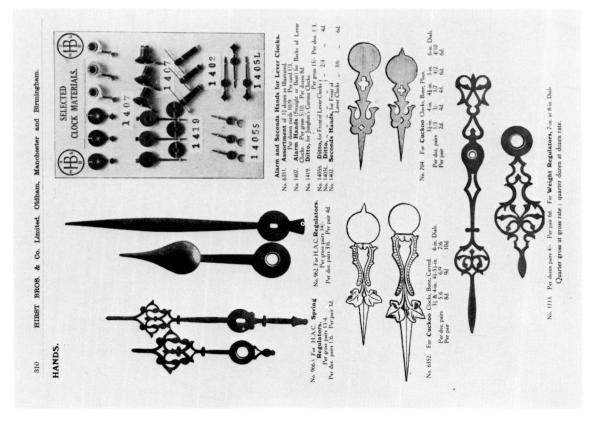

HIRST BROS. & Co., Limited, Oldham, Manchester and Birmingham.

HANDS.

SELECTED CLOCK MATERIALS.

1407   1402
1419   1405L
1405S   14405S

Alarm and Seconds Hands for Lever Clocks.
No. 6351. Assortment of 11 dozen as Illustrated. Per dozen cards 10/9. Per card 1/3.
No. 1407. **Alarm Hands** (Straight or Bent) for Backs of Lever Clocks. Per gross 5/10. Per dozen 8d.
No. 1419. **Ditto**, for Junghan's Corona Clocks. Per gross 13/-. Per doz. 1/3.
No. 1405S. **Ditto**, for Front of Lever Clocks } Per gross 2/4. " 4d.
No. 1405L. **Ditto**, " " 2/4 " 4d.
No. 1402. **Seconds Hands**, for Front of Lever Clocks. " 3/6. " 6d.

No. 966. For H.A.C. **Spring Regulators.**
Per gross pairs 13/4.
Per doz. pairs 1/6. Per pair 3d.

No. 962. For H.A.C. **Regulators.**
Per gross pairs 34/-.
Per doz. pairs 3/6. Per pair 4d.

No. 6352. For **Cuckoo** Clocks, Bone, Carved.
Per doz. pairs 7/6.
Per pair 8d. 9d.
6-in. Dial.
7/6 10d.

No. 704. For **Cuckoo** Clocks, Bone, Plain.
	3½-in.	4-in.	4½-in.	5-in.	6-in. Dial.
Per doz. pairs	1/3	3/-	3/7	4/2	4/10
Per pair	2d.	4d.	4d.	6d.	6d.

No. 1113. Per dozen pairs 4/-. Per pair 6d. For **Weight Regulators**, 7-in. or 8-in. Dials.
Quarter gross at gross rate : quarter dozen at dozen rate.

HIRST BROS. & Co. Limited, Oldham, Manchester and Birmingham.

KEYS.

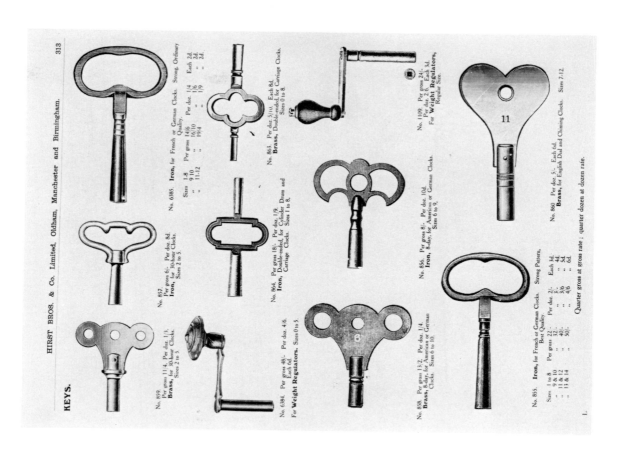

No. 859. Per gross 11/4. Per doz 1/3. **Brass**, for 30-hour Clocks. Sizes 2 to 5.

No. 857. Per gross 6/-. Per doz. 8d. **Iron**, for 30-hour Clocks. Sizes 2 to 5.

No. 6385. **Iron**, for French or German Clocks. Strong. Ordinary Quality.

Per doz. 1/4 Each 2d.
16/10 1/6 .. 2d.
19/4 1/9 .. 2d.

Sizes 1-8
,, 9-10
,, 11-12

No. 863. Per doz 5/10. Each 8d. **Brass**, Double-ended, for Carriage Clocks. Sizes 0 to 8.

No. 1109. Per gross 24/-. Per doz. 2/6 Each 3d. For **Weight Regulators**, Regular Size.

No. 6384. Per gross 48/-. Per doz. 4/6. Each 6d. For **Weight Regulators**. Sizes 0 to 5.

No. 864. Per gross 18/-. Per doz. 1/9. **Iron**, Double-ended, for Cylinder Drum and Carriage Clocks. Sizes 1 to 8.

No. 856. Per gross 8/-. Per doz. 10d. **Iron**, 8-day, for American or German Clocks. Sizes 6 to 9.

No. 860. Per doz. 5/-. Each 6d. **Brass**, for English Dial and Chiming Clocks. Sizes 7-12.

No. 858. Per gross 13/2. Per doz. 1/4. **Brass**, 8-day, for American or German Clocks. Sizes 6 to 10.

No. 853. **Iron**, for French or German Clocks. Strong Pattern.

Best Quality.

Sizes	Per gross	Per doz.		Strong Pattern	
1 to 8	22/-	32/-	2/-	Each 3d.	4d.
9 & 10		40/-	3/-	.. 4d.	5d.
11 & 12		50/-	4/6	.. 5d.	6d.
13 & 14					

Quarter gross at gross rate; quarter dozen at dozen rate.

HIRST BROS. & Co. Limited, Oldham, Manchester and Birmingham.

HANDS FOR FRENCH CLOCKS.

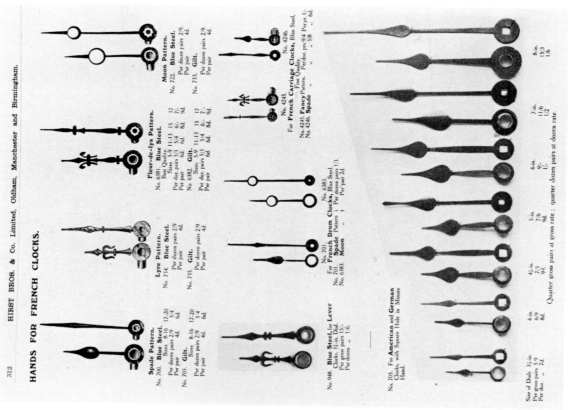

Spade Pattern.
No. 700. **Blue Steel.**
Sizes 8-16 17-20
Per dozen pairs 2/9 3/4
Per pair .. 4d.
No. 703. **Gilt.**
Sizes 8-16 17-20
Per dozen pairs 2/9 3/4
Per pair .. 4d. 6d.

Lyre Pattern.
No. 714. **Blue Steel.**
Per dozen pairs 2/9
Per pair .. 4d.
No. 735. **Gilt.**
Per dozen pairs 2/9
Per pair .. 4d.

Fleur-de-lys Pattern.
No. 6381. **Blue Steel.**
Best Quality.
Sizes 3-9 11-13 15 17
Per doz. pairs 5/1 5/4 6/- 8d.
Per pair .. 6d. 8d. 9d.
No. 6382. **Gilt.**
Sizes 3-9 11-13 15 17
Per doz. pairs 5/1 5/4 6/- 7/-
Per pair .. 6d. 8d. 9d.

Moon Pattern.
No. 722. **Blue Steel.**
Per dozen pairs 2/9
Per pair .. 4d.
No. 733. **Gilt.**
Per dozen pairs 2/9
Per pair .. 4d.

No. 4246.
For French Carriage Clocks, Blue Steel.
Fine Quality.
No. 4245. **Fancy** Pattern. Per doz. prs 9/4 Per pr. 1/-
No. 4246. **Spade** ,, ,, 5/8 ,, 6d.

No. 948. **Blue Steel**, for Lever Clocks, 3½-in. Dial.
Per gross pairs 15/-
Per doz. .. 1/6.

No. 6383.
For **French Drum Clocks**, Blue Steel.
No. 701. **Spade** Pattern. Per dozen pairs 1/3.
No. 6383. **Moon** ,, ,, Per pair 2d.

No. 705. For **American and German** Clocks, with Square Hole in Minute Hand.

Size of Dial	3½-in.	4-in.	4½-in.	5-in.	6-in.	7-in.	8-in.
Per gross pair	5/9	6/9	7/3	7/6	9/-	11/6	15/3
Per doz. ..	7d.	8d.	9d.	9d.	1/-	1/2	1/6

Quarter gross pairs at gross rate; quarter dozen pairs at dozen rate.

HIRST BROS. & Co. Limited, Oldham, Manchester and Birmingham.

LINES, GUT, CORDS, &c.

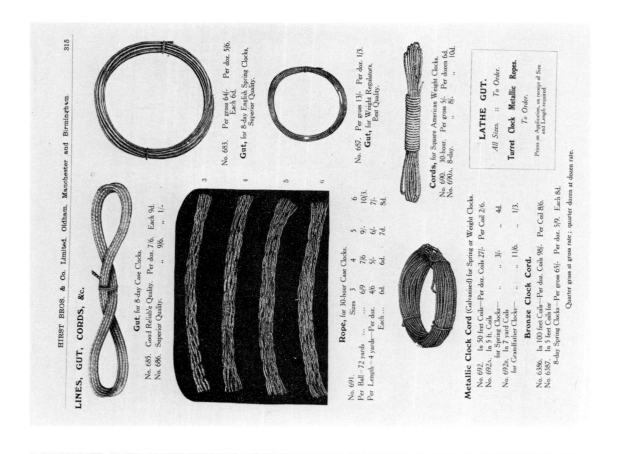

Gut, for 8-day Case Clocks.
No. 685. Good Reliable Quality. Per doz. 7/6. Each 9d.
No. 686. Superior Quality.   „ 9/6.   „ 1/-

No. 683. Per gross 64/-. Each 6d. **Gut**, for 8-day English Spring Clocks, Superior Quality.

No. 687. Per gross 13/-. Per doz. 1/3. **Gut**, for Weight Regulators, Best Quality.

Cords, for Square American Weight Clocks.
No. 690. 30-hour. Per gross 5/-. Per dozen 6d.
No. 690A. 8-day.   „ 8/-.   „ 10d.

Rope, for 30-hour Case Clocks.

Sizes	3	4	5	6
No. 691.				
Per Ball — 72 yards ...	6/9	7/6	9/-	10/3.
Per Length — 4 yards—Per doz.	4/6	5/-	6/-	7/-
Each ...	6d.	6d.	7d.	9d.

LATHE GUT.
All Sizes. :: To Order.
Turret Clock Metallic Ropes.
To Order.
Prices on Application, on receipt of Size and Length required.

Metallic Clock Cord (Galvanised) for Spring or Weight Clocks.
No. 692. In 50 feet Coil—Per doz. Coils 27/-. Per Coil 2/6.
No. 692A. In 5 ft. Coils   „ 3/-.   „ 4d.
No. 692H. In 7 yard Coils for Spring Clocks „ 11/6.   „ 1/3.
for Grandfather Clocks—

Bronze Clock Cord.
No. 6386. In 100 feet Coils—Per doz. Coils 98/-. Per Coil 8/6.
No. 6387. In 5 feet Coils for 8-day Spring Clocks—Per gross 65/-. Each 8d.

Quarter gross at gross rate ; quarter dozen at dozen rate.

KEYS—Continued.

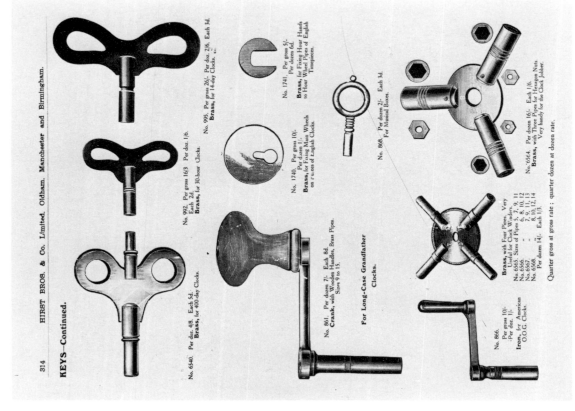

No. 6540. Per doz. 4/8. Each 5d. **Brass**, for 400-day Clocks.

No. 992. Per gross 16/3. Per doz. 1/6. Each 2d. **Brass**, for 30-hour Clocks.

No. 995. Per gross 26/-. Per doz. 2/6. Each 3d. **Brass**, for 14-day Clocks. &c.

No. 1740. Per gross 10/-. Per dozen 1/-. **Brass**, for Fixing Main Wheels on arbors of English Clocks.

No. 1741. Per gross 5/-. Per dozen 6d. **Brass**, for Fixing Hour Hands to Hour Wheel Pipes of English Timepieces.

No. 868. Per dozen 2/-. Each 3d. For Musical Boxes.

No. 861. Per dozen 7/-. Each 8d. **Crank**, with Wooden Handles, Brass Pipes. Sizes 9 to 15.

For Long-Case Grandfather Clocks.

No. 866. Per gross 10/-. Per doz. 1/-. **Iron**, for American O.O.G. Clocks.

No. 6564. Per dozen 16/-. Each 1/6. **Brass**, with Three Pipes for Hexagon Nuts. Very handy for the Clock Jobber.

Brass, with Four Pipes. Very Useful for Clock Winders.
No. 6565. Sizes of Pipes 5, 7, 9, 11
No. 6566.   „ 6, 8, 10, 13
No. 6567.   „ 7, 9, 11, 13
No. 6568.   „ 8, 10, 12, 14
Per dozen 14/-. Each 1/3.

Quarter gross at gross rate ; quarter dozen at dozen rate.

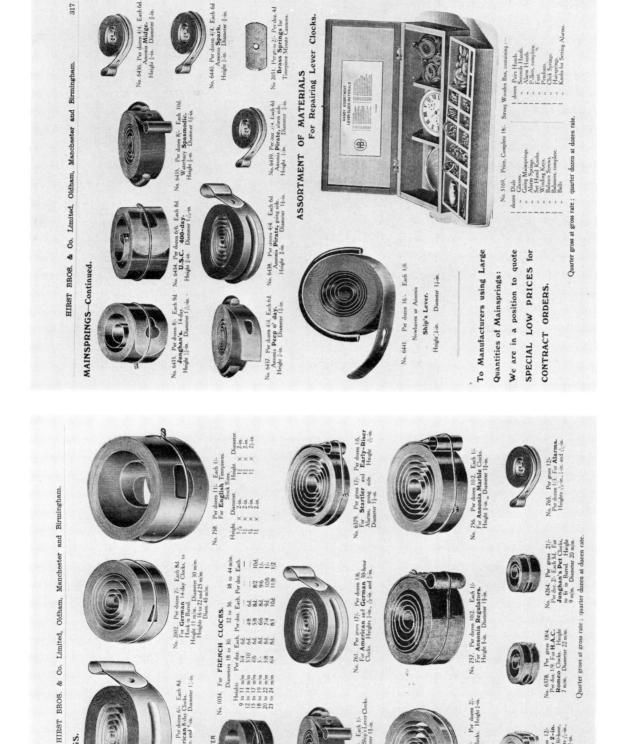

HIRST BROS. & Co. Limited, Oldham, Manchester and Birmingham.

MAINSPRINGS—Continued.

No. 6436. Per dozen 4/4. Each 6d. Ansonia **Midge.** Height ⅜-in. Diameter ⅞-in.

No. 6440. Per dozen 4/4. Each 6d. Ansonia **Spark.** Height ⅜-in. Diameter ⅞-in.

No. 2051. Per gross 2/-. Per doz. 4d. **Brass Springs** for Timepiece Minute Cannons.

No. 6433. Per dozen 8/-. Each 9d. **Junghan's.** 14-day. Height ⅝-in. Diameter 1⅛-in.

No. 6434. Per dozen 6/6. Each 8d. **U.S.C.** 400-day. Height ⅜-in. Diameter 1⅛-in.

No. 6435. Per dozen 8/-. Each 10d. Waterbury **Spasmodic.** Height ⅜-in. Diameter 1⅛-in.

No. 6439. Per doz. 2/4. Each 4d. Ansonia **Pirate,** alarm side. Height ⅜-in. Diameter 1¼-in.

No. 6437. Per dozen 4/4. Each 6d. Ansonia **Peep o' day.** Height ⅜-in. Diameter 1¼-in.

No. 6438. Per dozen 4/4. Each 6d. Ansonia **Pirate,** going side. Height ⅜-in. Diameter 1¼-in.

No. 6441. Per dozen 16/-. Each 1/6. Newhaven or Ansonia **Ship's Lever.** Height ⅜-in. Diameter 1¼-in.

ASSORTMENT OF MATERIALS
For Repairing Lever Clocks.

No. 5169. Price, Complete 18/-. Strong Wooden Box, containing:—

1 dozen Pairs Hands.
" " Seconds Hands.
" " Alarm Hands.
" " Pillars, complete.
" " Feet.
" " Pendants.
" " Click Springs.
" " Hairpins.
" " Knobs for Setting Alarms.

1 dozen Dials
" " Glasses.
" " Going Mainsprings.
" " Alarm Springs.
" " Set Hand Knobs.
" " Winding Keys.
" " Balance Screws.
" " Balances, complete.
" " Bells.

Quarter gross at gross rate : quarter dozen at dozen rate.

To Manufacturers using Large Quantities of Mainsprings:

We are in a position to quote

SPECIAL LOW PRICES for

CONTRACT ORDERS.

HIRST BROS. & Co. Limited, Oldham, Manchester and Birmingham.

MAINSPRINGS.

No. 759. Per gross 70/-. Per dozen 6/-. Each 8d. For **American** 8-day Clocks. Heights ⅜-in., 1¼-in. and ⅝-in. Diameter 1¼-in.

No. 2002. Per dozen 7/-. Each 8d. For **German** 14-day Clocks, to Hook in Barrel. Height 11 m/m. Diameter 30 m/m. Heights 16 to 22 and 25 m/m. Diam. 45 m/m.

No. 758. Per dozen 11/-. Each 1/-. For **English Timepieces.**
Stock Sizes.

Height.	Diameter.		Diameter.
1⅛	× 2-in.	1⅜ ×	2-in.
1⅛	× 2-in.	1½ ×	2-in.
1⅛	× 2-in.	1½ ×	2¼-in.

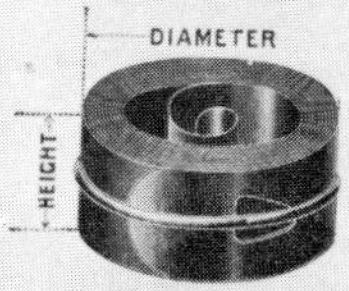

No. 1034. For **FRENCH CLOCKS.**

Diameters	18 to 30.	32 to 36.	38 to 44 m/m.
Heights	Each. Per doz.	Each.	Per doz. Each.
9 to 11 m/m	3/4	6d.	4/8 6d.
12 to 14 m/m	3/10	6d.	5/8 8d.
15 to 17 m/m	4/6	6d.	6/6 8d. 8/2 10d.
18 to 19 m/m	5/-	8d.	7/4 9d. 9/6 1/-
20 to 22 m/m	5/8	8d.	8/1 10d. 10/6 1/-
23 to 24 m/m	6/4	8d.	8/1 10d. 11/8 1/2

No. 6379. Per gross 17/-. Per dozen 1/6. For **Starticr** and **Early-Riser** Alarms, gong side. Height ½-in. Diameter 1¼-in.

No. 761. Per dozen 17/-. Each 1/6. For **American** and **German** 30-hour Clocks. Heights ⅜-in., 1¼-in. and ⅝-in.

No. 736. Per dozen 10/2. Each 1/-. For **Ansonia Marble** Clocks. Height ⅝-in. Diameter 1¼-in.

No. 765. Per gross 12/-. Per dozen 1/3. For **Alarms.** Heights ¹¹⁄₁₆-in., ⅝-in. and ⅝-in.

No. 754. Per dozen 10/8. Each 1/-. For **Ansonia** 8-day Ship's Lever Clocks. Height ½-in. Diameter 1¼-in.

No. 757. Per dozen 10/2. Each 1/-. For **Ansonia Regulators.** Height ⅝-in. Diameter 1¼-in.

No. 4264. Per gross 21/-. Per doz. 2/-. Each 3d. For **Junghan's Pet** Clocks, to fit in Barrel. Height 9 m/m. Diameter 20 m/m.

No. 764. Per dozen 22/-. Per doz. 2/-. Each 3d. For **Bee** Clocks. Height 1-in. Diameter 1⅝-in.

No. 6380. Per gross 12/-. Per doz. 1/3. For **2-in.** Drum Clocks, 30-hour, gong side. Heights ⅞-in., ⅝-in. and ⅝-in.

No. 6378. Per gross 18/4. Per doz. 1/9. For **H.A.C. Romeo** Clocks. Height 7 m/m. Diameter 22 m/m.

No. 6380. Per gross 12/-. Per doz. 1/3. For **Drum** Clocks, 30-hour, gong side.

Quarter gross at gross rate : quarter dozen at dozen rate.

HIRST BROS. & Co. Limited. Oldham, Manchester and Birmingham.

ORNAMENTS FOR CLOCK CASES.

COMPOSITION.

WOODEN KNOBS.

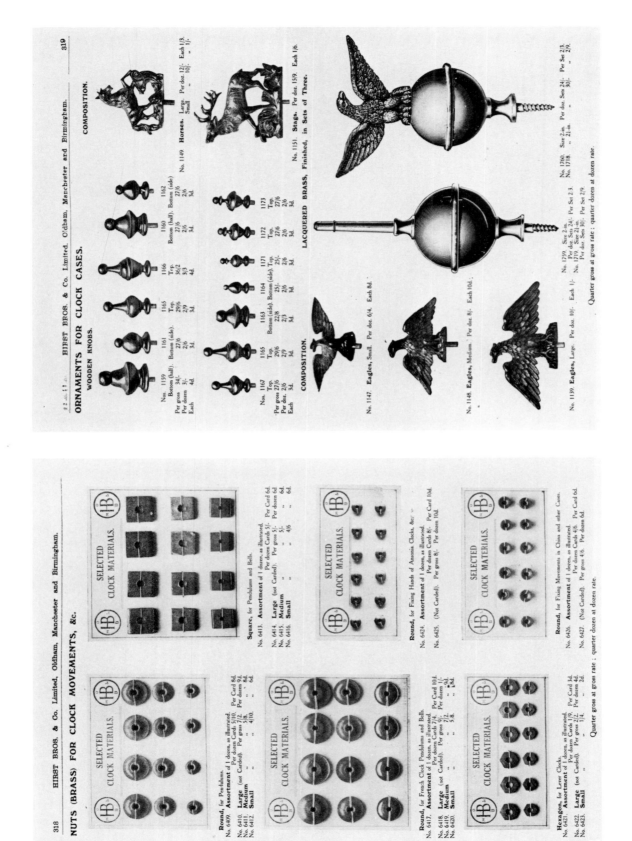

| Horses. | Large. | Per doz 12/-. | Each 1/3. |
| | Small. | „ 10/-. | „ 1/- |

No. 1149 Horses.

No. 1151. **Stags.** Per doz. 159. Each 1/6.

LACQUERED BRASS, Finished, in Sets of Three.

Nos.	1159 Bottom (half).	1161 Bottom (side).	1165 Top.	1166 Top.	1160 Bottom (half).	1162 Bottom (side).
Per gross	34/-	27/6	29/6	36/2	27/6	27/6
Per dozen	3/-	2/6	2/9	3/3	2/6	2/6
Each	4d.	3d.	3d.	4d.	3d.	3d.

Nos.	1167 Top.	1165 Top.	1163 Bottom (side).	1164 Bottom (side).	1171 Top.	1172 Top.	1173 Top.
Per gross	27/6	29/6	22/8	25/-	25/-	27/6	27/6
Per doz.	2/6	2/9	2/3	2/6	2/6	2/6	2/6
Each	3d.	3d.	3d.	3d.	3d.	3d.	3d.

COMPOSITION.

No. 1147. **Eagles,** Small. Per doz. 6/4. Each 8d.

No. 1148. **Eagles,** Medium. Per doz. 8/-. Each 10d.

No. 1139. **Eagles,** Large. Per doz. 10/-. Each 1/-.

No. 1759.	Size 2-in.	Per doz. Sets 24/-.	Per Set 2/3.
No. 1719.	Size 2½-in.	Per doz. Sets 24/-.	Per Set 2/3.
No. 1760.	Size 2-in.	Per doz. Sets 24/-.	Per Set 2/3.
No. 1718.	Size 2½-in.	Per doz. Sets 30/-.	Per Set 2/9.

Quarter gross at gross rate; quarter dozen at dozen rate.

---

318   HIRST BROS. & Co., Limited, Oldham, Manchester and Birmingham.

NUTS (BRASS) FOR CLOCK MOVEMENTS, &c.

SELECTED CLOCK MATERIALS.

Round, for Pendulums.

No. 6409. Assortment of 1 dozen, as illustrated.   Per Card 8d.
No. 6410. **Large** (not Carded).   Per dozen 5/10.   Per gross 7/2.
No. 6411. **Medium**   „   „ 5/8.   „ „ 9¼.
No. 6412. **Small**   „   „ 4/10.   „ „ 6d.

Square, for Pendulums and Bells.

No. 6413. Assortment of 1 dozen, as illustrated.   Per Card 6d.
No. 6414. **Large** (not Carded).   Per dozen 5/-.   Per gross 5/-.
No. 6415. **Medium**   „   „   „ 5/-.
No. 6416. **Small**   „   „   „ 4/6.

Round, for French Clock Pendulums and Bells.

No. 6417. Assortment of 1 dozen, as illustrated.   Per Card 10d.
No. 6418. **Large** (not Carded).   Per dozen 7/4.   Per gross 9/-.
No. 6419. **Medium**   „   „ 7/2.   „ „ 2/2.
No. 6420. **Small**   „   „ 5/8.   „ „ 1/8.

Round, for Fixing Hands of Ansonia Clocks, &c.

No. 6424. Assortment of 1 dozen, as illustrated.   Per Card 10d.
No. 6425. (Not Carded).   Per gross 8/-.   Per dozen 10d.

Hexagon, for Lever Clock.

No. 6421. Assortment of 1 dozen, as illustrated.   Per Card 3d.
No. 6422. **Large** (not Carded).   Per dozen 1/9.   Per gross 2/2.
No. 6423. **Small**   „   „ 1/4.   „ „ 2d.

Round, for Fixing Movements in China and other Cases.

No. 6426. Assortment of 1 dozen, as illustrated.   Per Card 4/6.
No. 6427. (Not Carded).   Per gross 4/6.   Per dozen 6d.

Quarter gross at gross rate; quarter dozen at dozen rate.

HIRST BROS. & Co. Limited, Oldham, Manchester and Birmingham.

PINIONS.

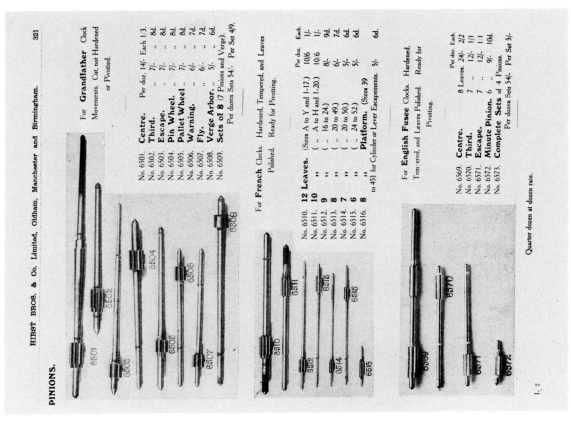

For Grandfather Clock Movements. Cut, not Hardened or Pivoted.

		Per doz.	Each.
No. 6501.	Centre.	14/-	1/3.
No. 6502.	Third.	7/-	8d.
No. 6503.	Escape.	7/-	8d.
No. 6504.	Pin Wheel.	7/-	8d.
No. 6505.	Pallet Wheel.	7/-	8d.
No. 6506.	Warning.	6/-	7d.
No. 6507.	Fly.	6/-	7d.
No. 6508.	Verge Arbor.	5/-	6d.
No. 6509.	Sets of 8 (7 Pinions and Verge).	Per dozen Sets 54/-	Per Set 4/9.

For French Clocks. Hardened, Tempered, and Leaves Polished. Ready for Pivoting.

			Per doz.	Each.
No. 6510.	12 Leaves.	(Sizes A to Y and 1-17.)	10/6	1/-
No. 6511.	10 "	(" A to H and 1-20.)	10/6	1/-
No. 6512.	9 "	(" 16 to 24.)	8/-	9d.
No. 6513.	8 "	(" 20 to 49.)	6/-	7d.
No. 6514.	7 "	(" 20 to 50.)	5/-	6d.
No. 6515.	6 "	(" 24 to 52.)	5/-	6d.
No. 6516.	8 "	Platform. (Sizes 39 to 45) for Cylinder or Lever Escapements.	5/-	6d.

For English Fusee Clocks. Hardened, Tempered, and Leaves Polished. Ready for Pivoting.

			Per doz.	Each.
No. 6569.	Centre.	8 Leaves.	24/-	2/2
No. 6570.	Third.	7 "	12/-	1/1
No. 6571.	Escape.	7 "	12/-	1/1
No. 6572.	Minute Pinion.	6 "	9/-	10d.
No. 6573.	Complete Sets of 4 Pinions.	Per dozen Sets 54/-	Per Set 5/-	

Quarter dozen at dozen rate.

HIRST BROS. & Co. Limited, Oldham, Manchester and Birmingham.

ORNAMENTS FOR CLOCKS (Continued), &c.

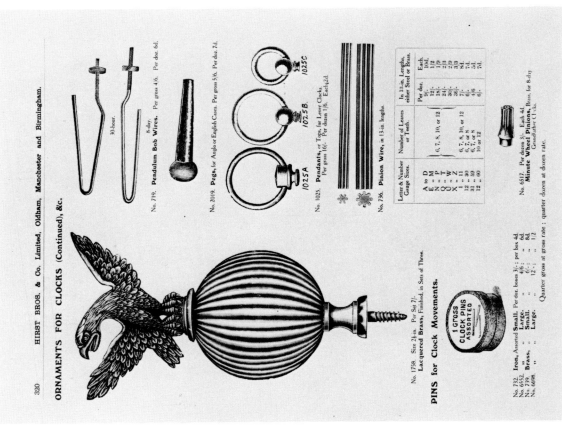

90-hour. 8-day.
No. 719. **Pendulum Bob Wires.** Per gross 4/6. Per doz. 6d.

No. 2019. **Pegs,** for Anglo or English Cases. Per gross 5/6. Per doz. 7d.

1025A 102.5B 1025C
No. 1023. **Pendants,** or Tops, for Lever Clocks. Per gross 16/- Per dozen 1/6. Each 2d.

No. 736. **Pinion Wire,** in 13-in. length.

Letter & Number Gauge Sizes.	Number of Leaves or Teeth.	In 13-in. Lengths, either Steel or Brass.	
		Per doz.	Each.
A to D		9/-	10d.
E " M		12/-	1/2
N " P	6, 7, 8, 10, or 12	13/-	1/3
Q " T		24/-	2/3
U " W		30/-	3/2
X " Z		30/-	3/8
11	6, 7, 8, 10, or 12	7/-	8d.
12 " 30	6, 7, or 8	6/-	7d.
31 " 59	4, 7, or 8	4/6	6d.
12 " 60	10 or 12	6/-	7d.

No. 6517. Per dozen 3/-. Each 4d. **Minute Wheel Pinions,** Brass, for 8-day Grandfather Clocks.

No. 1758. Size 2¼-in. Per Set 7/-. Lacquered Brass, Finished, in Sets of Three.

PINS for Clock Movements.

1 GROSS CLOCK PINS ASSORTED

No. 732.	Iron, Assorted Small. Per doz. boxes 3/-;	per box	4d.
No. 6152.	" Large.	4/6;	6d.
No. 739.	Brass, Small.	6/-;	8d.
No. 6698.	" Large.	12/-;	1/2

Quarter gross at gross rate; quarter dozen at dozen rate.

HIRST BROS. & Co. Limited, Oldham, Manchester and Birmingham.

PENDULUMS, &c.

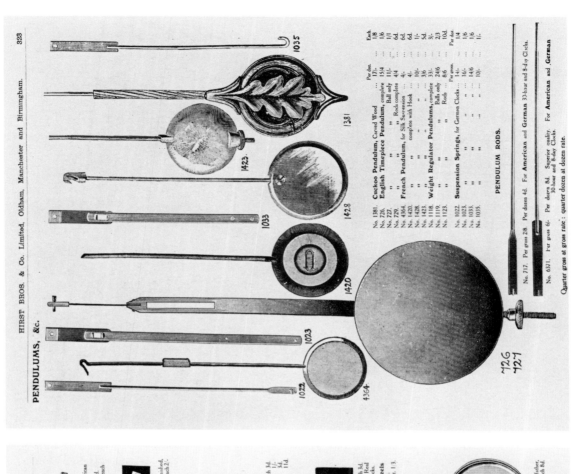

		Per doz.
No. 1381.	Cuckoo Pendulum, Carved Wood	17/-
No. 726.	English Timepiece Pendulum, complete	15/4
No. 727.	" " Ball only	11/-
No. 729.	" " Rods complete	4/4
No. 4364.	French Pendulum, for Silk Suspension	4/-
No. 1420.	" " complete with Hook	4/-
No. 1428.	" "	10/6
No. 1423.	" "	3/6
No. 1118.	Weight Regulator Pendulums, complete	33/-
No. 1119.	" " Ball only	24/6
No. 1123.	" " Rods	8/6

		Per doz.
		Each
	1/8
	1/6
	1/1
	6d.
	6d.
	6d.
	1/-
	5d.
	3/-
	2/3
	10d.

		Per gross
No. 1022.	Suspension Springs, for German Clocks	14/-
No. 1023.	" "	16/-
No. 1033.	" "	14/6
No. 1035.	" "	10/-

PENDULUM RODS.

No. 717. Per gross 2/8. Per dozen 4d. For American and German 30-hour and 8-day Clocks.

No. 6371. Per gross 6/- Per dozen 8d. Superior quality. For American and German 30-hour and 8-day Clocks.

Quarter gross at gross rate; quarter dozen at dozen rate.

HIRST BROS. & Co. Limited, Oldham, Manchester and Birmingham.

PALLETS, PALLET CRUTCHES AND PULLEYS.

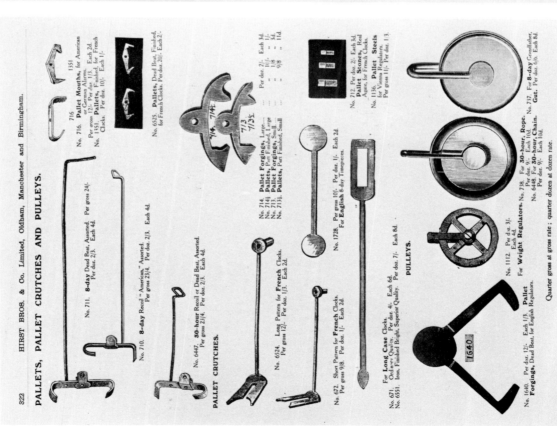

No. 716. Pallet Mouths, for American or German Alarms. Per gross 12/-. Per doz. 1/3. Each 2d.
No. 1351. Pallets, Finished, for French Clocks. Per doz. 10/-. Each 1/-

No. 6525. Pallets, Dead Beat, Finished, for French Clocks. Per doz. 20/-. Each 2/-

No. 711. 8-day Dead Beat, Assorted. Per gross 24/-

No. 710. 8-day Recoil "American," Assorted. Per gross 23/4. Per doz. 2/3. Each 4d.

No. 6447. 30-hour Recoil or Dead Beat, Assorted. Per gross 23/4. Per doz. 2/3. Each 4d.

PALLET CRUTCHES.

No. 6524. Long Pattern for French Clocks. Per gross 12/-. Per doz. 1/3. Each 2d.

No. 672. Short Pattern for French Clocks. Per gross 9/8. Per doz. 1/-. Each 2d.

	Per doz.
No. 714. Pallet Forgings, Large	2/-
No. 714½. Pallets, Part Finished, Large	10/-
No. 713. Pallet Forgings, Small	1/8
No. 713½. Pallets, Part Finished, Small	9/8

	Each
	3d.
	1/-
	3d.
	11d.

No. 1728. Per gross 10/-. Per doz. 1/-. Each 2d. For English 8-day Timepieces.

No. 712. Per doz. 2/-. Each 3d. Pallet Stones, Real Agate, for French Clocks.

No. 1136. Pallet Steels for Vienna Regulators. Per gross 11/-. Per doz. 1/3.

For Long Case Clocks.
No. 671. Ordinary Quality. Per doz. 4/-. Each 6d.
No. 6551. Iron, Finished Bright, Superior Quality. Per doz. 7/-. Each 8d.

PULLEYS.

No. 738. For 30-hour, Rope. Per doz. 9/-. Each 10d.
No. 6448. For 30-hour, Chain. Per doz. 9/-. Each 10d.
No. 737. For 8-day Grandfather, Gut. Per doz. 6/6. Each 8d.

No. 1112. Per doz. 3/-. Each 4d. For Weight Regulators.

No. 1640. Per doz. 12/-. Each 1/3. Pallet Forgings, Dead Beat, for English Regulators.

Quarter gross at gross rate; quarter dozen at dozen rate.

HIRST BROS. & Co. Limited, Oldham, Manchester and Birmingham.

PENDULUMS (Continued) and SCREWS.

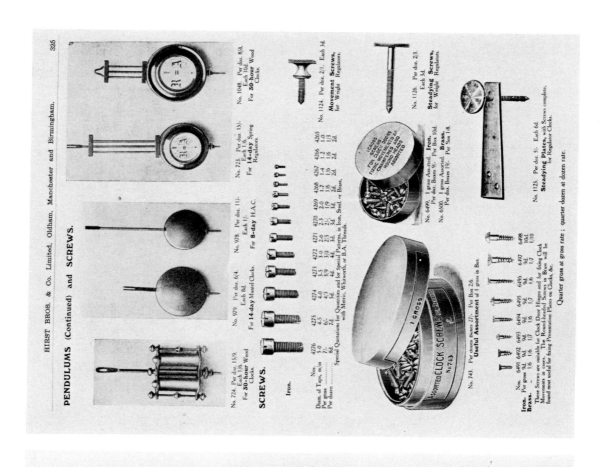

No. 724. Per doz. 15/9. Each 1/6. For **30-hour Wood Clocks.**

No. 979. Per doz. 6/4. Each 8d. For **14-day Mantel Clocks.**

No. 978. Per doz. 11/-. Each 1/-. For **8-day H.A.C.**

No. 723. Per doz. 15/-. Each 1/6. For **14-day Spring Regulators.**

No. 1048. Per doz. 8/3. Each 10d. For **30-hour Wood Clocks.**

No. 1124. Per doz. 2/1. Each 3d. **Movement Screws,** for Weight Regulators.

No. 1126. Per doz. 23. Each 3d. **Steadying Screws,** for Weight Regulators.

No. 1125. Per doz. 5/-. Each 6d. **Steadying Plates,** with Screws complete. Quarter gross at gross rate; quarter dozen at dozen rate.

SCREWS.

Iron.

Nos.	4276	4275	4274	4273	4272	4271	4270	4269	4268	4267	4266	4265
Diam. of Taps, m/m	5.0	4.5	4.0	3.5	3.0	2.6	2.3	2.0	1.7	1.4	1.2	1.0
Per gross	7/-	6/-	5/-	4/-	3/9	3/3	2/6	2/3	2/-	1/7	1/4	1/3
Per dozen	8d	7d	5d	4d	4d	4d	3d	3d	2d	2d	2d	2d

Special Quotations for Quantities and for Special Patterns, in Iron, Steel, or Brass, with Metric, Whitworth, or B.A. Threads.

No. 743. Per dozen Boxes 27. Per Box 26. **Useful Assortment** of 1 gross in Box.

No. 6499. 1 gross Assorted **Iron.** Per doz. Boxes 9/-. Per Box 10d.

No. 6500. 1 gross Assorted **Brass.** Per doz. Boxes 19/-. Per Box 18.

Nos.	6491	6492	6493	6494	6495	6496	6497	6498
Iron. Per gross	9d	9d	9d	9d	1/-	1/-	9d	10d
Brass. Per doz.	1/6	1/6	1/7	1/7	1/7	1/6	1/7	1/10

These Screws are suitable for Clock Door Hinges and for fixing Clock Movements in cases. The Round-headed Screws in Brass will be found most useful for fixing Presentation Plates on Clocks, &c.

Quarter gross at gross rate; quarter dozen at dozen rate.

HIRST BROS. & Co. Limited, Oldham, Manchester and Birmingham.

PENDULUMS, &c.—Continued.

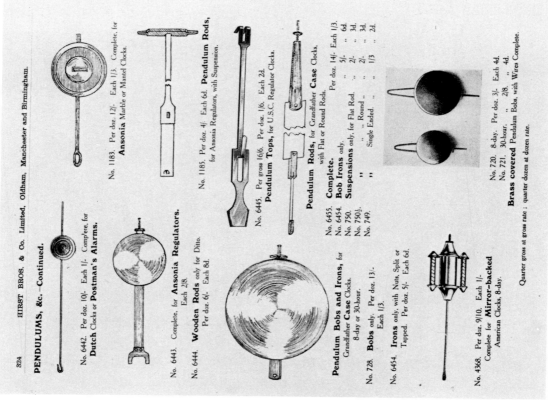

No. 6442. Per doz. 10/-. Each 1/-. Complete, for **Dutch Clocks or Postman's Alarms,**

No. 1183. Per doz. 12/-. Each 1/3. Complete, for **Ansonia Marble or Mantel Clocks.**

No. 6443. Complete, for **Ansonia Regulators.** Each 2/8.

No. 644. **Wooden Rods** only for Ditto. Per doz. 6/-. Each 8d.

No. 1185. Per doz. 4/. Each 6d. **Pendulum Rods,** for Ansonia Regulators, with Suspension.

No. 6445. Per gross 16/6. Per doz. 1/6. Each 2d. **Pendulum Tops,** for U.S.C. Regulator Clocks.

Pendulum Rods, for Grandfather **Case** Clocks, with Flat or Round Rods. **Complete.**

	Per doz.	Each
Complete.	14/-	1/3.
Bob Irons only.	5/-	6d.
Suspensions only, for Flat Rod.	2/-	3d.
" " Round.	2/-	3d.
" Single Ended.	1/3	2d.

No. 6455.
No. 6454.
No. 750.
No. 750½.
No. 749.

Pendulum Bobs and Irons, for Grandfather **Case** Clocks. 8-day or 30-hour.

No. 728. **Bobs** only. Per doz. 13/-. Each 1/3.

No. 6454. **Irons** only, with Nuts, Split or Tapped. Per doz. 5/-. Each 6d.

No. 4368. Per doz. 9/10. Each 1/-. Complete for **Mirror-backed** American Clocks, 8-day.

No. 720. 8-day. Per doz. 3/-. Each 4d.
No. 721. 30-hour. " 2/8. " 4d.
Brass covered Pendulum Bobs, with **Wire Complete.**

Quarter gross at gross rate; quarter dozen at dozen rate.

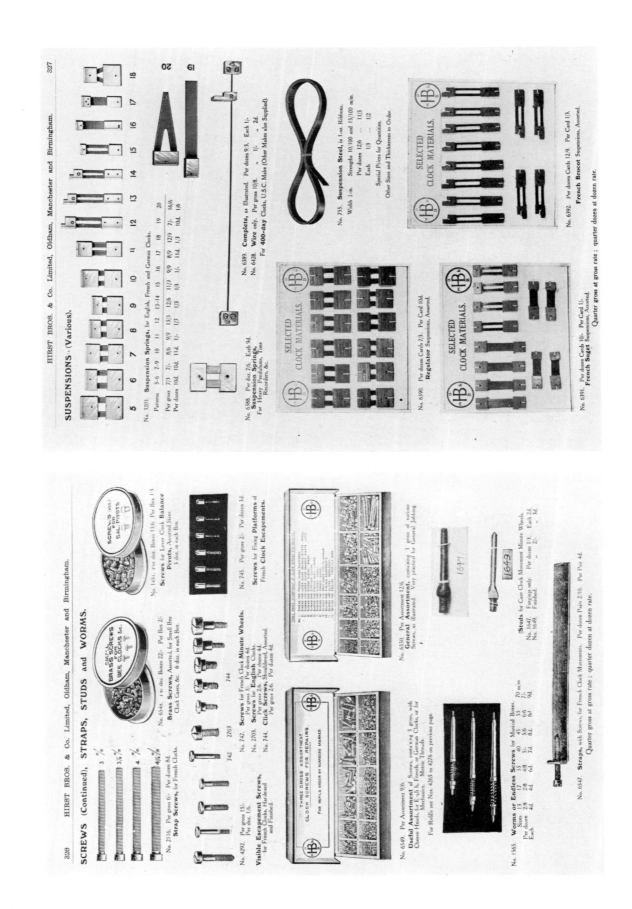

SUSPENSIONS - (Various).

No. 3203. **Suspension Springs**, for English, French and German Clocks.

Patterns	5–6	7–9	10	11	12	13–14	15	16	17	18	19	20
Per gross	7/3	7/–	8/6	9/9	13/3	12/6	11/9	9/9	8/9	12/9	7/–	16/6
Per dozen	10d.	10d.	11d.	1/–	1/3	1/3	1/3	1/3	11d.	1/3	10d.	1/6

No. 6388. Per doz. 7/6. Each 9d.
Suspension Springs, For Heavy Pendulums, Time Recorders, &c.

No. 6389. **Complete**, as illustrated. Per dozen 9/3. Each 1/-.
No. 6428. **Wire** only. Per gross 10/8. " 1/- " 2d.
For **400-day** Clocks, U.S.C. Make (Other Makes also Supplied).

No. 755. **Suspension Steel**, in 1-oz. Ribbons.
Width 1-in. Strengths 10/100 and 15/100 m/m.
Per dozen 12/6 ... 11/3
Each 1/3 ... 1/2
Special Prices for Quantities.
Other Sizes and Thickness to Order.

SELECTED CLOCK MATERIALS.

No. 6392. Per dozen Cards 12/9. Per Card 1/3.
French Brocot Suspension, Assorted.

No. 6390. Per dozen Cards 7/3. Per Card 10d.
Regulator Suspensions, Assorted.

No. 6391. Per dozen Cards 10/- Per Card 1/-.
French Suget Suspensions, Assorted.

SELECTED CLOCK MATERIALS.

SELECTED CLOCK MATERIALS.

Quarter gross at gross rate ; quarter dozen at dozen rate.

SCREWS (Continued), STRAPS, STUDS and WORMS.

No. 1451. Per doz. Boxes 13/6. Per Box 1/3.
Screws for Lever Clock **Balance Pivots**, Assorted Sizes.
3 doz. in each Box.

No. 6346. Per doz. Boxes 22/-. Per Box 2/-.
Brass Screws, Assorted, for Small Bee Clock Cases, &c. 6 doz. in each Box.

No. 4292. Per gross 15/-.
Per doz. 1/6.
Visible Escapement Screws, for French Clocks, Hardened and Finished.

No. 2116. Per gross 6/-. Per dozen 8d.
Strap Screws, for French Clocks.

No. 742. **Screws** for French Clock Minute Wheels. Per gross 3/-. Per dozen 4d.
No. 2703. **Screws** for **English** Clocks. Per gross 2/6. Per dozen 4d.
No. 744. **Click Screws**, Shouldered, Assorted. Per gross 2/6. Per dozen 4d.

No. 745. Per gross 2/-. Per dozen 3d.
Screws for Fixing **Platforms** of French **Clock Escapements**.

THREE GROSS ASSORTMENT
CLOCK SCREWS FOR REPAIRS.
FOR REFILLS ORDER BY NUMBERS MARKED.

No. 6549. Per Assortment 9/6.
Useful Assortment of Screws, containing 3 gross, with Cheese Heads, for English, French, or German Clocks, or for Mechanics. Metric Thread.
For Refills see Nos. 4365 to 4276 on previous page.

No. 6550. Per Assortment 12/6.
General Assortment, containing 3 gross of various Screws, as illustrated. Very practical for General Jobbing.

Studs for Case Clock Movement Minute Wheels.
No. 1647. Forging only. Per dozen 1/3. Each 2d.
No. 1649. Finished. " 2/-. " 3d.

No. 1565. **Worms or Endless Screws** for Musical Boxes.

Sizes	13	17	40	45	55	70 m/m
Per dozen	28	28	5/-	5/6	6/6	7/-
Each	4d.	4d.	6d.	8d.	8d.	9d.

No. 6547. **Straps**, with Screws, for French Clock Movements. Per dozen Pairs 2/10. Per Pair 4d.

Quarter gross at gross rate ; quarter dozen at dozen rate.

HIRST BROS. & Co. Limited, Oldham, Manchester and Birmingham.

WHEELS and PINIONS, Finished, for 8-day Timepieces.

		Per doz.
No. 6307.	Barrels, with Arbors	28/10
No. 6308.	Fusees, Complete	89/-
No. 6309.	Centre Wheels and Pinions	12/-
No. 6310.	Intermediate Wheels & Pinions	10/10
No. 6311.	Pallets	20/-
No. 6312.	Minute Cannons	4/-

		Each.
		2/6
		7/9
		1/3
		1/-
		1/10
		6d.

		Per doz.	Each.
No. 6313.	Minute Wheels and Pinions	4/6	6d.
No. 6314.	Hour Wheels	6/-	8d.
No. 6315.	Escape Wheels and Pinions	12/-	1/3
No. 6316.	Crutches	8/-	10d.
No. 6317.	Pallet Arbors	8/-	10d.

Only Stock Sizes and Counts supplied at above prices.

		Per doz.	Each.
No. 4285.	Escape Wheels for French Clocks	8/-	10d.
No. 1144.	" " Regulators	5/3	6d.
No. 769.	" " American and German Clocks		4d.
No. 768.	Pallet Alarm Wheels Nos. of Teeth 32, 35, 36, 40, 42, 45	3/6	4d.
No. 4514.	Escape Wheels for Timepieces	5/-	6d.
No. 4515.	" " Case Clocks	5/-	6d.

No. 1566. Per dozen 5/- ; Each 6d.   No. 1567. Per dozen 5/- ; Each 6d.   No. 1568. Per dozen 3/8 ; Each 6d.
For Dot Clocks.   For Gnat Clocks.   For Bee Clocks.
Quarter dozen at dozen rate.

No. 2047. Brass Wheel Blanks. Price on Application, stating Size, Thickness and Quantity.

HIRST BROS. & Co. Limited, Oldham, Manchester and Birmingham.

SUSPENSIONS—Continued.

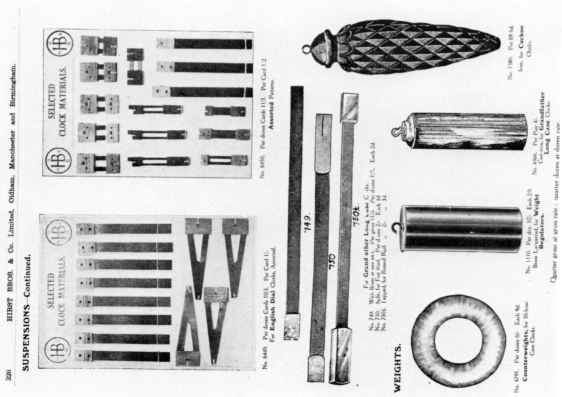

SELECTED CLOCK MATERIALS.

No. 6449. Per dozen Cards 10/3. Per Card 1/-. For English Dial Clocks, Assorted.

SELECTED CLOCK MATERIALS.

No. 6450. Per dozen Cards 11/3. Per Card 1/2. Assorted Patterns.

No. 749. For Grandfather Long Case C cks. Per gross 11/3. Per dozen 1/2. Each 2d.
No. 750. With Brass at one en l. Per gross 2/- Per dozen 3d. Each 3d.
No. 750½. Split, for Full Rod. " dozen 2/- Each 3d.
Tapped, for Round Rod. " 2/- " 3d.

WEIGHTS.

No. 1380. Per lb. 6d. Iron, for Cuckoo Clocks.

No. 1110. Per doz. 32/-. Each 2/9. Brass Lacquered, for Weight Regulators.

No. 4566. Per Pair 4/-. Cast-iron, for Grandfather Long Case Clocks. Quarter gross at gross rate ; quarter dozen at dozen rate.

No. 4791. Per dozen 6/-. Each 8d. Counterweights, for 30-hour Case Clocks.

WHEELS, &c.—Continued.

No. 6519. Per doz. 6/- ;
Each 7d.
Main Wheels for
Ansonia Bee Clocks.

No. 6521. Per doz. 5/10 ; Each 7d.
Centre Wheels for Ansonia
Bee Clocks.

No 6520. Per doz. 7/- ; Each 8d.
Escape Wheels and Pinions
complete, for U.S.C. Weight
Regulators.

No. 5126. Per Set 6/-
Sprocket Chain Wheels for 30-Hour
Grandfather Clocks. Set includes 2, or 1
pair Sprockets, with Click and Chain complete.

WINDERS FOR LEVER CLOCKS.

Nos. 6526/7. Right or Left, and sizes 7, 8 & 9
Nos. 6528/9.   ,,   ,,   ,,   ,,   ,,   5 & 6
No. 1200. For Pirate Alarms.

Nos.   6526,
Per gross 18/-
Per doz. 1/9

6527.
16/-
1/6

6528.
12/-
1/3

6529.
12/-
1/3

1200.
20/-
1/9

No. 1222. Per dozen pairs 6/- ; per pair 7d.
Wings, Polished Hardwood, for Anglo Cases,
Light or Dark.

VARIOUS MATERIAL.

No. 1192. **Balances,** Complete
Per dozen 3/8 ; Each 5d.
No. 1188. **Balance Staffs**
Per gross 9/9 ; per doz. 1/-
No. 1194. **Click Springs**
Per gross 8/- ; per doz. 9d.
No. 1195. **Click Springs**
Per gross 8/- ; per doz. 9d.
No. 1196. **Click Springs** for Main
Wheels. Per gross 3/3 ; per doz. 6d.
No. 1189. **Escape Wheels and
Pinions.** Per doz. 2/- ; Each 3d.
No. 1186. **Handsetters**
Per doz. 2/- ; Each 3d.
No. 1187. **Handsetters**
Per doz. 1/6 ; Each 2d.
No. 1215. **Pallets**
Per doz. 2/4 ; Each 3d.
No. 1200. **Winders**
Per doz. 1/9 ; Each 2d.
No. 1190. **Winding Wheels**
Per doz. 2/- ; Each 3d.
No. 1191. **Winding Wheels**
Per doz. 1/6 ; Each 2d.

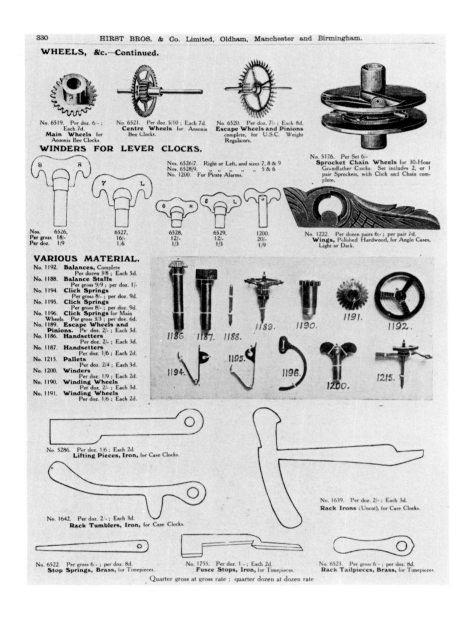

1186.   1187.   1188.   1189.   1190.   1191.   1192.

1194.   1195.   1196.   1200.   1215.

No. 5286. Per doz. 1/6 ; Each 2d.
Lifting Pieces, Iron, for Case Clocks.

No. 1639. Per doz. 2/- ; Each 3d.
Rack Irons (Uncut), for Case Clocks.

No. 1642. Per doz. 2/- ; Each 3d.
Rack Tumblers, Iron, for Case Clocks.

No. 6522. Per gross 6/- ; per doz. 8d.
Stop Springs, Brass, for Timepieces.

No. 1755. Per doz. 1/- ; Each 2d.
Fusee Stops, Iron, for Timepieces.

No. 6523. Per gross 6/- ; per doz. 8d.
Rack Tailpieces, Brass, for Timepieces.

Quarter gross at gross rate ; quarter dozen at dozen rate

APPENDIX II

The British Clock Manufacturer's supplement for March 1940.

This publication includes line drawings of the mechanical English movements
still in production in 1940, and is reproduced courtesy of NAG Press Ltd.

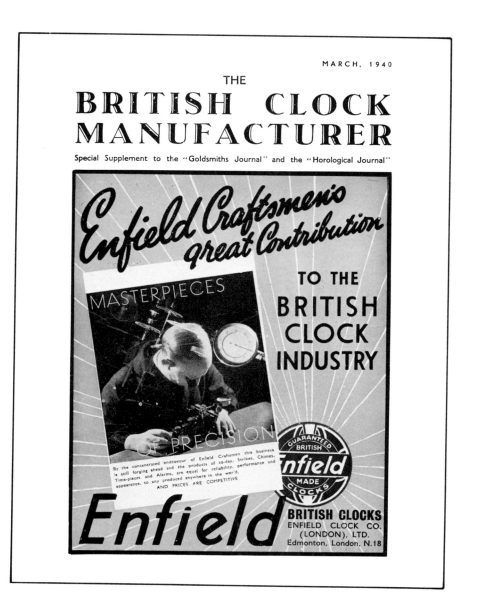

THE

BRITISH CLOCK MANUFACTURER

Special Supplement to the "Goldsmiths Journal" and the "Horological Journal"

Editor: ARTHUR TREMAYNE

EIGHTH YEAR   MARCH, 1940

THE FUTURE OF THE BRITISH CLOCK MANUFACTURING INDUSTRY

This Important Article is by

D. W. BARRETT

Director and Manager of Smith's English Clocks Ltd.

AT a recent Horological Trade Meeting someone reminded those present that Hitler said "Germany must export or die." The person in question went on to say that "we in Great Britain must import if we are to export."

Really, I have been unable to get the inference of this remark out of my mind. Believe me, what was being suggested was that we must increase our imports if we hope to increase our exports, but, more serious still, we must increase our imports *of manufactured goods.*

Well, I wonder if that person is aware that we import more than any other country in the world, that for the past two years our visible adverse trade balance has been over 400 million pounds and that this year it might reach the figure of 600 millions or more.

We are now and may always be the largest importers in the world of raw materials and food stuffs. Therefore, it is in our vital interests to keep down our imports of other things, which in the main means manufactured goods.

These remarks and all that they engender may seem far removed from "The Future of the British Clock Industry," but actually they bring us to the crux of the subject.

For years now certain thinking people have realised that, with practically every country in the world aiming at economic self-sufficiency, our exports of manufactured goods are bound to continue to decline from the pinnacle of 1913. Other countries do not want our manufactures in such abundance because they are making similar things for themselves. Therefore, if we are to survive we in turn must make for ourselves things we have never made before or things we stopped making because we could exchange other manufactured goods and coal for them.

Our coal exports over the last twenty-five years have been almost halved and our exports of manufactures and re-exports are down by over 33½ per cent. This being the case, what are we to do? I suggest that as world rationalisation appears to be beyond human compass, the only thing left for us to do is to make for ourselves things which we have previously imported and we must also grow more of our own food.

The net result of making more goods for ourselves will be that we shall import more and more raw materials and, if our suppliers of these are to be compensated, then they must take our manufactures.

In just a few words I have endeavoured to give a rough outline of a picture which one's imagination can fill in with detail.

[Imports for 1939 were valued at £886 millions and exports at £418¾ millions. For 1938 the figures were—imports £919¼ millions and exports £470¾ millions.]

A FINE OLD INDUSTRY FADES AWAY

It was not just chance which caused clockmaking to be attempted once again on a large scale in this country. Let me remind readers that once we were the largest clockmaking country in the world. Think what is implied by mention of the names Tompion, Graham, Harrison, Mudge and other giants of the past. Such names simply echo the greatness of British horological achievements in by-gone days.

Why, then, did an industry—indeed, a craft—with such great old masters as its foundation dwindle until it barely existed? Simply because British watch and clock makers were such fine craftsmen that they did not feel the need for machinery and did not use it when it became available. Others did, notably in Germany, Switzerland and France, and in consequence the inevitable happened—we were swamped and under we went. But did we go completely under? No! Upon reflection I am reminded that we have never lost our art of making the really good "stuff" such as chronometers, astronomical clocks and master clock systems. How many foreign chronometers and master clock systems have been or are being sold in this country? Precious few, as the names of Mercer, Synchronome, and Gent, to mention only three, of many, stand here and everywhere for all that is best in their spheres.

The dawn of the machine age was the dawn of the demise of British domestic clockmaking but, at the same time, the machine was responsible for Britain's industrial millennium and the foundation of an export trade as yet unsurpassed by any other country. Did not our exports in 1913 reach the value of 635 million pounds, and did we not build up overseas investments totalling the enormous value of 4,000 million pounds?

THE NEW START

The machine virtually made us what we are and with it we made endless things, including instruments of all descriptions, and eventually it was the instrument and the gramophone manufacturers of this country who had two thoughts. In effect they said to themselves, why should approximately 6,000,000 foreign clocks be sold in this country annually, and, secondly, why should we not start making clocks? One could add to this that, like every other substantial British manufacturer, they felt it was altogether wrong that foreign manufactured goods should be allowed to come into this country in large quantities without being challenged by a British product. The sequel was that domestic clockmaking re-started here early in 1931, and I can assure my readers that very large sums of money would have been invested in the enterprise had it not been for the fact that, almost from the day it started, the prices of foreign clocks came tumbling down. Actually within the space of two years the prices of foreign "strikes" and "chimes" were halved, and we have unquestionable evidence that they were reduced to this extent simply with the idea of putting the newly-formed British clock industry out of existence.

OUR STRUGGLE

Much has been written about the struggle the British manufacturers have had in their efforts to keep their heads above water, especially in competition with the products from Germany, which for many years were lavishly subsidised by the German Government.

Hard times there have been, almost heart-breaking times. However, the industry still exists, and it is stronger to-day than ever, although it would have been much stronger had it not been faced with the uneconomic

competition referred to. Had it not been for this, the industry would to-day have been in a position not only to meet the whole demand for this market but for all the Empire markets, from which, of course, as from Britain, the enemy-made article is barred.

AFTER THIS WAR

The eight to nine years of life of the newly-formed British clock industry is a story which has often been told, so now let us turn to the future. What does it hold for us? As a clockmaker my first consideration is, will the industry after the war have to compete once again with foreign subsidised competition? I think not, as I am confident that a non-Nazi Government would not adopt a similar practice. Relieved of such unfair competition, there is no reason why the British industry should not go ahead rapidly.

After this war there will be available many new factories with fine equipment and highly skilled labour which can be switched over to clockmaking. I suggest clockmaking will not be a side-line as it has been to several big firms, but it will be of prime importance to them, and in consequence, each and every one of its needs will be carefully studied. In addition a power of influence, skill and ability will be brought to bear in an effort to provide the right type of clock for every room in every British home both here and overseas—British homes in Australia, New Zealand, South Africa and the Empire at large.

Later, increasing production and, in consequence, lower costs will enable the industry to compete with its foreign rivals for the business in foreign non-clock manufacturing countries.

I can visualise—in fact, I am greatly looking forward to—the time when the British factories' output will be numbered in millions of pieces, and when the trade and public will not buy British clocks simply because they are British but because they are as good as or better than foreign-made equivalents and are as competitive in price.

TARIFF PROTECTION ESSENTIAL

One should bear in mind that, if it is to succeed, the industry must have adequate tariff protection, as we have a high standard of living and expensive social services in this country and, therefore, cannot compete with countries where lower standards prevail.

I am confident that the day is not far distant when—everything being equal—British clock manufacturers will be able to produce as cheaply as the foreigners.

You will agree that a clockmaking industry would not exist in the U.S.A. if it were not for the high tariff duties. In the U.S.A. the duty ranges as high as 165 per cent., which, of course, is prohibitive. It is significant that no country in the world has been successful in building up a large scale clockmaking industry without the aid of high duties. Even Japan imposes a 100 per cent. duty on foreign-made electric clocks!

PROUD OF ITS WAR EFFORT

The existing British clock industry can be proud of the fact that by making intricate instruments which could not have been made here had it not been for our clockmaking machinery and skilled clockmakers, it is playing a tremendously important part in the National war effort. Although the British clock factories and the instrument concerns with which they are associated are at present so very busy with work of national importance, I can assure readers that the principals of these concerns are not losing sight of the future, and they are not overlooking the fact that one day the war will end. They are, therefore, carrying on—in fact speeding up—their development work so that when peace does come they will immediately be in a position to launch big manufacturing programmes. New tools will be ready, the wheels of the machines will start turning once again after the briefest possible lapse; thus the British clockmaking industry will be "all set" for rapid expansion—and why not for development into the largest industry of its kind in the world?

450

BRITISH CLOCKS FOR EVERY HOME

A CLAIM that the British clock manufacturer is as well able to supply any kind of clock which may be required may seem a bold assertion, but that it is nothing more than the simple truth is proved by even the most casual survey of the products of British factories. In all but the markets which demand the cheapest, there are clocks of British make which can compete on an equal footing with any made elsewhere, and the wide range of models and prices available is a remarkable achievement.

Every kind of timekeeper for domestic use is represented, from the low-priced thirty-hour timepiece to the dignified long-case clock with alternative chimes and "super" finish. In the following pages, no less than twenty-nine British movements are illustrated and described, and every possible requirement is covered by at least one model. Timepieces, striking and chiming clocks, all are produced by British craftsmen, and constructed along lines which ensure that they will give good service and prove sound investments to their purchasers.

The single exception is, perhaps, the humble alarm clock, but for this omission the British maker must not be blamed, for the Continental makers of this type of clock produced them at prices which did not encourage our own factories to compete with them. All the same, British alarm clocks of the synchronous electric type are to be had, where owners do not mind paying a little more for a clock that is undeniably a high-class article. So even there the British maker has an answer to the critic.

In an examination of the movements shown, nothing is more impressive than the determination of makers to provide just what is wanted by the buying public. Until recently this keen appreciation of the taste and needs of the purchaser was often spoken of as a characteristic of the foreign manufac-

turer. "They study your wants and cater for them," was a favourite phrase. In the following pages it is shown conclusively that this appreciation of the buyer's needs is no more a monopoly than is the making of good clocks. The British maker is quite as capable of giving the clock owner what he wants, and, indeed, can often do it better than his competitors.

When it is considered that little more than eight years has elapsed since the revival of British clockmaking, the technical advances made and the high standard of quality attained are remarkable enough, but these factors alone are not enough to enable a newly re-founded industry to capture markets already well supplied from other sources.

The answer to the question "How did the British manufacturer do it?" is shown in this issue of the BRITISH CLOCK MANUFACTURER. The kinds of clocks needed for various purposes were studied, and care was taken to cover every type and to meet particular requirements.

Narrow models for modern mantels, alternative chimes for people who like a change of time, small timepieces of both the cheap and the more expensive kinds for bedrooms and similar places, synchronous electric clocks of particularly high quality, and even synchronous chiming and striking models, all are in the British range at prices which represent excellent value.

Nor is this all, for the novelties have not been by any means neglected. The "Mystery" clock, for instance, with its fascinating concealed method of operation, and the "Teesmade" clock, which provides a cup of tea at any set time, not to mention the more business-like "Cooker Time Switch," which is a robot clock for the kitchen, and almost takes the responsibility of cooking.

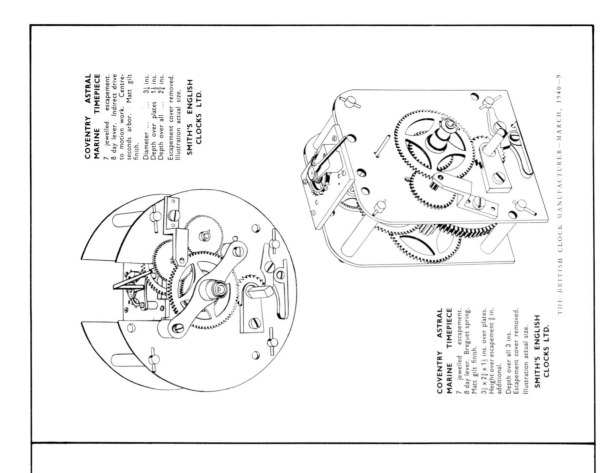

COVENTRY ASTRAL MARINE TIMEPIECE

7 jewelled escapement. 8 day lever. Indirect drive to motion work. Centre-seconds arbor. Matt gilt finish.

Diameter ... ... 3⅛ ins.
Depth over plates 1⅛ ins.
Depth over all ... 2¼ ins.
Escapement cover removed. Illustration actual size.

SMITH'S ENGLISH CLOCKS LTD.

COVENTRY ASTRAL MARINE TIMEPIECE

7 jewelled escapement. 8 day lever. Breguet spring. Matt gilt finish.

3⅛ x 2¼ x 1⅛ ins. over plates. Height over escapement ⅜ in. additional.

Depth over all 3 ins.
Escapement cover removed. Illustration actual size.

SMITH'S ENGLISH CLOCKS LTD.

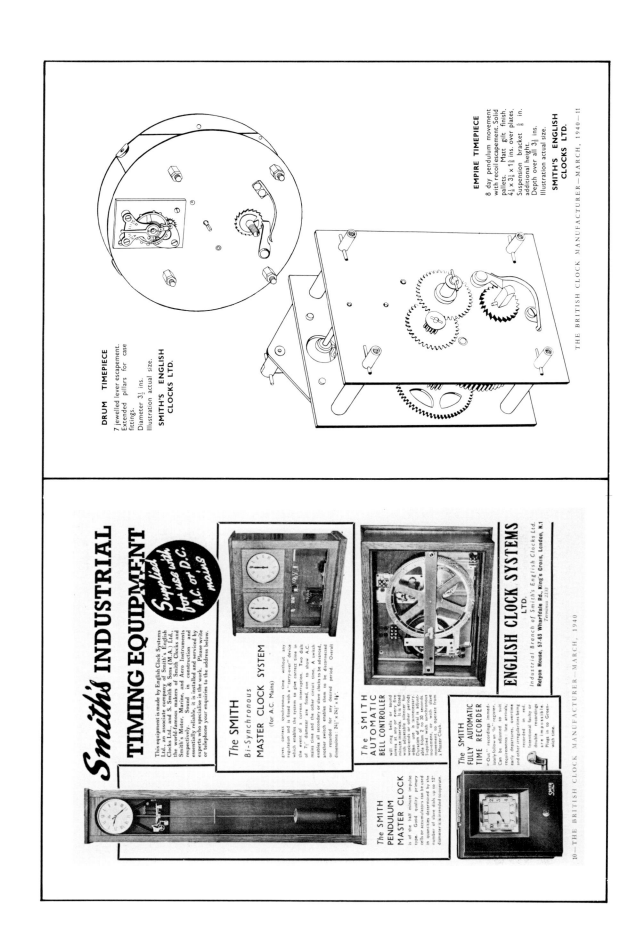

DRUM TIMEPIECE

7 jewelled lever escapement.
Extended pillars for case
fittings.
Diameter 3½ ins.
Illustration actual size.

**SMITH'S ENGLISH
CLOCKS LTD.**

EMPIRE TIMEPIECE

8 day pendulum movement
with recoil escapement. Solid
pallets. Matt gilt finish.
4½ x 3½ x 1⅜ ins. over plates.
Suspension bracket ⅝ in.
additional height.
Depth over all 3½ ins.
Illustration actual size.

**SMITH'S ENGLISH
CLOCKS LTD.**

THE BRITISH CLOCK MANUFACTURER—MARCH, 1940—11

Smith's INDUSTRIAL
TIMING EQUIPMENT

Supplied for use with A.C. or D.C. mains

This equipment is made by English Clock Systems
Ltd., an associate company of Smith's English
Clocks Ltd., and S. Smith & Sons (M.A.) Ltd.,
the world-famous makers of Smith Clocks and
Smith's Motor, Marine, and Aero Instruments
respectively. Sound in construction and
essentially reliable, it is installed and serviced by
experts who specialise in the work. Please write
or telephone your enquiries to the address below.

The SMITH
Bi-Synchronous
MASTER CLOCK SYSTEM
(for A.C. Mains)

gives correct synchronous time without any
regulation and is fitted with a "carry-over" device
which enables the system to give correct time in
the event of a current interruption. Two dials
of 7½" diameter are fitted, one to show A.C.
mains time and the other circuit time. A switch
enables all secondary or slave clocks to be advanced,
another switch enables them to be disconnected
or retarded for any desired period. Overall
dimensions: 24½" x 26½" x 4½".

The SMITH
AUTOMATIC
BELL CONTROLLER

will ring bells or sound
sirens at any or every five-
minute intervals. It is fitted
with adjustable silencer for
week-ends or other periods
when use is unnecessary.
Duration of signal is adjustable from 2 to 30 seconds.
Supplied with synchronous
movement, or with slave
movement to operate from
a Master Clock.

The SMITH
PENDULUM
MASTER CLOCK

is of the half minute impulse
type. Good quality primary
cells or accumulators can be used
in quantities determined by the
number of slave clocks up to 12"
diameter it is intended to operate.

The SMITH
FULLY AUTOMATIC
TIME RECORDER

"Out" recordings immed-
iately follow an "In" register.
Can be adjusted to suit
requirements, late arrivals,
early departures, overtime
and other irregularities being
recorded in red.
Intentional faulty or
double recordings
are impossible.
Plugs in to Green-
wich time.

ENGLISH CLOCK SYSTEMS
LTD.

Industrial Branch of Smith's English Clocks Ltd.

Relyon House, 57-63 Wharfdale Rd., King's Cross, London, N.1

Terminus 2233

10—THE BRITISH CLOCK MANUFACTURER—MARCH, 1940

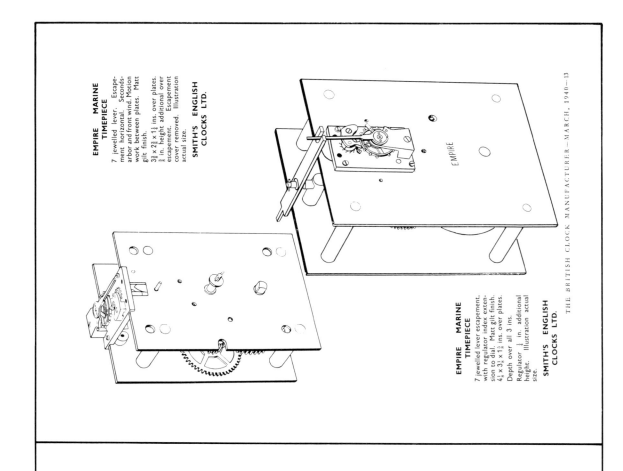

EMPIRE MARINE TIMEPIECE

7 jewelled lever. Escapement horizontal. Seconds-arbor and front wind. Motion work between plates. Matt gilt finish.

3⅞ x 2¾ x 1¼ ins. over plates. ¾ in. height additional over escapement. Escapement cover removed. Illustration actual size.

SMITH'S ENGLISH CLOCKS LTD.

EMPIRE MARINE TIMEPIECE

7 jewelled lever escapement, with regulator index extension to dial. Matt gilt finish. 4¼ x 3⅜ x 1⅝ ins. over plates. Depth over all 3 ins. Regulator ¼ in. additional height. Illustration actual size.

SMITH'S ENGLISH CLOCKS LTD.

454

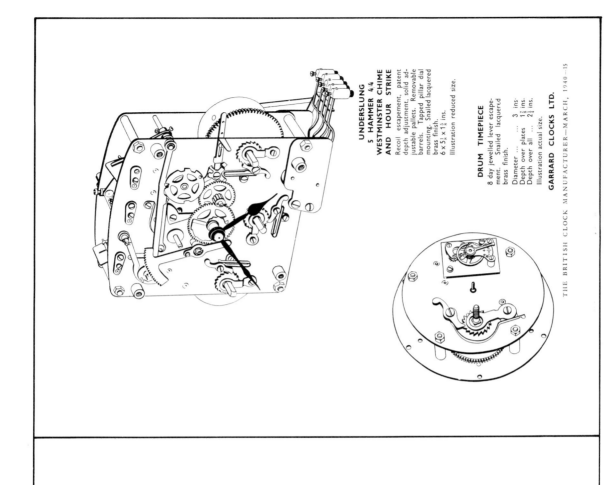

GARRARD
Clocks

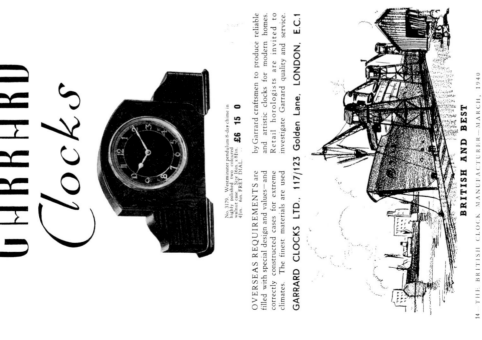

No. 3179. Westminster pendulum 8-day chime in highly polished two coloured walnut case. Size 16in. x 8½in. x 4½in. 6in. FRET DIAL. **£6 15 0**

OVERSEAS REQUIREMENTS are filled with special design and values—and correctly constructed cases for extreme climates. The finest materials are used by Garrard craftsmen to produce reliable and artistic clocks for modern homes. Retail horologists are invited to investigate Garrard quality and service.

GARRARD CLOCKS LTD., 117/123 Golden Lane, LONDON. E.C.1

BRITISH AND BEST

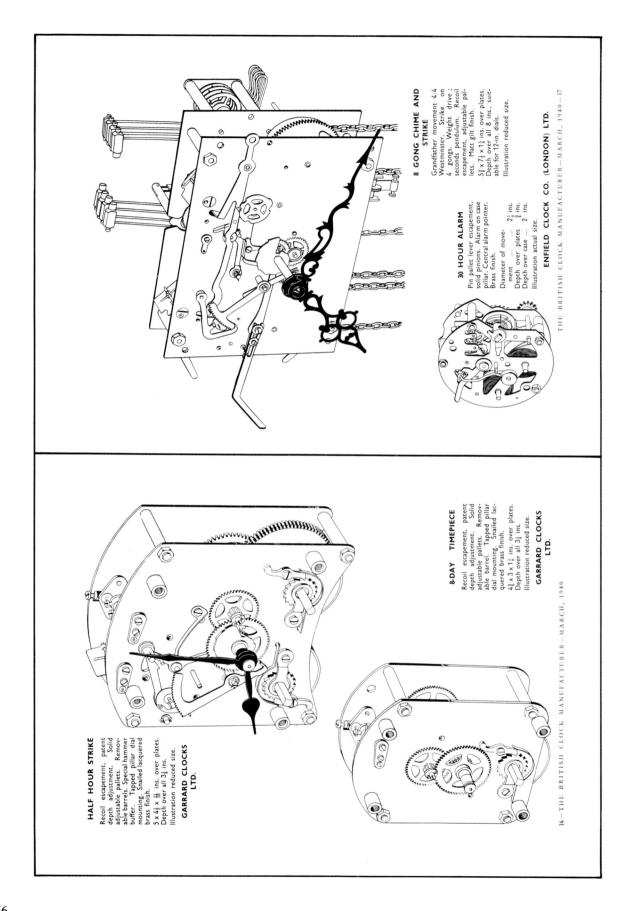

8 GONG CHIME AND STRIKE

Grandfather movement 4/4 Westminster. Strike on 4 gongs. Weight drive ; seconds pendulum. Recoil escapement, adjustable pallets. Matt gilt finish.
5¾ x 7½ x 1⅝ ins. over plates. Depth over all 8 ins., suitable for 12-in. dials.
Illustration reduced size.

30 HOUR ALARM

Pin pallet lever escapement, solid pinions. Alarm on case pillar. Central alarm pointer. Brass finish.

Diameter of movement ... ... 2¼ ins.
Depth over plates ... 1⅞ ins.
Depth over case ... ... 2 ins.
Illustration actual size.

ENFIELD CLOCK CO. (LONDON) LTD.

THE BRITISH CLOCK MANUFACTURER—MARCH, 1940—17

HALF HOUR STRIKE

Recoil escapement, patent depth adjustment. Solid adjustable pallets. Removable barrels. Special hammer buffer. Tapped pillar dial mounting. Snailed lacquered brass finish.
5 x 4¼ x ⅞ ins. over plates.
Depth over all 3¼ ins.
Illustration reduced size.

GARRARD CLOCKS LTD.

8-DAY TIMEPIECE

Recoil escapement, patent depth adjustment. Solid adjustable pallets. Removable barrel. Tapped pillar dial mounting. Snailed lacquered brass finish.
4¼ x 3 x 1⅛ ins. over plates.
Depth over all 3¼ ins.
Illustration reduced size.

GARRARD CLOCKS LTD.

16—THE BRITISH CLOCK MANUFACTURER—MARCH, 1940

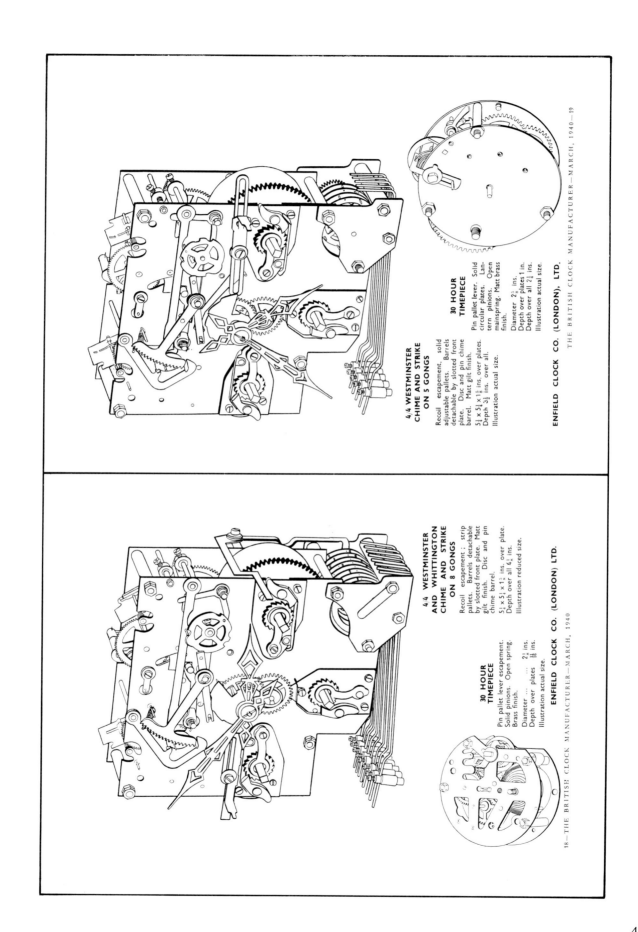

4/4 WESTMINSTER CHIME AND STRIKE ON 5 GONGS

Recoil escapement, solid adjustable pallets. Barrels detachable by slotted front plate. Disc and pin chime barrel. Matt gilt finish.

5¼ x 5¼ x 1⅝ ins. over plates. Depth 3¼ ins. over all.

Illustration actual size.

30 HOUR TIMEPIECE

Pin pallet lever. Solid circular plates. Lantern pinions. Open mainspring. Matt brass finish.

Diameter 2⅝ ins.
Depth over plates 1 in.
Depth over all 2¼ ins.

Illustration actual size.

ENFIELD CLOCK CO. (LONDON), LTD.

4/4 WESTMINSTER AND WHITTINGTON CHIME AND STRIKE ON 8 GONGS

Recoil escapement; strip pallets. Barrels detachable by slotted front plate. Matt gilt finish. Disc and pin chime barrel.

5¼ x 5¼ x 1¾ ins. over plate. Depth over all 4¼ ins.

Illustration reduced size.

30 HOUR TIMEPIECE

Pin pallet lever escapement. Solid pinions. Open spring. Brass finish.

Diameter ... ... 2¼ ins.
Depth over plates 1⅛ ins.

Illustration actual size.

ENFIELD CLOCK CO. (LONDON) LTD.

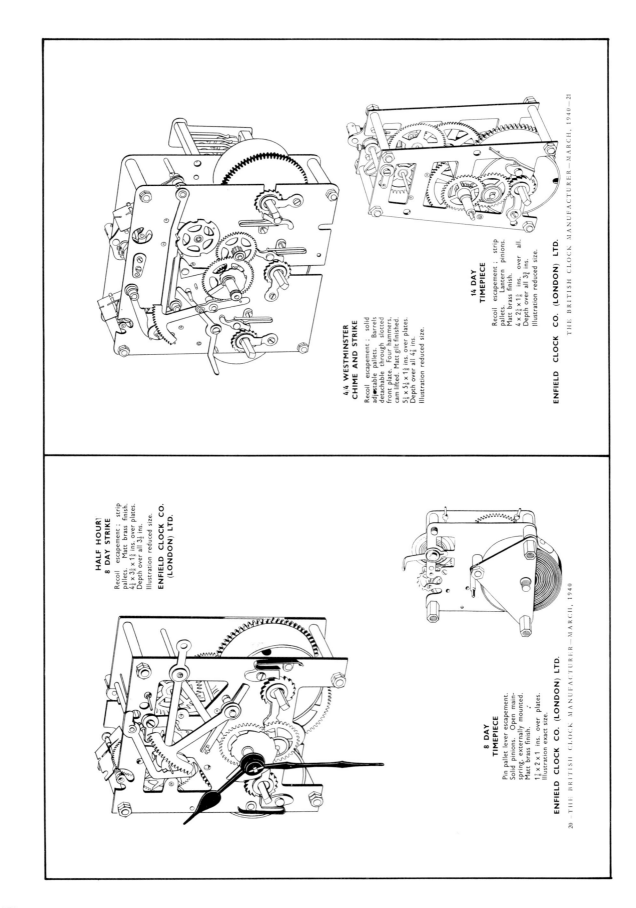

4/4 WESTMINSTER CHIME AND STRIKE

Recoil escapement ; solid adjustable pallets. Barrels detachable through slotted front plate. Four hammers, cam lifted. Matt gilt finished.

5⅛ x 5¼ x 1⅝ ins. over plates. Depth over all 4⅞ ins.

Illustration reduced size.

14 DAY TIMEPIECE

Recoil escapement ; strip pallets. Lantern pinions. Matt brass finish.

4 x 2⅝ x 1⅛ ins. over all. Depth over all 3⅜ ins.

Illustration reduced size.

ENFIELD CLOCK CO. (LONDON) LTD.

THE BRITISH CLOCK MANUFACTURER—MARCH, 1940—21

**HALF HOUR\
8 DAY STRIKE**

Recoil escapement ; strip pallets. Matt brass finish.

4¼ x 3¾ x 1⅜ ins. over plates. Depth over all 3⅝ ins.

Illustration reduced size.

ENFIELD CLOCK CO. (LONDON) LTD.

8 DAY TIMEPIECE

Pin pallet lever escapement. Solid pinions. Open mainspring, externally mounted. Matt brass finish.

1⅞ x 2 x 1 ins. over plates. Illustration exact size.

ENFIELD CLOCK CO. (LONDON) LTD.

20—THE BRITISH CLOCK MANUFACTURER—MARCH, 1940

459

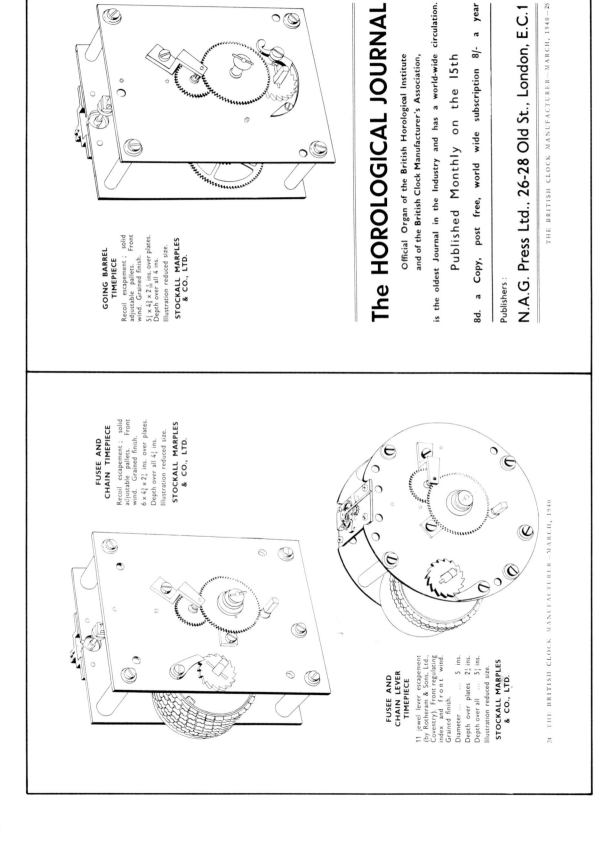

GOING BARREL TIMEPIECE

Recoil escapement; solid adjustable pallets. Front wind. Grained finish.
5¼ x 4⅜ x 2⅛ ins. over plates.
Depth over all 4 ins.
Illustration reduced size.

STOCKALL MARPLES & CO., LTD.

FUSEE AND CHAIN TIMEPIECE

Recoil escapement; solid adjustable pallets. Front wind. Grained finish.
6 x 4⅜ x 2¼ ins. over plates.
Depth over all 4¼ ins.
Illustration reduced size.

STOCKALL MARPLES & CO., LTD.

FUSEE AND CHAIN LEVER TIMEPIECE

11 jewel lever escapement (by Rotheram & Sons, Ltd., Coventry). Front regulating index and front wind. Grained finish.

Diameter ... ... 5 ins.
Depth over plates 2⅜ ins.
Depth over all ... 5¼ ins.
Illustration reduced size.

STOCKALL MARPLES & CO., LTD.

PERIVALE RULES the HOURS

The Perivale range of all-British clocks incorporating fine reliable movements of first-class materials in well-designed cases.

Every effort is being made to increase the production of Perivale British Clocks.

The British Clock now rules the world's markets.

EXPORT ENQUIRIES INVITED

THE BENTIMA COMPANY LIMITED, 18 Bury Street, LONDON, E.C.3

26—THE BRITISH CLOCK MANUFACTURER—MARCH, 1940

4/4 WESTMINSTER CHIME AND STRIKE

5 gongs. Recoil escapement. Solid adjustable pallets. Detachable barrels. Disc-cam chiming barrel. Matt gilt finish front plate, smailed finish back plate.

$5\frac{3}{8} \times 5\frac{1}{4} \times 1\frac{7}{8}$ ins. over plates. Depth over all $4\frac{1}{2}$ ins.

Illustration reduced size.

30 HOUR TIMEPIECE

Pin pallet lever escapement. Solid pinions. Matt brass finish.

Diameter ... ... $2\frac{1}{4}$ ins.
Depth over plates $1\frac{1}{8}$ ins.
Depth over all ... $1\frac{1}{4}$ ins.

Illustration actual size.

PERIVALE MANUFACTURING CO., LTD.

THE BRITISH CLOCK MANUFACTURER—MARCH, 1940—27

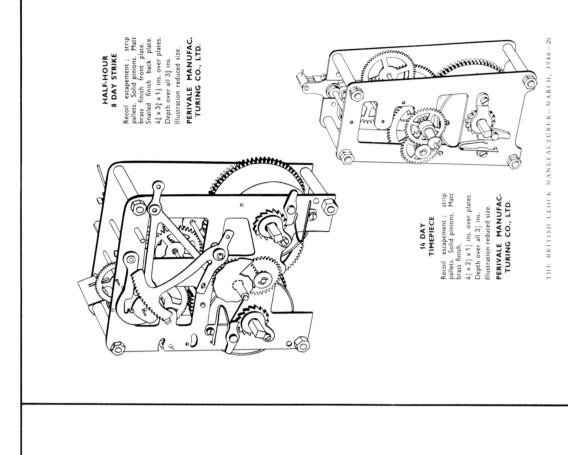

462

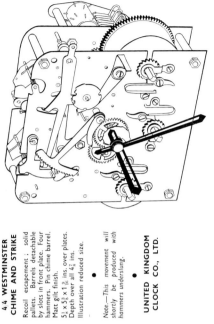

4/4 WESTMINSTER CHIME AND STRIKE

Recoil escapement; solid pallets. Barrels detachable by slots in front plate. Four hammers. Pin chime barrel. Matt gilt finish.

5¼ x 5¾ x 1 9/16 ins. over plates. Depth over all 4⅞ ins. Illustration reduced size.

Note.—This movement will shortly be produced with hammers underslung.

● UNITED KINGDOM CLOCK CO., LTD.

INDEX TO MOVEMENT DIAGRAMS

THE BRITISH CLOCK MANUFACTURER—MARCH, 1940—31

INDEX

Note: The index is arranged under eight main headings: Clockmakers and Manufacturers; Clocks — some distinctive types; Escapements; Materials; Patents; Retailers and Distributors; Trade Names, Trade Marks. Page numbers appear first in medium type, followed by Figure numbers and Colour Plate numbers in bold type.

CLOCKS —
some distinctive types

ESCAPEMENTS

MATERIALS (this is not an exhaustive list and does not include items in Appendices)